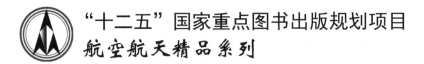

"十二五"国家重点图书出版规划项目
航空航天精品系列

AUTOMATIC CONTROL PRINCIPLE
自动控制原理
（中英文对照）

● 李道根　主编　　● 王建华　主审

哈尔滨工业大学出版社
HARBIN INSTITUTE OF TECHNOLOGY PRESS

Brief of contents

This textbook states the basic concepts, principles and various analysis techniques of classical control theory. The textbook consists of seven chapters, among them the former six chapters describe the analysis and synthesis of linear time-invariant systems, and the last chapter describes non-linear systems.

The book is printed in a bilingual form to make convenience for reader′s reading and studying. This text book could be used as a professinal teaching material for the speciallies of automation, electrical engineering and automation, and so on, as well as a self-studying material or a reference book for graduated students of other relevaut specialties and teachers.

内 容 提 要

本书阐述了经典控制理论的基本概念、原理和分析方法。全书共7章，前6章介绍线性时不变系统的分析和综合，最后一章介绍非线性系统。

本书以双语形式出版，以方便读者的阅读和学习。本书是工科院校自动控制及自动化类专业教材，也是专业英语教材，同时也可作为其他相关专业研究生和教师的参考书。

图书在版编目(CIP)数据

自动控制原理(中英文对照)/李道根主编．—哈尔滨：哈尔滨工业大学出版社，2007.8(2023.7重印)

ISBN 978-7-5603-2542-2

Ⅰ.自… Ⅱ.李… Ⅲ.自动控制理论 Ⅳ.TP13

中国版本图书馆 CIP 数据核字(2007)第 116159 号

策划编辑　张秀华　杨　桦
责任编辑　康云霞
封面设计　卞秉利
出版发行　哈尔滨工业大学出版社
社　　址　哈尔滨市南岗区复华四道街10号　邮编150006
传　　真　0451－86414749
网　　址　http://hitpress.hit.edu.cn
印　　刷　哈尔滨市颉升高印刷有限公司
开　　本　787mm×1092mm　1/16　印张 25.5　字数 588 千字
版　　次　2007年8月第1版　2023年7月第6次印刷
书　　号　ISBN 978-7-5603-2542-2
定　　价　38.00元

(如因印装质量问题影响阅读,我社负责调换)

Preface

With the coming of economic globalization and knowledge-based economy, higher demand of the quality of higher education is put forward. In 2001, National Ministry of Education introduced "On the strengthening of Undergraduate Education in improving the quality of teaching some of the views", in which twelve measurements to enhance undergraduate teaching and improve teaching quality were brought forward to answer the new situation of higher education, and "bilingual teaching" was one of them. National Ministry of Education calls for actively promotion to develop bilingual teaching in colleges and universities, and exert much effort to open up 5% ~ 10% of the bilingual courses in three or five years.

Early of 2007, National Ministry of Education and Ministry of Finance jointly introduced the "Opinion on the implementation of quality of undergraduate education and teaching reform", encouraging colleges and universities to build 500 bilingual teaching model curriculums within five years. The complier of this book conforms to the trend of the development of higher education, carried through bilingual teaching of "Automatic control principle" in automation and electrician engineering and automation specialties, actively explored for the bilingual teaching models and methodologies adapting to ordinary colleges. This book is one of the fruits of their bilingual teaching practice.

The compliers take two years to accomplish this bilingual teaching material based on the long-term accumulation of teaching, bilingual teaching experience and the reference to the large number of original material. Two academic years' bilingual teaching practice indicates that this book is satisfactory for teaching demand of automation and relevant speciality.

The whole book is well-structured with concise content which not only considers the integrity of the classical control theory, but also strives to give prominence to key concepts and connects theory with practice. This book is guidebook-structured, explains the profound things in a simple way, and concisely expatiates basic concepts avoiding complicated mathematic derivation, that provides strong readability.

This book can be used as professional teaching materials for the specialties of automation, electrical engineering and automation and so on, as well as reference books for relevant technical staff.

Li Wenxiu
July, 2007

序

经济全球化进程的加快和知识经济时代的到来,对高等教育质量提出了更高的要求。教育部2001年出台了《关于加强高等学校本科教学工作提高教学质量的若干意见》,针对我国高等教育面对的新形势提出了十二条加强本科教学工作、提高教学质量的措施和意见,其中之一就是"开展双语教学"。教育部要求各高校积极推动使用英语等外语进行公共课和专业课的教学,力争在三年内,开出5%~10%的双语课程。

2007年初,教育部、财政部又联合出台了《关于实施高等学校本科教学质量与教学改革工程的意见》,提出在五年内要建设500门双语教学示范课程。编者顺应高等教育的发展潮流,对所在学校的自动化和电气工程及其自动化专业的"自动控制原理"课程进行了双语教学,积极探索适合普通高等院校(理工类专业)的双语教学模式和方法,本书正是编者开展双语教学研究的成果之一。

编者基于长期的教学积累和开展双语教学的心得体会,在参考了大量原版教材的基础上,编写了这本《自动控制原理》双语教材,并经过了两个学年的双语教学实践,教学效果表明该教材基本能够满足自动化及相关专业双语教学的需要。

全书内容精练,结构严谨,既考虑了经典控制理论体系的系统性和完整性,又力求做到重点突出,理论联系实际。全书层次分明、深入浅出,对基本概念的阐述简明扼要,避免了繁杂的数学推导,可读性较强,便于学生掌握经典控制理论的基本知识。

本书适用于工程与应用类自动化专业、电气工程及其自动化专业或相近专业的教学,也可作为相关专业工程技术人员的参考书。

李文秀

2007年7月

Foreword

Automatic control theory is widely applied in so many fields, such as industry, agriculture, shipbuilding and aviation, that greatly contributes to social development and economic construction. In recent years the compliers carried through bilingual teaching of "Automatic control principle" in the specialties of automation and electrical engineering and automation. This book is a product of our bilingual teaching practice and experience. The book is printed in a bilingual form to provide convenience for reader.

This book tries our best to explain the profound things in a simple way, apply rigorous theory and give prominence to system organization and give balance to engineering. We expect this book would play an active role in improving english application ability of reader, cultivating correction ideation and the ability to connect theory with practice of students.

This book can be used as a professional teaching material for the speciallies of automation, electricial engineering and automation and so on, as well as a reference book for graduated students of other relevant speciaties and teachers.

This book is edited by Li Daogen. The chief revisor is Li Wenxiu. Dr. Zhu Zhiyu compiles Chapter two, three, Dr. Liu Weiting compiles Chapter four, and Li Daogen writes the rest. We would especially like to thank Hao Peng, Wang Fang and Zong Yang for drawing the figures and tables in this book. And we would like to thank all the teachers and students for their help in the production and editing of this book.

It is inevitable there are some mistakes in this book because of our limited level, any criticism and correction will be appreciated.

Compiler Li Daogen
June, 2007

前　言

"自动控制理论"广泛应用于工业、农业、船舶、航空航天等诸多领域，在促进社会发展和经济建设中做出了重要贡献。近几年来，编者对自动化和电气工程及其自动化等专业的"自动控制原理"课程进行了双语教学尝试，本书就是编者在开展"双语"教学实践的基础上，结合多年讲授"自动控制原理"课程的心得和体会编写而成的。全书采用中英文对照的形式出版，以方便读者阅读和学习。

全书力求做到深入浅出，理论严谨，突出系统性，并兼顾工程性。本书对提高读者的英语应用能力，培养学生的辨正思维能力和理论联系实践的能力，都具有一定的作用。

本书可作为工科自动化、电气工程及其自动化专业教材，也可作为相关专业研究生和教师的参考书目。

本书由李道根主编，王建华主审。本书第1章、第5~7章由李道根编写，第2、3章由朱志宇编写，第4章由刘维亭编写。郝鹏、王芳、宗阳等绘制了书中的图表。在此，仅向参与和关心本书编写工作的各位教师和同学表示感谢。

由于作者水平有限，书中存在的错误和不妥之处，敬请读者批评指正。

编　者
2007年6月

Contents

Chapter 1 Introduction to Control Systems ·············· (1)

   1.1 Introduction ·············· (1)
   1.2 Open-and Closed-loop Control ·············· (4)

Chapter 2 Mathematic Models of Control Systems ·············· (8)

   2.1 Introduction ·············· (8)
   2.2 Differential Equation and Transfer Function ·············· (9)
   2.3 Linear Approximation of Nonlinear Systems ·············· (16)
   2.4 Block Diagram ·············· (19)
   2.5 Signal Flow Graph ·············· (26)
   2.6 Transfer Functions of Linear System ·············· (31)
   2.7 Impulse Response of Linear Systems ·············· (33)
   Problems ·············· (35)

Chapter 3 Time-Domain Analysis of Control Systems ·············· (40)

   3.1 Introduction ·············· (40)
   3.2 Time Response of First-Order system ·············· (45)
   3.3 Time Response of Second-Order System ·············· (49)
   3.4 Time Response of Higher-Order System ·············· (57)
   3.5 Stability of Linear Systems ·············· (63)
   3.6 Steady-State Error ·············· (72)
   3.7 Disturbance Rejection ·············· (81)
   Problems ·············· (86)

Chapter 4 Root Locus Method ·············· (92)

   4.1 Root Locus of Feedback System ·············· (92)
   4.2 Rules for Plotting Root Locus ·············· (94)
   4.3 Other Configuration of Root Locus ·············· (104)
   4.4 Application of the Root Locus Method ·············· (107)
   Problems ·············· (109)

Chapter 5 Frequency Response Method (111)

5.1 Introduction (111)
5.2 Bode Diagrams of Elementary Factors (115)
5.3 Open-Loop Frequency Response (121)
5.4 Nyquist Stability Criterion (130)
5.5 Relative Stability (140)
5.6 Closed-Loop Frequency-Domain Analysis (143)
5.7 Opened-Loop Frequency-Domain Analysis (149)
Problems (152)

Chapter 6 Compensation of Control System (156)

6.1 Introduction (156)
6.2 Phase-Lead Compensation (160)
6.3 Phase-Lag Compensation (164)
6.4 Phase Lag-Lead Compensation (168)
6.5 PID Controller (171)
6.6 Feedback Compensation (173)
Problems (177)

Chapter 7 Nonlinear System Analysis (181)

7.1 Introduction (181)
7.2 Describing Function Method (186)
7.3 Phase-plane Method (198)
Problems (207)

References (209)

目　录

第 1 章　控制系统概述 ·· (210)

　1.1　概　述 ·· (210)

　1.2　开环控制和闭环控制 ·· (212)

第 2 章　控制系统的数学模型 ·· (216)

　2.1　引　言 ·· (216)

　2.2　微分方程和传递函数 ·· (216)

　2.3　非线性系统的线性近似 ··· (223)

　2.4　方框图 ·· (225)

　2.5　信号流图 ··· (232)

　2.6　线性系统的传递函数 ·· (236)

　2.7　线性系统的脉冲响应 ·· (238)

　习　题 ··· (239)

第 3 章　控制系统的时域分析 ·· (244)

　3.1　引　言 ·· (244)

　3.2　一阶系统的时间响应 ·· (248)

　3.3　二阶系统的时间响应 ·· (252)

　3.4　高阶系统的时间响应 ·· (259)

　3.5　线性系统的稳定性 ··· (264)

　3.6　稳态误差 ··· (272)

　3.7　扰动的抑制 ··· (281)

　习　题 ··· (284)

第 4 章　根轨迹法 ·· (290)

　4.1　反馈系统的根轨迹 ··· (290)

　4.2　绘制根轨迹的法则 ··· (292)

　4.3　其他形式的根轨迹 ··· (301)

　4.4　根轨迹法的应用 ··· (304)

　习　题 ··· (306)

第5章 频率响应法 ……………………………………………… (308)

5.1 引 言 …………………………………………………… (308)
5.2 基本环节的伯德图 ……………………………………… (311)
5.3 开环频率响应 …………………………………………… (317)
5.4 奈奎斯特稳定性判据 …………………………………… (325)
5.5 相对稳定性 ……………………………………………… (333)
5.6 闭环频域分析 …………………………………………… (336)
5.7 开环频域分析 …………………………………………… (342)
习 题 ………………………………………………………… (344)

第6章 控制系统的校正 ……………………………………… (348)

6.1 引 言 …………………………………………………… (348)
6.2 相位超前校正 …………………………………………… (351)
6.3 相位滞后校正 …………………………………………… (355)
6.4 滞后-超前校正 ………………………………………… (358)
6.5 比例-积分-微分(PID)调节器 ………………………… (361)
6.6 反馈校正 ………………………………………………… (363)
习 题 ………………………………………………………… (367)

第7章 非线性系统分析 ……………………………………… (370)

7.1 引 言 …………………………………………………… (370)
7.2 描述函数法 ……………………………………………… (374)
7.3 相平面法 ………………………………………………… (385)
习 题 ………………………………………………………… (393)

参考文献 ……………………………………………………… (396)

Chapter 1　Introduction to Control System

1.1　Introduction

1.1.1　Control Engineering and Automation

As an application of scientific and mathematical principles to the design, manufacture, and operation of systems, such as machines, plants, processes, and etc., control engineering is concerned with understanding and controlling the material and forces of nature for the benefit of humankind. Control system engineers are concerned with understanding and controlling systems to provide useful economic products for society. Control engineering is based on the foundations of feedback theory and linear system analysis, and it integrates the concepts of network theory and communication theory. Therefore, control engineering is not limited to any engineering discipline but is equally applicable to aeronautical, chemical, mechanical, environmental, and electrical engineering. For example, quite often, a control system includes electrical, mechanical, and chemical components. Furthermore, as the understanding of the dynamics of business, social, and political systems increases, the ability to control these systems will increase also.

The control of a plant or process by automatic rather than manual means is called automation; or we can say that automation is an automatic technology of machines, plants, processes, and etc. Automation is used to improve productivity and obtain high-quality products.

1.1.2　History of Automatic Control

The simplest way to automate the control of a plant or process is through conventional feedback control. The use of feedback to control a system has had a long history. Historically, a key step forward in the development of control occurred during the industrial revolution. At that time, machines were developed that greatly enhanced the capacity to turn raw materials into products of benefit to society. The associated machines, especially steam engines, involved large amounts of power, and it was soon realized that this power needed to be controlled in an organized fashion if the systems were to operate safely and efficiently. A major development at that time was Watt's fly-ball governor. This device regulates the speed of a steam engine by throttling the flow of steam.

The World Wars also led to many developments in control engineering. Some of these were associated with guidance systems and anti-aircraft systems while others were connected with the enhanced manufacturing requirements necessitated by the war effort.

The push into space in the 1960's and 1970's also depended on control development. These developments then flowed back into consumer goods, as well as into commercial, environmental, and medical applications.

By the end of the twentieth century, control has become a ubiquitous (but largely unseen) element of modern society. Virtually every system we come in contact with is underpinned by sophisticated control systems. Examples range from simple household products (temperature regulation in air-conditioners, thermostats in hot-water heaters, etc.), to large-scale systems (such as chemical plants, aircraft, and manufacturing processes).

Beyond these industrial examples, feedback regulatory mechanisms are central to the operation of biological systems, communication networks, national economies, and even human interactions. Indeed, if one thinks carefully, control in one form or another can be found in every aspect of life.

Thus, control engineering is an exciting multidisciplinary subject with an enormously large range of practical applications. Moreover, interest in control is unlikely to diminish in the foreseeable future. On the contrary, it is likely to become ever more important, because of the increasing globalization of markets and environmental concerns.

Market globalization is increasingly occurring, and this situation means that manufacturing industries are necessarily placing increasing emphasis on issues of quality and efficiency. This focuses attention on the development of improved control systems, so that the plants and processes operate in the best possible way. In particular, improved control is a key enabling technology underpinning.

Again, control engineering is a core enabling technology in reaching the goals of respecting finite natural resources and preserving our fragile environment.

1.1.3 Historical Periods of Control Theory

We have seen above that control engineering has taken several major steps forward at crucial events in history. Each of these steps has been marched by a corresponding burst of development in the underlying theory of control.

Early on, when the concept of feedback was applied, engineers sometimes encountered unexpected results. These then became catalysts for rigorous analysis.

The developments around the period of the Second World War were also marched by significant developments in control theory. This resulted in simple graphical means for analyzing single-variable feedback control problems. These methods are now generally known by the generic term classical control theory.

The 1960's saw the development of an alternative state-space approach to control. This followed the publication of work by Wiener, Kalman (and others) on optimal estimation and control. This work allowed multivariable problem to be treated in a unified fashion. This had been difficult, if not impossible, in the classical framework. This set of developments is termed modern control theory.

The theory of automatic control is a large, exciting, and extremely useful engineering discipline. One can readily understand the motivation for the study of automatic control theory.

1.1.4 Future Evolution of Control Systems

The continuing goal of control systems is to provide extensive flexibility and high level of autonomy. Today's industrial robot is perceived as quite autonomous: once it is programmed, further intervention is not normally required. Because of sensory limitations, these robotic systems have limited flexibility in adapting to work environment changes, which is the motivation of computer vision research. The control system is very adaptable, but it relies on human supervision. Advanced robotic systems are striving for task adaptability through enhanced sensory feedback. Research areas concentrating on artificial intelligence, sensor integration, computer vision, and off-line CAD/CAM programming will make systems more universal and economical. Control systems are moving toward autonomous operation as an enhancement to human control. Research in supervisory control, human-machine interface methods to reduce operator burden, and computer database management is intended to improve operator efficiency. Many research activities are common to robotics and control systems and are aimed toward to reducing implementation cost and expanding the realm of application. These include improved communications methods and advanced programming languages.

1.1.5 Control System

A control system is an organized collection of interacting units designed to achieve some specified objectives by manipulation and control of materials, energy, and information. In studying automatic control theory we are concerned with the signal flow within a system rather than the material flow and energy flow.

A control system is dynamic in nature and there exists cause-and-effect relationship among the components of a system. Hence, a system and its components can be represented graphically with a so-called block diagram, as shown in Fig. 1.1. Within the block there is the name, function, or mathematic model of the corresponding system or its component; the lines with arrow indicate the direction of

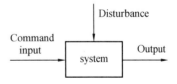

Figure 1.1 Block representing a system

signal flow. In a control system, the plant or process to be controlled is called the controlled object, and the quantity to be chosen to characterize the system behavior is called the controlled variable, or output for short. Corresponding to the controlled variable, a desired value or command input is established. The input-output relationship represents a cause-and-effect relationship of the system. A disturbance is an external action (other than the command input) that tends to drive the controlled variable away from its desired value. The disturbance-output relationship is another cause-and-effect relationship of the system.

1.2 Open-and Closed-Loop Control

Usually, an automatic control system consists of the controlled object and its controller. Based on how the control action is generated, i. e. , whether the generation of control action is depended on the actual output, the control systems may be classified as open-loop control systems and closed-loop control systems.

1.2.1 Open-Loop Control

An open-loop control system is a system without feedback, and in which the signal flow from input to the output is unidirectional. In the case of open-loop control, the generation of control action is independent of the actual output, i. e. , the control action is depended only on the command input and/or disturbance.

1. Manipulation according to command input

One mode of open-loop control is the manipulation according to command input, as shown in Fig. 1. 2. The controller accepts the command input and manipulates the plant to obtain a desired output. The desired value of the output may be changed and then the input will need to be changed to adjust the plant operation.

Figure 1.2　Functional block diagram of manipulation according to input

An example is a turntable used to rotate a disk at a constant speed, as shown in Fig. 1. 3. This system uses a battery source to provide a reference voltage that is proportional to the desired speed. This voltage is amplified and applied to the dc motor, which provides a speed proportional to the applied voltage. In this

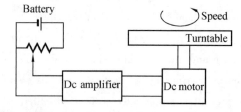

Figure 1.3　Open-loop control of turntable speed

system, a variation of the speed from the desired value, due to some reason, can in no way cause a change of voltage applied on the dc motor to maintain the desired speed. In this case, it can also be said that the output has no influence on the input.

Another example of manipulation according to command input is a generator-load system, as shown in Fig. 1. 4. The voltage provided by the generator is proportional to the exciting voltage of exciter, which can be adjusted with a potentiometer.

Obviously, the voltage across the load will be affected by the disturbance, for example the fluctuation of load or the parameter variation of certain components, the actual output will be away from the desired value and cannot return to the original condition by itself.

2. Compensation according to disturbance

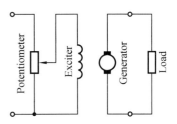

Figure 1.4   Open-loop control of a generator-load system

Another mode of open-loop control is the compensation according to disturbance, as shown in Fig. 1.5. In this case, the controller accepts the disturbance signal, if it is measurable, and generates an additional control action to compensate the effect of disturbance on the system.

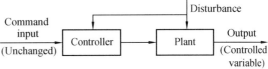

Figure 1.5   Compensation according to disturbance

An example of this open-loop control fashion is the generator-load system, as shown in Fig. 1.6, but a part of the exciter winding is connected in the generator-load loop. Now, if the load voltage is decreased due to a increment of load, the current flowing through the load and exciting winding will be increased, which will result in increasing the terminal voltage of generator.

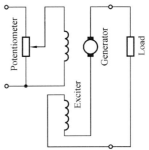

Although an open-loop system is simpler and less expensive to construct, it requires detailed knowledge of each component in order to determine the input value for a required output. Moreover, the variation of system parameters and/or the external disturbance may have bad influences on the control accuracy.

Figure 1.6   A generator-load system with compensation in terms of disturbance

1.2.2   Closed-Loop Control

In contrast to an open-loop control system, a closed-loop control system utilizes an additional measure of the actual output to compare with the desired value of output, as shown in Fig. 1.7. The measure of the actual output is called feedback signal. A feedback control system is a control system that tends to maintain a prescribed relationship of one system variable to another by comparing functions of these variables and using the difference as a means of control.

Fig. 1.8 shows a closed-loop control of the speed of a turntable, where a tachometer is used to measure the actual speed and output a voltage proportional to the speed. This system uses a function of the output and input to control the plant. The difference between the input and feedback signal is used to control the system so that the difference is continuously reduced. The feedback concept has been the foundation for control system analysis and design.

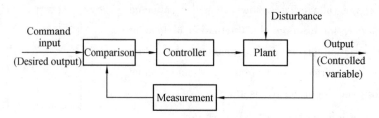

Figure 1.7 Function block diagram of a closed-loop control system

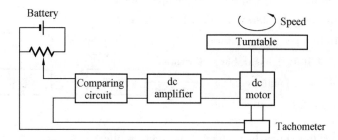

Figure 1.8 Closed-loop control of the speed of a turntable

One of the noteworthy features of the closed-loop system is its ability faithfully to reproduce the command input owing to feedback. Since the system is adjusted based on the deviation of the output from the input, the control action persists in generating sufficient additional output to bring the two into correspondence. Unfortunately, feedback is also responsible for the tendency to oscillate in the system. Another important feature of closed-loop system is that, in direct contrast to the open-loop system, it usually performs accurately even in the presence of nonlinearities.

1.2.3 General Functional Structure of Feedback System

Every feedback control system consists of components that perform specific functions. A convenient and useful method of representing this functional characteristic of a control system is the block diagram. Basically this is a means of representing the operations performed in the system and the manner in which signal flows throughout the system. The block diagram is concerned not with the physical characteristics of any specific system but only with the functional relationship among various parts in the system. The general functional layout and structure of a feedback control system is shown in Fig. 1.9.

We can draw the functional block diagram of a control system from its operational schematic diagram as follows. At first, we can determine the controlled object, controlled variable, and possible disturbances according to the task of the system to be completed. Then, each functional block can be determined according to the signal transfer sequence.

Since the operation of a feedback control system is based on the deviation, i.e., the actuating signal just is the difference between the actual output and desired output, there must be a comparing element, i.e. a comparator, in every feedback control system.

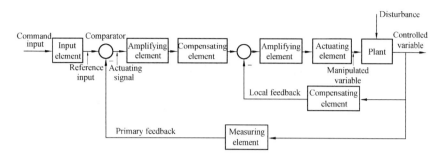

Figure 1.9 Functional layout of a feedback loop

The measuring element measures the actual controlled variable, directly or indirectly, and provides the comparator a feedback signal. The input element provides a functional conversion of the command input into a suitable mode for comparing with the feedback signal and results in reference input of the system.

The amplifying element is used to amplify the deviation into a power signal and, if it is necessary, to transform the deviation into a suitable physical form.

Finally, the actuating elements executes the controlling action to manipulate the plant.

Besides above functional blocks, almost all control systems need the compensating element to improve the system performances.

1.2.4 Performance Requirements on Control System

A suitable control system should have some of the following properties:

① It should be stable and present acceptable response to input command, i. e., the controlled variable should follow the changes in the reference input at a suitable speed without unduly large oscillations or overshoots.

② It should operate with as little error as possible.

③ It should be able to mitigate the effect of undesirable disturbances.

Chapter 2 Mathematic Models of Control Systems

2.1 Introduction

One of the most important steps in the analysis of control systems is the mathematical description and modeling of the systems. The term mathematical model means the mathematical representations describing the relationships among the variables of a system. In classical control theory, it usually means the mathematical relationships relating the output of a system, or a component, to its input. A mathematical model of a control system is essential because it allows us to gain a clear understanding of the system in terms of cause-and-effect relationship among the system components.

Because the systems under consideration are dynamic in nature, the descriptive equations are usually differential equations. Furthermore, if these equations are linear or can be linearized, then the Laplace transform can be utilized to simplify the method of solution and transfer functions become valuable tools for analysis as well as for design. In practice, the complexity of systems and our ignorance of all the relevant factors require the introduction of assumptions concerning the system operation. Therefore, we will often find it useful to consider the physical system, introduce some necessary assumptions, and linearize the system. Then, by using the physical laws describing the linear equivalent system, we can obtain a set of linear differential equations. Finally, utilizing the mathematical tools, such as the Laplace transform, we obtain a solution describing the operation of the system.

In general, a physical system also can be represented by a schematic diagram, which portrays the relationship and interconnections among the system components. In control system theory, the block diagram is often used to portray systems of all types.

In this chapter we develop models for some common physical systems. It should be pointed that this chapter serves only as an introduction to system modeling and is not intended to be a complete treatise on the subject.

It should be understood that no mathematical model of a physical system is absolutely exact. We may be able to increase the accuracy of a model by increasing the complexity of the equations, but we never achieve exactness. We generally strive to develop a model that is adequate for the problem at hand without making the model overly complex.

2.2 Differential Equation and Transfer Function

2.2.1 Introduction

From the mathematical standpoint, differential equations can be used to describe the dynamic behavior of a system. The components of a control system are diverse in nature and may include electrical, mechanical, thermal, and fluidic devices and etc. The differential equations for these devices are obtained using the basic laws of physics. These include balancing forces, energy and mass.

Consider that a linear time-invariant continuous system is described by the following n th-order differential equation

$$a_0 \frac{d^n c(t)}{dt^n} + a_1 \frac{d^{n-1} c(t)}{dt^{n-1}} + \cdots + a_{n-1} \frac{dc(t)}{dt} + a_n c(t) =$$
$$b_0 \frac{d^m r(t)}{dt^m} + b_1 \frac{d^{m-1} r(t)}{dt^{m-1}} + \cdots + b_{m-1} \frac{dr(t)}{dt} + b_m r(t) \tag{2.1}$$

where $c(t)$ is the output variable and $r(t)$ is the input variable, the coefficients $a_0, a_1, \ldots, a_n$ and $b_0, b_1, \ldots, b_m$ are constants, and $n \geq m$. This differential equation represents a complete description of the system between the input $r(t)$ and the output $c(t)$. Once the input and the initial conditions of the system are specified, the output response may be obtained by solving the equation. However, it is apparent that the differential equation method of describing a system is, although essential, a rather cumbersome one, and the higher-order differential equation is of little practical use in design. More important is the fact that although efficient software is available on digital computer for solution of a high-order differential equation, the important development in linear control theory relies on analysis and design techniques without actual solutions of the system differential equations.

A convenient way of describing linear system is made possible by the use of transfer function. To obtain the transfer function of the linear system represented by Eq. (2.1), we take the Laplace transform on both sides of the equation, assuming zero initial conditions, then we have

$$(a_0 s^n + a_1 s^{n-1} + \ldots + a_{n-1} s + a_n) C(s) =$$
$$(b_0 s^m + b_1 s^{m-1} + \ldots + b_{m-1} s + b_m) R(s) \tag{2.2}$$

The transfer function of the system is defined as the ratio of $C(s)$ to $R(s)$, i.e.,

$$\frac{C(s)}{R(s)} = \frac{b_0 s^m + b_1 s^{m-1} + \ldots + b_{m-1} s + b_m}{a_0 s^n + a_1 s^{n-1} + \ldots + a_{n-1} s + a_n} \tag{2.3}$$

Summarizing over the properties of transfer function we have:

① A transfer function is defined only for a linear time-invariant system.

② A transfer function between an input and an output of a system is defined as the ratio of the Laplace transform of the output to the Laplace transform of the input. Meanwhile, all initial conditions of the system are assumed to be zero.

③ A transfer function is independent of input excitation.

Transfer function plays an important role in the characterization of linear time-invariant systems. Together with block diagram, transfer function forms the basis of representing the input-output relationships of a linear time-invariant system in the classical control theory.

Now the mathematical models of some simple physical systems are developed by using the relative physical laws.

2.2.2 Simple Networks

1. Electrical circuit

The classical way of writing differential equations of an electrical network is the loop method and node method, which are formulated from the two laws of Kirchhoff.

Example 2.1 As the first example, consider the simple electrical circuit in Fig. 2.1, where an input voltage $v_i(t)$ is applied to a RC circuit.

According to Kirchhoff's voltage law, the output voltage $v_o(t)$ is related to the input through the differential equation

$$v_i(t) = RC \frac{dv_o(t)}{dt} + v_o(t)$$

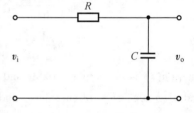

Figure 2.1 RC circuit

which can be rearranged in a normalized form

$$RC \frac{dv_o(t)}{dt} + v_o(t) = v_i(t) \tag{2.4}$$

Taking Laplace transform, assuming zero initial condition, we have

$$(RCs+1) V_o(s) = V_i(s)$$

From above equation we get the transfer function

$$\frac{V_o(s)}{V_i(s)} = \frac{1}{RCs+1} \tag{2.5}$$

Example 2.2 As another example, consider the circuit in Fig. 2.2, where $v_i(t)$ is the input voltage and $v_o(t)$ is the output voltage.

By Kirchhoff's voltage law, assuming zero initial conditions, we get

$$v_i(t) = Ri(t) + L \frac{di(t)}{dt} + v_o(t)$$

$$v_o(t) = \frac{1}{C} \int i(t) \, dt$$

where loop current i is an intermediate variable. Eliminating the variable i results in

$$LC \frac{d^2 v_o(t)}{dt^2} + RC \frac{dv_o(t)}{dt} + v_o(t) = v_i(t) \tag{2.6}$$

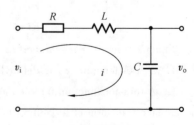

Figure 2.2 RLC circuit

Taking Laplace transform of this equation, we get the transfer function

$$\frac{V_o(s)}{V_i(s)} = \frac{1}{LCs^2 + RCs + 1} \tag{2.7}$$

2. Mechanical systems

(1) Translational system

The motion of translation is defined as a motion that takes place along a straight line. The variables that are used to describe translational motion are force, displacement, velocity, and acceleration. The elements of a translational mechanical system may be mass, damper, and spring. The symbols for these elements are given in Fig. 2.3.

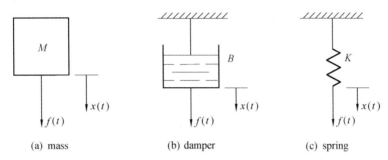

(a) mass (b) damper (c) spring

Figure 2.3 Basic elements of translational movement

In Fig. 2.3(a), $f(t)$ represents the applied force, $x(t)$ represents the displacement, and M represents the mass. Then, by Newton's second law, we have

$$f(t) = M \frac{d^2 x(t)}{dt^2} \tag{2.8}$$

It is assumed that the mass is rigid.

For the remaining two elements, the top connection point can move relative to the bottom connection point. First consider damper shown in Fig. 2.3(b). A physical realization of damping phenomenon is the viscous friction associated with oil, air and so on. A physical device is shock absorber on an automobile. The mechanical model of damper is given by

$$f(t) = B \frac{dx(t)}{dt} \tag{2.9}$$

where B is the viscous friction coefficient.

The defining equation of spring shown in Fig. 2.3(c), from Hooke's law, is given by

$$f(t) = K x(t) \tag{2.10}$$

where K is the spring constant.

Example 2.3 Fig. 2.4 shows a simple spring-mass-damper mechanical system, where an external force $f(t)$ is applied and the motion is indicated by $x(t)$.

In this example we model the wall viscous friction as a

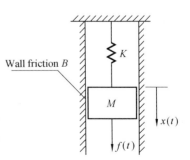

Figure 2.4 Spring-mass-damper system

damper, i. e. , the friction force is linearly proportional to the velocity of the mass. Using Newton's second law yields

$$f(t) - K x(t) - B \frac{dx(t)}{dt} = M \frac{d^2 x(t)}{dt^2}$$

i. e. ,

$$M \frac{d^2 x(t)}{dt^2} + B \frac{dx}{dt} + K x(t) = f(t) \tag{2.11}$$

Taking Laplace transforms, assuming zero initial conditions, we get

$$(Ms^2 + Bs + K) X(s) = F(s)$$

The result in the transfer function is

$$\frac{X(s)}{F(s)} = \frac{1}{Ms^2 + Bs + K} \tag{2.12}$$

(2) Rotational system

The rotational motion of a body may be defined as motion about a fixed axis, and the variables generally used to describe the motion of rotation are torque, angular displacement, angular velocity and angular acceleration. The linear rotational systems are analogues to the linear translational systems (equations are of the same form), and the same procedure used for writing the equations for the linear translational systems may be used for the rotational systems.

The three elements of the linear rotational systems are depicted in Fig. 2.5.

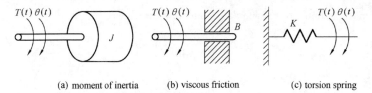

(a) moment of inertia (b) viscous friction (c) torsion spring

Figure 2.5 Basic elements of rotational movement

The first element in Fig. 2.5(a), moment of inertia, is defined by the relationship

$$T(t) = J \frac{d^2 \theta(t)}{dt^2} \tag{2.13}$$

where $T(t)$ is the applied torque, J is the moment of inertia, and $\theta(t)$ is the angle of rotation. This equation is analogous to that for the mass in translational system.

The defining equation for the second element in Fig. 2.5(b), viscous friction, is

$$T(t) = B \frac{d\theta(t)}{dt} \tag{2.14}$$

where B is the damping coefficient. As for the torsion spring shown in Fig. 2.5(c) the defining equation is

$$T(t) = K \theta(t) \tag{2.15}$$

where K is the torsion spring coefficient.

Example 2.4 Consider the torsional pendulum in Fig. 2.6. The moment inertia of the pendulum bob is represented by J, the friction between the bob and air by B, and the elastance of the suspension strip by K. It is assumed that the torque is applied at the bob.

Summing torques at the pendulum bob, we have

$$J\frac{d^2\theta(t)}{dt^2}+B\frac{d\theta(t)}{dt}+K\theta(t)=T(t) \quad (2.16)$$

The transfer function is then derived easily

$$\frac{\theta(s)}{T(s)}=\frac{1}{Js^2+Bs+K} \quad (2.17)$$

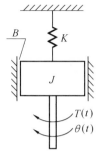

Figure 2.6 Torsional pendulum

(3) Analogous systems

We can easily note the equivalence of Eq. (2.6), (2.11) and (2.16), where voltage $v_i(t)$, force $f(t)$ and torque $T(t)$ are equivalent variables, usually called the analogous variables, and the systems are analogous systems.

Analogous systems with similar solutions exist for electrical, mechanical, thermal, and fluid systems. The existence of analogous systems provides the analyst with the ability to extend the solution of one system to all analogous systems with the same describing differential equations. Therefore what one learns about the analysis and design of electrical systems is immediately extended to an understanding of mechanical, thermal, and fluid systems.

2.2.3 Gear Train

A gear train is a mechanical device that transmits energy from one part of a system to another in such a way that force, torque, speed, and displacement are altered. These devices may also be regarded as matching devices used to attain maximum power transfer.

An equivalent representation of a gear train is shown in Fig. 2.7, where $T(t)$ is the applied torque, $\theta_1(t)$ and $\theta_2(t)$ are the angular

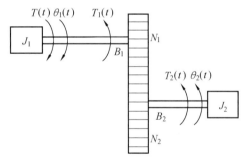

Figure 2.7 Gear train

displacements, $T_1(t)$ and $T_2(t)$ are the torques transmitted to gears, J_1 and J_2 are the moments of inertial of gears, N_1 and N_2 are the numbers of teeth, B_1 and B_2 are the viscous frictional coefficients.

The torque equation for the driven gear is written

$$T_2(t)=J_2\frac{d^2\theta_2(t)}{dt^2}+B_2\frac{d\theta_2(t)}{dt} \quad (2.18)$$

The torque equation on the side of driving gear is

$$T(t)=J_1\frac{d^2\theta_1(t)}{dt^2}+B_1\frac{d\theta_1(t)}{dt}+T_1(t) \quad (2.19)$$

Since the relation between the torques T_1 and T_2 is given by

$$\frac{T_1}{T_2} = \frac{N_1}{N_2} \tag{2.20}$$

and the relation between angular speeds ω_1 and ω_2 is

$$\frac{\omega_1}{\omega_2} = \frac{d\theta_1/dt}{d\theta_2/dt} = \frac{N_2}{N_1} \tag{2.21}$$

hence we have

$$T_1(t) = \frac{N_1}{N_2} T_2(t) = \left(\frac{N_1}{N_2}\right)^2 J_2 \frac{d^2\theta_1(t)}{dt^2} + \left(\frac{N_1}{N_2}\right)^2 B_2 \frac{d\theta_1(t)}{dt} \tag{2.22}$$

This equation indicates that it is possible to reflect the moment of inertial, torque, speed, and displacement from one side of a gear train to the other. Substituting Eq. (2.22) into (2.19) yields

$$T(t) = J_{1e} \frac{d^2\theta_1(t)}{dt^2} + B_{1e} \frac{d\theta_1(t)}{dt} \tag{2.23}$$

where $J_{1e} = J_1 + \left(\frac{N_1}{N_2}\right)^2 J_2$, $B_{1e} = B_1 + \left(\frac{N_1}{N_2}\right)^2 B_2$.

2.2.4 Armature-Controlled dc Motor

The dc motor is a power actuator device that converts dc electrical energy into rotational energy to delivers energy to a load. Fig. 2.8 shows a schematic diagram of an armature-controlled dc motor. In this case armature voltage $v_a(t)$ is the system input while the motor shaft angle $\theta_m(t)$ is considered to be the output. The resistance and inductance of the armature coil are R_a and L_a respectively.

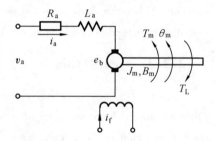

Figure 2.8 Armature-controlled dc motor

The armature current is related to the input voltage applied to the armature as

$$v_a(t) = L_a \frac{di_a(t)}{dt} + R_a i_a(t) + e_b(t) \tag{2.24}$$

where $e_b(t)$ is the back electromotive force proportional to the motor speed $\omega_m(t)$. Hence we have

$$e_b(t) = C_e \omega_m(t) = C_e \frac{d\theta_m(t)}{dt} \tag{2.25}$$

where C_e is the motor back emf (**e**lectro **m**otive **f**orce) constant.

The electromagnetic torque produced by the motor is given by

$$T_m(t) = C_m i_a(t) \tag{2.26}$$

where C_m is the motor torque constant. From summing torque on the motor armature it is derived

$$T_m(t) = J_m \frac{d^2\theta_m(t)}{dt^2} + B_m \frac{d\theta_m(t)}{dt} + T_L(t) \qquad (2.27)$$

where J_m is the moment inertia of the motor, B_m is the damping coefficient due to friction, and T_L is the load torque.

From above equations we have

$$J_m L_a \frac{d^3\theta_m}{dt^3} + (B_m L_a + J_m R_a) \frac{d^2\theta_m}{dt^2} + (B_m R_a + C_e C_m) \frac{d\theta_m}{dt} =$$

$$C_m v_a - L_a \frac{dT_L}{dt} - R_a T_L \qquad (2.28)$$

If the armature inductance L_a is negligible, then Eq. (2.28) becomes

$$J_m R_a \frac{d^2\theta_m}{dt^2} + (B_m R_a + C_e C_m) \frac{d\theta_m}{dt} = C_m v_a - R_a T_L \qquad (2.29)$$

Taking Laplace transform, we get the transfer functions

$$\frac{\theta_m(s)}{V_a(s)} = \frac{C_m}{s(J_m R_a s + B_m R_a + C_e C_m)} = \frac{K_m}{s(T_m s + 1)} \qquad (2.30)$$

$$\frac{\theta_m(s)}{T_L(s)} = -\frac{R_a}{s(J_m R_a s + B_m R_a + C_e C_m)} = -\frac{K_L}{s(T_m s + 1)} \qquad (2.31)$$

where $T_m = \frac{JR_a}{fR_a + C_e C_m}$, $K_m = \frac{C_m}{fR_a + C_e C_m}$ and $K_L = \frac{R_a}{fR_a + C_e C_m}$.

2.2.5 Position Control System

Example 2.5 A dc position-control system or servomechanism is shown in Fig. 2.9, where the position of the output shaft is required to follow that of the input shaft. An error voltage v_s, proportional to the difference between the input shaft position θ_i and the output shaft position θ_o, is applied to the amplifier. The output of the amplifier is connected to the armature of a dc servomotor as shown. The shaft of the motor is connected to the output shaft through a reduction gear of ratio $i = Z_2/Z_1$.

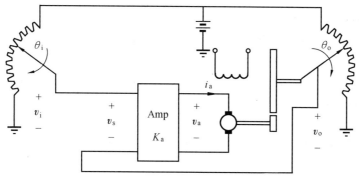

Figure 2.9　dc position-control system

The following relations for error detector and amplifier are easily derived

$$\begin{aligned}v_s(t) &= v_i(t) - v_o(t) \\ v_i(t) &= K_1 \theta_i(t) \\ v_o(t) &= K_1 \theta_o(t) \\ v_a(t) &= K_a v_s(t)\end{aligned} \qquad (2.32)$$

where K_1 is the constant of the transducer (the potentiometer in this case) and K_a is the voltage amplifying coefficient of amplifier.

Assuming the load torque be zero, the relation between the armature voltage v_a and the angular displacement θ_m of output shaft is given by

$$JR_a \frac{d^2 \theta_m}{dt^2} + (BR_a + C_e C_m) \frac{d\theta_m}{dt} = C_m v_a \qquad (2.33)$$

where J is the effective inertia and B is the effective damping. As for the reduction gear, we have

$$\theta_o(t) = \frac{\theta_m(t)}{i} \qquad (2.34)$$

Finally we have

$$J \frac{d^2 \theta_o(t)}{dt^2} + F \frac{d\theta_o(t)}{dt} + K \theta_o(t) = K \theta_i(t) \qquad (2.35)$$

where $F = \dfrac{BR_a + C_m C_e}{i R_a}$ and $K = \dfrac{C_m K_a K_1}{i R_a}$. Taking Laplace transform yields

$$\frac{\theta_o(s)}{\theta_i(s)} = \frac{K}{Js^2 + Fs + K} \qquad (2.36)$$

2.3 Linear Approximation of Nonlinear Systems

It is useful for us to model physical systems with linear time-invariant differential equations whenever possible. In this chapter we've presented linear models of several types of physical systems.

A great majority of physical systems are linear within some range of the variables. For example, the spring-mass-damper system of Example 2.3 is linear and described by a linear differential equation as long as the mass is subjected to small deflections $x(t)$. However, if $x(t)$ were continually increased, eventually the spring would be overextended and break. Therefore the question of linearity and the range of applicability must be considered for each system.

A system characterized by the relation $y = x^2$ is not linear, because the superposition property is not satisfied. A system represented by the relation $y = ax + b$ is not linear, because it does not satisfy the homogeneity property. However, the second system may be considered linear about an operating point (x_0, y_0) for small changes Δx and Δy. When $x = x_0 + \Delta x$ and $y = y_0 + \Delta y$, we have

$$y_0+\Delta y=a(x_0+\Delta x)+b \qquad (2.37)$$

and therefore

$$\Delta y=a\Delta x \qquad (2.38)$$

which satisfies the properties of superposition and homogeneity for the linear systems.

Strictly speaking, all physical systems are inherently nonlinear. Thus, when we use a linear model of a physical system, we are employing some form of linearization. The linearity of many mechanical and electrical systems can be assumed over a reasonable range of the variables. This is not usually the case for thermal and fluid elements, which are more frequently nonlinear in character. Fortunately, one can often linearize nonlinear elements assuming small-signal condition. This is the normal approach used to obtain a linear equivalent circuit for electric circuits and transistors. In many cases, the linearized models yield accurate results, i.e., the linear models accurately model the physical systems.

In other cases the linearized model is a very poor approximation to the physical system and cannot be used with any confidence. In these cases other techniques of analysis should be employed. For some systems, no valid analysis techniques have been found, and the system characteristics must be determined through simulation.

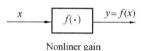

Nonliner gain

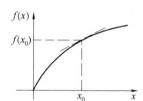

Figure 2.10   Nonlinear gain characteristic

In order to introduce linearization, we consider the nonlinear gain characteristic of Fig. 2.10, which has an input of x and an output of $y=f(x)$. It is assumed that $y=f(x)$ is a smooth function over the range of interest and the operating point on the gain curve is at the input value x_0, as shown.

Suppose that a small change Δx occurs in x. Now the input to the nonlinearity is $x=x_0+\Delta x$. The Taylor series expansion of $f(x)$ about the operating point (x_0, y_0) is

$$y=f(x_0)+\frac{df(x)}{dx}\bigg|_{x=x_0}(x-x_0)+\frac{1}{2!}\frac{d^2f(x)}{dx^2}\bigg|_{x=x_0}(x-x_0)^2+\ldots \qquad (2.39)$$

This expansion can be rewritten, with $\Delta x=x-x_0$, as

$$y-y_0=\Delta y=\frac{df(x)}{dx}\bigg|_{x=x_0}\cdot \Delta x+\frac{1}{2!}\frac{d^2f(x)}{dx^2}\bigg|_{x=x_0}\cdot(\Delta x)^2+\ldots \qquad (2.40)$$

The tangent slope at the operating point is

$$\frac{df(x)}{dx}\bigg|_{x=x_0}$$

which is a good approximation to the curve over a small range of $(x-x_0)$, the deviation from the operating point. Then, as a reasonable approximation, neglecting the higher-order derivative terms in the expansion of $f(x)$, Eq. (2.40) becomes

$$\Delta y\approx \frac{df(x)}{dx}\bigg|_{x=x_0}\cdot \Delta x \qquad (2.41)$$

For convenience, Eq. (2.41) is often rewritten as

$$y = K x \tag{2.42}$$

where

$$K = \frac{\mathrm{d} f(x)}{\mathrm{d} x}\bigg|_{x=x_0}$$

It should be noted, however, both variables x and y in this equation are the increments.

The relation between Δy and Δx is linear, since the derivative of a function evaluated at a point is a constant. The nonlinear gain of Fig. 2.10 in the vicinity of the operating point (x_0, y_0) is linearized. The accuracy of Eq. (2.41) is a function of the magnitude of Δx and the smoothness of $y = f(x)$ in the vicinity of the operating point (x_0, y_0).

For the smooth nonlinear function with two inputs, $y = f(x_1, x_2)$, the linearization has a similar procedure. The Taylor series expansion of $y = f(x_1, x_2)$ about the operating point $y_0 = f(x_{10}, x_{20})$ is

$$y = f(x_{10}, x_{20}) + \left[\frac{\partial f}{\partial x_1}\bigg|_{\substack{x_1=x_{10}\\x_2=x_{20}}} \cdot (x_1 - x_{10}) + \frac{\partial f}{\partial x_2}\bigg|_{\substack{x_1=x_{10}\\x_2=x_{20}}} \cdot (x_2 - x_{20})\right] +$$

$$\frac{1}{2!}\left[\frac{\partial^2 f}{\partial x_1^2}\bigg|_{\substack{x_1=x_{10}\\x_2=x_{20}}} \cdot (x - x_{10})^2 + 2\frac{\partial^2 f}{\partial x_1 x_2}\bigg|_{\substack{x_1=x_{10}\\x_2=x_{20}}} \cdot (x - x_{10})(x - x_{20}) + \frac{\partial^2 f}{\partial x_2^2}\bigg|_{\substack{x_1=x_{10}\\x_2=x_{20}}} \cdot (x - x_{20})^2\right] + \ldots \tag{2.43}$$

Neglecting the higher-order derivative terms in the expansion of $y = f(x_1, x_2)$ yields

$$y = f(x_{10}, x_{20}) + \left[\frac{\partial f}{\partial x_1}\bigg|_{\substack{x_1=x_{10}\\x_2=x_{20}}}(x_1 - x_{10}) + \frac{\partial f}{\partial x_2}\bigg|_{\substack{x_1=x_{10}\\x_2=x_{20}}}(x_2 - x_{20})\right] \tag{2.44}$$

that is

$$y - y_0 = K_1(x_1 - x_{10}) + K_2(x_2 - x_{20}) \tag{2.45}$$

or for brief

$$y = K_1 x_1 = K_2 x_2 \tag{2.46}$$

where

$$K_1 = \frac{\partial f}{\partial x_1}\bigg|_{\substack{x_1=x_{10}\\x_2=x_{20}}}, \quad K_2 = \frac{\partial f}{\partial x_2}\bigg|_{\substack{x_1=x_{10}\\x_2=x_{20}}}$$

Example 2.6 The characteristic curve of a SCR three-phase bridge rectifier is shown in Fig. 2.11, where the input is the control angle α, the output is the rectified voltage E_d, and the normal operating point is A, i.e. $(E_d)_0 = E_{d0}\cos\alpha_0$, where E_{d0} is the rectified voltage as $\alpha = 0$. Determine the linearized mathematical model of this SCR rectifier.

Solution The relation between E_d and α is given by

$$E_d = 2.34 E_2 \cos\alpha = E_{d0}\cos\alpha \tag{2.47}$$

where E_2 is the effective value of ac phase voltage.

If the control angle α is varied in a small range,

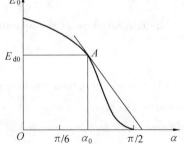

Figure 2.11 Characteristic of a SCR rectifier

this nonlinear characteristic can be linearized. From Eq. (2.47) we have

$$E_d - E_{d0} \cos \alpha_0 = K_s(\alpha - \alpha_0) \tag{2.48}$$

or

$$\Delta E_d = K_s \Delta \alpha \tag{2.49}$$

where $K_s = \dfrac{dE_d}{d\alpha}\bigg|_{\alpha = \alpha_0} = -E_{d0} \sin \alpha_0$.

2.4 Block Diagram

Because of its simplicity and versatility, block-diagram is often used to portray control systems of all types. A block diagram can be used simply to represent the composition and interconnection of a system. Or, it can be used, together with transfer functions, to represent the dynamic cause-and-effect relationships throughout the system.

If the mathematical and functional relationships of all the system elements are known, the block diagram can be used as a reference for the analytical or the computer solution of the system. Furthermore, if all the system elements are assumed to be linear, the transfer function for the overall system can be obtained by means of the block-diagram algebra.

2.4.1 Concepts of Block Diagram

Block diagram consists of unidirectional, operational blocks that represent the transfer function of the variables of interest. By definition, for a linear time-invariant system or a component the transfer function is the ratio of the Laplace transform of the input variable to the Laplace transform of the output variable. Let $R(s)$ be the input variable, $C(s)$ be the output variable, and $G(s)$ be the transfer function. The block diagram shown in Fig. 2.12 is another method of graphically denoting the algebraic equation

Figure 2.12   Block diagram representation of input-output relation

$$C(s) = G(s)R(s) \tag{2.50}$$

For this block, the output is equal to the transfer function given in the block multiplied by the input. The input and the output are defined by the direction of the arrowheads of signal line as shown.

Sometimes, one additional element required to represent an algebraic equation by a block diagram is the summing junction, which is illustrated in Fig. 2.13 for Eq. (2.51)

$$C(s) = G_1(s)R_1(s) \pm R_2(s) \tag{2.51}$$

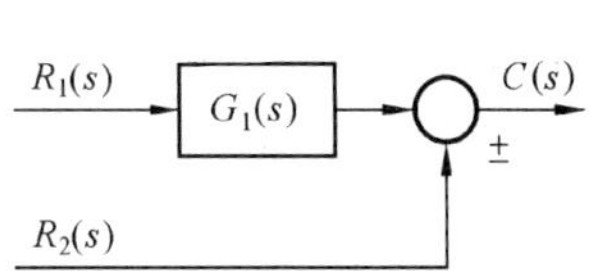

Figure 2.13   Block diagram with summing junction

In the block diagram, a summing junction is denoted with a circle, as in Fig. 2.13. By the definition, the signal out of the summing junction is equal to the sum of the signals into the junction with the sign of each component determined by the sign placed next to the arrowhead of the component. Note that whereas a summing junction can have any number of inputs, we show only one output.

To represent a more complex system, an interconnection of blocks is utilized. For example, the system shown in Fig. 2.14 has two input variables and two output variables. Using transfer function relations we can write the simultaneous equations for the output variables as

$$C_1(s) = G_{11}(s)R_1(s) + G_{12}(s)R_2(s)$$
$$C_2(s) = G_{21}(s)R_1(s) + G_{22}(s)R_2(s)$$
(2.52)

where $G_{ij}(s)$ is the transfer function relating the ith output variable to the jth input variable. The block diagram representing this set of equations is shown in Fig. 2.14, where another useful element denoted with " · " is pick-off point.

Usually, a block diagram is consisted of blocks, signal lines with arrowhead, summing junctions, and pick-off points.

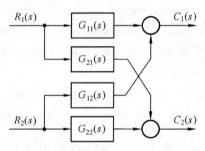

Figure 2.14 Block diagram representation of a two-input-two-output system

2.4.2 Examples of Establishing Block Diagram

Now, some simple examples are given to illustrate how to establish the block diagram of a physical system from its dynamic equation set.

Example 2.7 Consider the RC passive network shown in Fig. 2.15, where $v_i(t)$ is the input and $v_o(t)$ is the output.

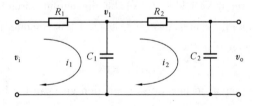

Figure 2.15 RC passive network

Using the loop currents and node voltages designated in the same figure, one set of independent equations is

$$v_i = R_1 i_1 + v_1$$

$$v_1 = \frac{1}{C_1}\int (i_1 - i_2)\,\mathrm{d}t$$

$$v_1 = R_2 i_2 + v_o$$

$$v_o = \frac{1}{C_2}\int i_2\,\mathrm{d}t$$

Taking laplace transform and writing the result in cause-and-effect relation form yield

$$I_1(s) = \frac{1}{R_1}[V_i(s) - V_1(s)]$$

$$V_1(s) = \frac{1}{C_1 s}[I_1(s) - I_2(s)]$$

$$I_2(s) = \frac{1}{R_2}[V_1(s) - V_o(s)]$$

$$V_o(s) = \frac{1}{C_2 s}I_2(s)$$

With the variables $V_i(s), I_1(s), V_1(s), I_2(s), V_o(s)$ arranged from left to right in order, the block diagram of the network is constructed as shown in Fig. 2.16.

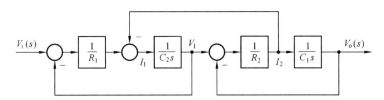

Figure 2.16 Block diagram of RC passive network

Example 2.8 From Eq. (2.24) ~ (2.27), the dynamic behavior of an armature-controlled dc motor can be described by the following equations

$$v_a(t) = L_a \frac{\mathrm{d}i_a(t)}{\mathrm{d}t} + R_a i_a(t) + e_b(t)$$

$$T_m(t) = C_m i_a(t)$$

$$e_b(t) = C_e \omega_m(t)$$

$$T_m(t) = J_m \frac{\mathrm{d}\omega_m(t)}{\mathrm{d}t^2} + B_m \omega_m(t) + T_L(t)$$

$$\omega_m(t) = \frac{\mathrm{d}\theta_m(t)}{\mathrm{d}t}$$

where v_a is the input armature voltage, θ_m is the output angular displacement, i_a is the armature current, e_b is the back emf, ω_m is the motor speed, T_m is the electromagnetic torque, and T_L is the load torque. Taking Laplace transform yields

$$V_a(s) = (L_a s + R_a) I_a(s) + E_b(s)$$
$$T_m(s) = C_m I_a(s)$$
$$E_b(s) = C_e \omega_m(s)$$
$$T_m(s) = (J_m s + B_m) \omega_m(s) + T_L(s)$$
$$\omega_m(s) = s \theta_m(s)$$

Rewriting above equations in cause-and-effect relation form, we get

$$I_a(s) = \frac{1}{L_a s + R_a} [V_a(s) - E_b(s)]$$
$$T_m(s) = C_m I_a(s)$$
$$E_b(s) = C_e \omega_m(s)$$
$$\omega_m(s) = \frac{1}{J_m s + B_m} [T_m(s) - T_L(s)]$$
$$\theta_m(s) = \frac{1}{s} \omega_m(s)$$

Considering $E_b(s)$ be a back emf, with the variables $V_a(s), I_a(s), T_m(s), \omega_m(s), \theta_m(s)$ arranged from left to right in order, the block diagram of the motor is constructed as shown in Fig. 2.17.

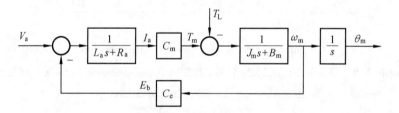

Figure 2.17 Block diagram of dc armature-controlled motor

Although a dc motor is an open-loop system, the block diagram in Fig. 2.17 shows that the dc armature-controlled motor has a "built-in" feedback loop caused by the back emf.

2.4.3 Simplification of Block Diagram

In order to analyze a complex feedback control system or obtain its overall transfer function, it is usual, through proper equivalent transform, to rearrange the block diagram so as to enable it to be easily analyzed or reduce its block diagram to one block.

1. Block diagram algebra

Applying block algebra, the block diagrams may be directly reduced to a single element in the three cases discussed as below.

ⅰ. Serial blocks

This configuration consists of two (or more) individual blocks as shown in Fig. 2.18, where each transfer function is known. In this case the output of the first block is the input of the second block. Two blocks connected in this manner are said to be serial, or cascaded. The overall transfer function is then

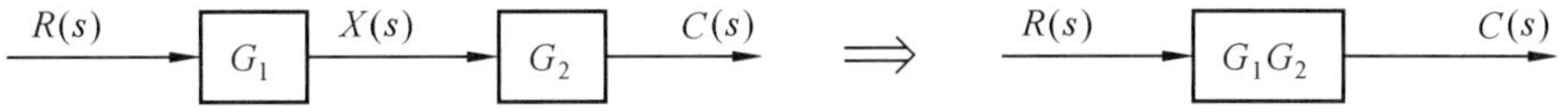

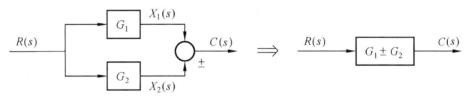

Figure 2.18 Reduction of serial blocks

$$\frac{C(s)}{R(s)} = \frac{C(s)}{X(s)} \cdot \frac{X(s)}{R(s)} = G_1(s) G_2(s) \qquad (2.53)$$

According to this rule the blocks in serial connection can be combined into one block and the overall transfer function is simply the product of the individual block transfer functions. This rule can extend to more than two blocks in series.

ⅱ. Parallel blocks

Another configuration to be considered is shown in Fig. 2.19. The input to each block is the same, and the system output is the sum of the outputs of the individual blocks. These blocks are said to be parallel. The overall transfer function is then

$$\frac{C(s)}{R(s)} = \frac{X_1(s) \pm X_2(s)}{R(s)} = G_1(s) \pm G_2(s) \qquad (2.54)$$

Figure 2.19 Reduction of parallel blocks

According to this rule the blocks in parallel connection can be combined into one block and the overall transfer function is then the sum of the individual block transfer functions. This rule is also seen to extend directly to more than two blocks in parallel.

ⅲ. Prototype feedback connection

As shown in Fig. 2.20, this system comprises a forward-path transfer function $G(s)$ and a feedback transfer function $H(s)$. From the diagram it is known that

$$E(s) = R(s) - H(s) C(s) \qquad (2.55)$$

$$C(s) = G(s) E(s) \qquad (2.56)$$

Substituting Eq. (2.55) into (2.56) yields

$$C(s) = G(s) [R(s) - H(s) C(s)] = G(s) R(s) - G(s) H(s) C(s) \qquad (2.57)$$

Rearranging this equation results in

$$[1 + (G(s) H(s))] C(s) = G(s) R(s) \qquad (2.58)$$

i.e.,

$$\frac{C(s)}{R(s)} = \frac{G(s)}{1+G(s)H(s)} \qquad (2.59)$$

Therefore, the diagram may be reduced to a single block as shown. This closed-loop transfer function is particular important because it represents many of the existing practical control systems.

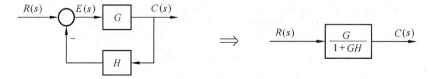

Figure 2.20 Reduction of prototype feedback case

Note that if the feedback loop is positive instead of negative, then

$$\frac{C(s)}{R(s)} = \frac{G(s)}{1-G(s)H(s)} \qquad (2.60)$$

2. Block manipulation rules

Sometimes, in order to simplify a block diagram it is wise to moving individual blocks across summing junctions or pick off points. In doing so, there are a number of other rules that will be useful. These rules are summarized in Fig. 2.21, and they are

iv. moving a summing junction ahead of a block,

v. moving a summing junction past a block,

vi. moving a pick-off point ahead of a block,

vii. moving a pick-off point past a block.

In each case, the equivalent block diagram has a same output equation as the original block diagram.

In general, the simplification of a complex block diagram may be accomplished by systematic modification of the diagram using the rules in Fig. 2.21 applied in the following order:

Step 1: Combine all serial blocks (rule i).

Step 2: Combine all parallel blocks (rule ii).

Step 3: Eliminate all inner loops (rule iii).

Step 4: Move summing junctions and/or pick off points left/right (rules iv-vii).

Chapter 2 Mathematic Models of Control Systems

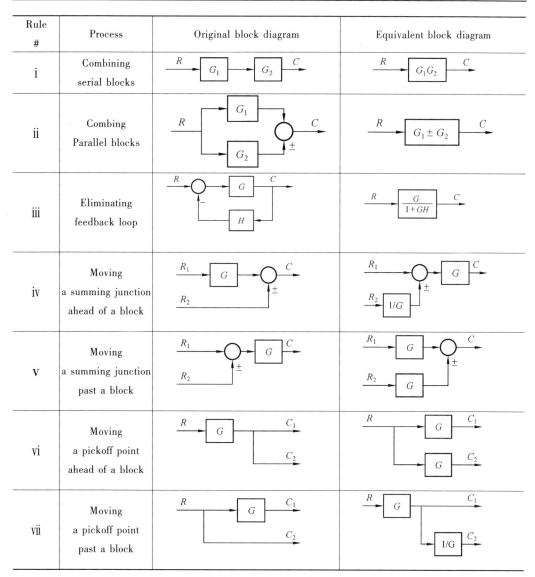

Figure 2.21 Block manipulation rules

Example 2.9 Simplify the block diagram in Fig. 2.22 and determine the transfer function $C(s)/R(s)$.

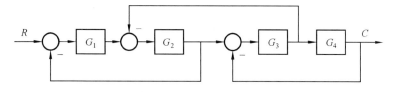

Figure 2.22

Solution Observing Fig. 2.22, it may be seen that there are no serial elements, no parallel elements, and no prototype feedback loop to be reduced. The simplification begins with step 4, as shown in Fig. 2.23(b), by moving summing junction s_2 ahead of its left block and exchanging its position with summing joint s_1. Then, using step 7, pick off point p_2 is move past its right block, after exchanging positions p_2 of and p_3.

The part formed by $G_1(s)$, $G_2(s)$ and the part formed by $G_3(s)$, $G_4(s)$ may be identified as a serial connection respectively, and may be reduced to the equivalent blocks as Fig. 2.23(c).

Then two simple feedback loops can be identified and reduced to the equivalent blocks as Fig. 2.23(d).

Thus, obviously, we can simplify the given block diagram to a simple block as shown in Fig. 2.23(e).

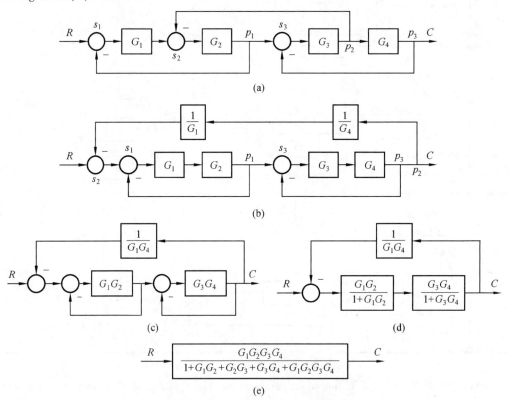

Figure 2.23 Simplification procedure of Example 2.9

Finally, the transfer function is obtained as

$$\frac{C(s)}{R(s)} = \frac{G_1 G_2 G_3 G_4}{1 + G_1 G_2 + G_2 G_3 + G_3 G_4 + G_1 G_2 G_3 G_4}$$

2.5 Signal Flow Graph

The block diagram is useful for analyzing and designing control systems. However, in order to determine the transfer function for a system with reasonably complex interrelationship, the block diagram reduction procedure is cumbersome and often quite difficult to complete. An alternative representation of control systems is the signal flow graph, which looks like a simplified block diagram. The main advantage of signal flow graph method is the availability of a flow graph gain formula, which provides the relation between system variables without requiring any reduction procedure or manipulation of the flow graph.

2.5.1 Definitions for Single Flow Graph

A signal flow graph is a diagram consisting of nodes that are connected by several directed branches, and is a graphical representation of a set of linear algebraic equations. For example, the diagram shown in Fig. 2.24 is a signal flow graph for the following set of equations

$$x_2 = a x_1 + c x_3$$
$$x_3 = b x_2 + d x_4$$
$$x_5 = x_3$$

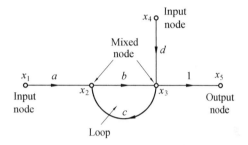

Figure 2.24 Signal flow graph

where x_1, x_2, x_3, x_4 and x_5 are variables.

Nodes represent the system variables and denoted by the symbol "○". A node performs two functions: (1) Addition of the signals of all incoming branches. (2) Transmission of the total node signal (the sum of all incoming signals) to all outgoing branches. A branch is a unidirectional line segment connected between two nodes and acts as a signal multiplier, the direction of signal flow is indicated by an arrow placed on the branch, and the multiplication factor (gain or transfer function) is placed near the arrow. In addition, the following terms are useful for using the flow-graph gain formula.

Source node (input node). This is a node that has only incoming branches. For example, in Fig. 2.24 nodes x_1 and x_4 are source nodes.

Sink node (output node). This is a node that has only outgoing branches. For example, in Fig. 2.24 node x_5 is a sink node. Any non-input node may be treated as a sink node by introducing a branch with unity gain.

Mixed node (general node). This is a node that has both incoming and outgoing branches.

Path. A path is any connected sequence of branches in the same direction.

Forward path. This is a path that started at an input node and ends at an output node and along which no node is traversed more than once. For example, in Fig. 2.24, there is a forward path from the node x_1 to x_5, and there is another forward path from the node x_4 to x_5.

Loop. A loop is a closed path that originates and terminates on the same node and along which no other node is encountered more than once. For example, in Fig. 2.24 there is only one loop.

Nontouching loop. Some loops are said to be nontouching if they do not have any common node.

Path gain. Path gain is the product of the branch gains encountered in traversing a path.
Forward-path gain. Forward-path gain is defined as the path gain of a forward path.
Loop gain. Loop gain is defined as the path gain of a loop.

2.5.2 Examples for Drawing Signal Flow Graph

The signal flow graph is simply a graphical method of writing a set of algebraic equations so as to indicate the interrelationship among the variables.

Example 2.10 Again consider the RC passive network shown in Example 2.7. The RC passive network shown in Fig. 2.15 can be described by the following set of equations in the cause-and-effect relation form

$$I_1(s) = \frac{1}{R_1}[V_i(s) - V_1(s)] = \frac{1}{R_1}V_i(s) - \frac{1}{R_1}V_1(s)$$

$$V_1(s) = \frac{1}{C_1 s}[I_1(s) - I_2(s)] = \frac{1}{C_1 s}I_1(s) - \frac{1}{C_1 s}I_2(s)$$

$$I_2(s) = \frac{1}{R_2}[V_1(s) - V_o(s)] = \frac{1}{R_2}V_1(s) - \frac{1}{R_2}V_o(s)$$

$$V_o(s) = \frac{1}{C_2 s}I_2(s)$$

With the variables $V_i(s), I_1(s), V_1(s), I_2(s), V_o(s)$ arranged from left to right in order, the signal flow graph of the network is drawn as shown in Fig. 2.25.

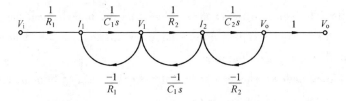

Figure 2.25 Signal flow graph for Example 2.10

2.5.3 Mason's Gain Formula

In general, the overall gain (transfer function) between a source node (input variable) and a sink node (output variable) can be calculated by Mason's gain formula

$$T = \sum_{k=1}^{n} \frac{T_k \Delta_k}{\Delta} \tag{2.61}$$

where

T = overall gain between the specified source node and sink node;

T_k = gain of the k th forward path;

n = total number of the forward paths;

Δ = determinant of the flow graph;

Δ_k = cofactor of the flow-graph determinant for the kth forward path.

The determinant

$$\Delta = 1 - \sum L_{m1} + \sum L_{m2} - \sum L_{m3} + \ldots \qquad (2.62)$$

where L_{mr} is the gain product of the mth possible combination of r non-touching loops, i.e.

$\Delta = 1 - ($ sum of all different loop gains$) +$

(sum of the gain products of all combination of two nontouching loops) −

(sum of the gain products of all combination of two nontouching loops) + ...

The cofactor Δ_k is the determinant with the loops touching the kth forward path removed.

Example 2.11 Consider the signal flow graph of Example 2.10. Determine the transfer function $V_o(s)/V_i(s)$ by use of the gain formula.

Solution From the flow graph in Fig. 2.25, the following conclusions are obtained by inspection:

① There are three loops in the signal flow graph

$$L_{11} = -\frac{1}{R_1 C_1 s}, \quad L_{21} = -\frac{1}{R_2 C_1 s}, \quad L_{31} = -\frac{1}{R_2 C_2 s}$$

② Loop L_{11} does not touch loop L_{31}. Thus the determinant of flow graph is

$$\Delta = 1 - \left(-\frac{1}{R_1 C_1 s} - \frac{1}{R_2 C_1 s} - \frac{1}{R_2 C_2 s}\right) + \left(-\frac{1}{R_1 C_1 s}\right)\left(-\frac{1}{R_2 C_2 s}\right) = \frac{R_1 R_2 C_1 C_2 s^2 + (R_1 C_1 + R_2 C_1 + R_2 C_2)s + 1}{R_1 R_2 C_1 C_2 s^2}$$

③ There is only one forward path from $V_i(s)$ to $V_o(s)$, and the forward path gain is

$$T_1 = \frac{1}{R_1} \cdot \frac{1}{C_1 s} \cdot \frac{1}{R_2} \cdot \frac{1}{C_2 s} = \frac{1}{R_1 R_2 C_1 C_2 s^2}$$

④ Since the forward path is in touch with all loops, we have

$$\Delta_1 = 1$$

Therefore, the transfer function of the network is

$$\frac{V_o(s)}{V_i(s)} = \frac{T_1 \Delta_1}{\Delta} = \frac{1}{R_1 R_2 C_1 C_2 s^2 + (R_1 C_1 + R_2 C_1 + R_2 C_2)s + 1}$$

Because of the similarity between the block diagram and the signal flow graph, Mason's gain formula can be used to determine the input-output relationship of either. Given a block diagram of a linear system, the forward path gains, loop gains, determinant and cofactors can be obtained directly from the block diagram.

Example 2.12 Determine the transfer functions $C(s)/R(s)$, $E(s)/R(s)$ and $X_1(s)/R(s)$ of the block diagram shown in Fig. 2.26.

Solution (1) Determine the transfer function $C(s)/R(s)$.

① There are five individual loops, and the loop transfer functions are, respectively,

$$L_{11} = -G_1 G_2 H_1, L_{21} = -G_2 G_3 H_2, L_{31} = -G_1 G_2 G_3, L_{41} = -G_4 H_2, L_{51} = -G_1 G_4$$

② There are no nontouching loops, thus the flow-graph determinant is

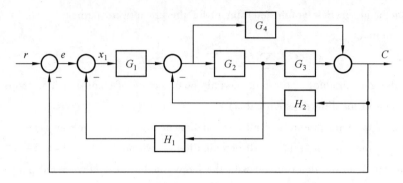

Figure 2.26 Block diagram of Example 2.12

$$\Delta = 1 + G_1 G_2 H_1 + G_2 G_3 H_2 + G_1 G_2 G_3 + G_4 H_2 + G_1 G_4$$

③ There are two forward paths from $R(s)$ to $C(s)$, thus we have

$$T_1 = G_1 G_2 G_3, \quad \Delta_1 = 1$$
$$T_2 = G_1 G_4, \quad \Delta_2 = 1$$

Therefore, the transfer function from $R(s)$ to $C(s)$ is

$$\frac{C(s)}{R(s)} = \sum_{k=1}^{2} \frac{T_k \Delta_k}{\Delta} = \frac{G_1 G_2 G_3 + G_1 G_4}{1 + G_1 G_2 H_1 + G_2 G_3 H_2 + G_1 G_2 G_3 + G_4 H_2 + G_1 G_4}$$

(2) Determine the transfer function $E(s)/R(s)$. Note that the determinant of a flow graph is unique. There is only one forward path from $R(s)$ to $E(s)$, and the path transfer function is

$$T_1 = 1$$

This forward path does not touch the loops L_{11}, L_{21} and L_{41}. Thus the cofactor of this forward path is

$$\Delta_1 = 1 + G_1 G_2 H_1 + G_2 G_3 H_2 + G_4 H_2$$

Therefore, the transfer function from $R(s)$ to $E(s)$ is

$$\frac{E(s)}{R(s)} = \frac{T_1 \Delta_1}{\Delta} = \frac{1 + G_1 G_2 H_1 + G_2 G_3 H_2 + G_4 H_2}{1 + G_1 G_2 H_1 + G_2 G_3 H_2 + G_1 G_2 G_3 + G_4 H_2 + G_1 G_4}$$

(3) Determine the transfer function $X_1(s)/R(s)$. There is only one forward path from $R(s)$ to $X_1(s)$, and the path transfer function is

$$T_1 = 1$$

In this case, the forward path does not touch the loops L_{21} and L_{41}. Thus the cofactor of this forward path is

$$\Delta_1 = 1 + G_2 G_3 H_2 + G_4 H_2$$

Finally, the transfer function from $R(s)$ to $X_1(s)$ is

$$\frac{X_1(s)}{R(s)} = \frac{1 + G_2 G_3 H_2 + G_4 H_2}{1 + G_1 G_2 H_1 + G_2 G_3 H_2 + G_1 G_2 G_3 + G_4 H_2 + G_1 G_4}$$

2.5.4 Drawing Signal Flow Graph from Block Diagram

Mason's formula can be directly used to determine the input-output relationship of a linear system from its block diagram. However, in order to be able to identify all the loops and nontouching parts clearly, sometimes it may be useful if an equivalent signal flow graph is drawn for a block diagram before applying the Mason's gain formula.

To illustrate how the signal flow graph and the block diagram are related, the block diagram of Example 2.12 and its equivalent signal flow graph are shown in Fig. 2.27.

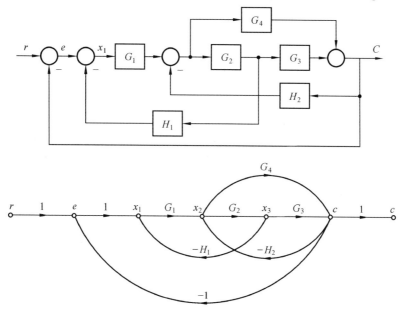

Figure 2.27 Block diagram and its equivalent signal flow graph

Since a node on the signal flow graph is interpreted as a summing point of all incoming signals, generally, it is recommended that,

① all input variables and disturbances are chosen as the source nodes;

② all output variables are chosen as the sink nodes; and

③ all other variables following the summing points and departing from the pick-off points are chosen as the mixed nodes.

2.6 Transfer Functions of Linear System

After familiarizing with mathematical modeling of physical systems, we can extend the ideas of modeling to include control system characteristics, such as transient response to input test signals, steady-state errors, disturbance rejection, and etc. Following transfer functions defined in the feedback control systems are very useful in our study.

Consider a control system with a standard construction depicted in Fig. 2.28, in which there are only one forward path from input to output and one loop. In the figure $R(s)$ is the input, $C(s)$ is

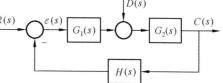

Figure 2.28 Control system with a standard construction

the output, $D(s)$ is the disturbance acting on the forward path, $\varepsilon(s)$ is the deviation, $G_1(s)$ and $G_2(s)$ are the transfer functions of the forward path ahead of and past the disturbance acting point respectively, and $H(s)$ is the feedback transfer function. Usually, the transfer function of forward path is denoted with $G(s)$. Obviously, for the system shown in Fig. 2.28 we have

$$G(s) = G_1(s) G_2(s) \tag{2.63}$$

In many cases, $H(s)$ is equal to 1 or a constant other than 1.

2.6.1 Closed-Loop Transfer Functions

Using Mason's formula, it is easy to obtain the following transfer functions of the system shown in Fig. 2.28:

$$\Phi(s) = \frac{C(s)}{R(s)} = \frac{G_1(s) G_2(s)}{1 + G_1(s) G_2(s) H(s)} = \frac{G(s)}{1 + G(s) H(s)} \tag{2.64}$$

$$\Phi_d(s) = \frac{C(s)}{D(s)} = \frac{G_2(s)}{1 + G_1(s) G_2(s) H(s)} = \frac{G_2(s)}{1 + G(s) H(s)} \tag{2.65}$$

$$\Phi_\varepsilon(s) = \frac{\varepsilon(s)}{R(s)} = \frac{1}{1 + G_1(s) G_2(s) H(s)} = \frac{1}{1 + G(s) H(s)} \tag{2.66}$$

$$\Phi_{\varepsilon,d}(s) = \frac{\varepsilon(s)}{D(s)} = \frac{-G_2(s) H(s)}{1 + G_1(s) G_2(s) H(s)} = \frac{-G_2(s) H(s)}{1 + G(s) H(s)} \tag{2.67}$$

where $\Phi(s)$ is called the closed-loop transfer function of output to input and often called the closed-loop transfer function for short, $\Phi_d(s)$ is called the closed-loop transfer function of output to disturbance, $\Phi_\varepsilon(s)$ is called the closed-loop transfer function of deviation to input, and $\Phi_{\varepsilon,d}(s)$ is called the closed-loop transfer function of deviation to disturbance.

If the control signal $R(s)$ and disturbance signal $D(s)$ are applied to the system simultaneously, then the output signal is given by

$$C(s) = \Phi(s) R(s) + \Phi_d(s) D(s) =$$
$$\frac{G_1(s) G_2(s)}{1 + G_1(s) G_2(s) H(s)} R(s) + \frac{G_2(s)}{1 + G_1(s) G_2(s) H(s)} D(s) \tag{2.68}$$

Similarly, in this case, the deviation signal is given by

$$\varepsilon(s) = \Phi_\varepsilon(s) R(s) + \Phi_{\varepsilon,d}(s) D(s) =$$
$$\frac{1}{1 + G_1(s) G_2(s) H(s)} R(s) - \frac{G_2(s) H(s)}{1 + G_1(s) G_2(s) H(s)} D(s) \tag{2.69}$$

Observing these four closed-loop transfer functions, we can find that they have the same denominator $1 + G(s) H(s)$, i.e., they have same closed-loop poles. It means the closed-loop poles of a system are depended only on the equation

$$1 + G(s) H(s) = 0 \tag{2.70}$$

which is called the characteristic equation of feedback control system.

2.6.2 Open-Loop Transfer Function

In the characteristic equation given by Eq. (2.70) the quantity $G(s)H(s)$ is defined as the open-loop transfer function. For the single-loop feedback system shown in Fig. 2.28 the open-loop transfer function just is the product of forward-path transfer function and feedback-path transfer function, i.e.,

$$G(s)H(s) = G_1(s)G_2(s)H(s) \qquad (2.71)$$

It should be noted that the open-loop transfer function of a closed-loop system is neither the transfer function of open-loop system nor the loop transfer function of closed-loop system. In Fig. 2.28 the transfer function of open-loop system is $G_1(s)G_2(s)$ and the loop transfer function is $-G_1(s)G_2(s)H(s)$.

The open-loop transfer function is very useful in many methods of analysis and design, and it is often written in one of the following forms. The first one is the polynomial form

$$G(s)H(s) = \frac{M(s)}{N(s)} \qquad (2.72)$$

where both $M(s)$ and $N(s)$ are polynomials in s. The second one is the pole-zero form

$$G(s)H(s) = \frac{k\prod_{i=1}^{m}(s-z_i)}{\prod_{j=1}^{n}(s-p_j)} \qquad (2.73)$$

where $z_i(i=1,2,\ldots,m)$ and $p_j(j=1,2,\ldots,n)$ are the zeros and poles of the open-loop transfer function, respectively. The third one is the standard factor form

$$G(s)H(s) = \frac{K\prod_{i=1}^{m}(\tau s+1)}{s^v\prod_{j=1}^{l}(Ts+1)\prod_{k=1}^{(n-v-l)/2}(\hat{T}^2s^2+2\hat{T}\xi s+1)} \qquad (2.74)$$

where K is called the open-loop gain and defined as

$$K = \lim_{s \to 0} s^v G(s)H(s) \qquad (2.75)$$

2.7 Impulse Response of Linear Systems

2.7.1 Definition of Impulse Response

The impulse response of a linear system is defined as the output response of the system when the input is a unit impulse function. Hence, for a system with transfer function $G(s)$, if $r(t) = \delta(t)$, the Laplace transform of the system output is simply the transfer function of the system, i.e.,

$$C(s) = G(s)L[\delta(t)] = G(s) \qquad (2.76)$$

since the Laplace transform of the unit impulse function is unity.

Taking the inverse Laplace transform on both sides of Eq. (2.76) yields

$$c(t) = g(t) \qquad (2.77)$$

where $g(t)$ is the inverse Laplace transform of $G(s)$ and is the impulse response (sometimes also called the weighing function) of a linear system. Therefore, we can state that the Laplace transform of the impulse response is the transfer function.

Since the transfer function is a powerful way of characterizing linear systems, this means that if a linear system has zero initial conditions, theoretically, the system can be described or identified by exciting it with a unit impulse and measuring the output. In practice, although a true impulse cannot be generated physically, a pulse with a very narrow pulse-width usually provides a suitable approximation.

2.7.2 Purpose of Impulse Response

The derivation of transfer function in Eq. (2.3) is based on the knowledge of the system differential equation, and the solution of $C(s)$ from Eq. (2.3) is also assumed that $R(s)$ and transfer function to be available in analytical forms. This is not always possible, for quite often the input signal $r(t)$ is not Laplace transformable or is available only in the form of experimental data. Under such conditions, to analyze the system we would have to work with the impulse response $g(t)$.

The evaluation of the impulse response of a linear system is sometimes an important step in the analysis and design of a class of systems known as the adaptive control systems. In real life the dynamic characteristics of most systems vary to some extent over an extended period of time. This may be caused by simple deterioration of components due to wear and tear, drift in operating environments, and etc. Some systems simply have parameters that vary with time in a predictable or unpredictable fashion. For instance, the transfer characteristic of a guided missile in flight will vary in time because of the change of mass of the missile and the change of the atmospheric conditions. On the other hand, for a simple mechanical system with mass and friction, the latter may be subject to unpredictable variation either due to "aging" or surface conditions. Thus the control system designed under the assumption of known and fixed parameters may fail to yield satisfactory response should the system parameters vary. In order that the system may also have the ability of self-correction or self-adjustment in accordance with varying parameters and environment, it is necessary that the system's transfer characteristics will be identified continuously or at appropriate intervals during the operation of the system. One of the methods of identification is to measure the impulse response of the system so that design parameters may be adjusted accordingly to attain optimal control at all times.

Thus, in the previous discussions, definitions of transfer function and impulse response of a linear system have been presented. The two functions are directly related through the Laplace transform, and they represent essentially the same information about the system. However, it must be emphasized that transfer function and impulse response are defined only for linear systems and that the initial conditions are assumed to be zero.

Problems

P2.1 The following differential equations represent linear time-invariant systems, where $r(t)$ denotes the input and $c(t)$ denotes the output. Find the transfer function of each of the systems.

(a) $\dfrac{d^3 c(t)}{dt^3} + 6 \dfrac{d^2 c(t)}{dt^2} + 11 \dfrac{dc(t)}{dt} + 6c(t) = r(t)$

(b) $\dfrac{d^3 c(t)}{dt^3} + 3 \dfrac{d^2 c(t)}{dt^2} + 4 \dfrac{dc(t)}{dt} + c(t) = 2 \dfrac{dr(t)}{dt} + r(t)$

P2.2 A mass-spring-damper system is shown in Fig. P2.2. Determine the differential equation between the input force f and the output displacement x.

P2.3 Determine the transfer function $X_2(s)/F(s)$ for the system shown in Fig. P2.3. Both masses slide on a frictionless surface, and $K = 1$ N/M.

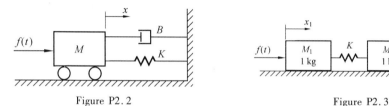

Figure P2.2   Figure P2.3

P2.4 A thermistor has a response to temperature represented by
$$R = R_0 e^{-0.1T}$$
where R is resistance, T is temperature in degrees Celsius, and $R_0 = 10\,000$ Ω. Find the linearized model for the thermistor operating at $T = 20\,°\mathrm{C}$ and for a small range of variation of temperature.

P2.5 Obtain the transfer functions, $V_o(s)/V_i(s)$, for the passive networks shown in Fig. P2.5.

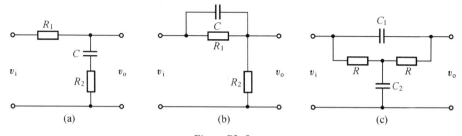

Figure P2.5

P2.6 Obtain the transfer functions, $V_o(s)/V_i(s)$, for the active networks shown in Fig. P2.6.

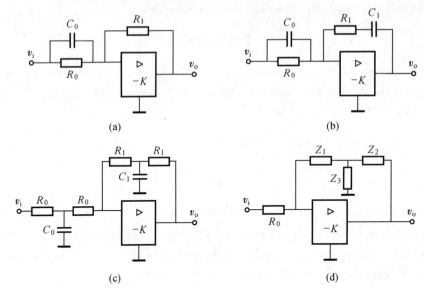

Figure P2.6

P2.7 Draw a block diagram showing all the variables for the system described by the following set of differential equations.

$$x_1(t) = r(t) - c(t) - n_1(t)$$
$$x_2(t) = K_1 x_1(t)$$
$$x_3(t) = x_2(t) - x_5(t)$$
$$T\dot{x}_4 = x_3(t)$$
$$x_5(t) = x_4(t) - K_2 n_2(t)$$
$$\ddot{c} + \dot{c}(t) = K_0 x_5(t)$$

where $r(t)$ is the input, $c(t)$ is the output, $n_1(t)$ and $n_2(t)$ are two disturbances, K_0, K_1, K_2, and T are constants.

P2.8 A system is described by the following set of equations:

$$X_1(s) = G_1(s)R(s) - G_1(s)[G_7(s) - G_8(s)]C(s)$$
$$X_2(s) = G_2(s)[X_1(s) - G_6(s)X_3(s)]$$
$$X_3(s) = G_3(s)[X_2(s) - G_5(s)C(s)]$$
$$Y(s) = G_4(s)X_3(s)$$

Draw the block diagram showing all the variables for this system and find the transfer function for $C(s)/R(s)$.

P2.9 A system with two inputs and two outputs is shown in Fig. P2.9. Determine the transfer functions $C_1(s)/R_1(s)$, $C_1(s)/R_2(s)$, $C_2(s)/R_1(s)$, and $C_2(s)/R_2(s)$. Write the expressions of $C_1(s)$ and $C_2(s)$.

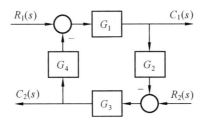

Figure P2.9

P2.10 By block diagram deduction find the transfer functions of following systems.

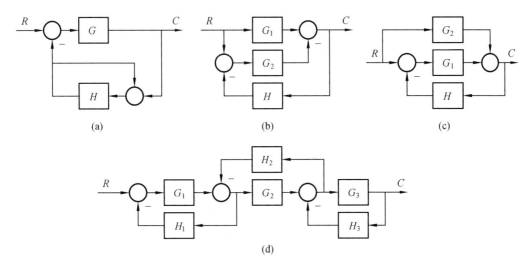

Figure P2.10

P2.11 For the following systems:

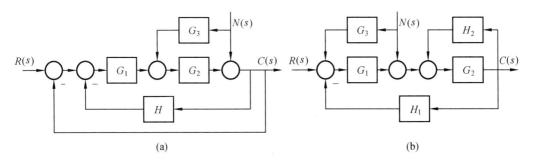

Figure P2.11

Derive the transfer functions for $C(s)/R(s)$ and $C(s)/N(s)$ by block diagram deduction.

P2.12 By block diagram deduction find the transfer functions of the systems shown in Fig. P2.12.

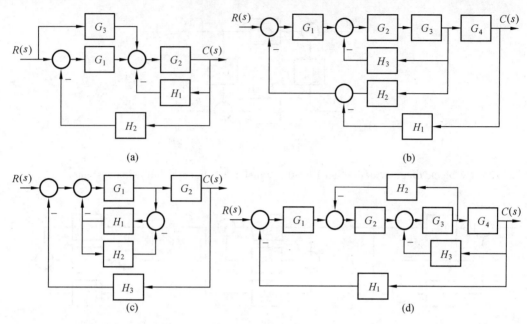

Figure P2.12

P2.13 For the system shown in the figure P2.13, determine K_1, K_2, and $H(s)$ if it is required that the feedforward transfer function of the system

$$\frac{C(s)}{R(s)} = \frac{100(s+10)}{s(s+5)(s+20)}$$

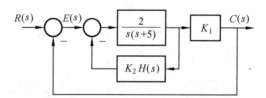

Figure P2.13

P2.14 Are the two systems shown in Fig. P2.14(a) and (b) equivalent? Explain.

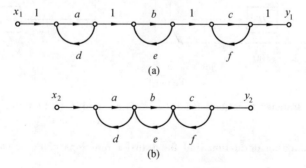

Figure P2.14

P2.15 Find the overall transfer functions for the systems shown in Fig. P2.15.

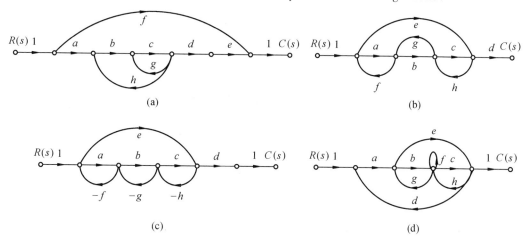

Figure P2.15

P2.16 Find the transfer function for the multi-loop crossing system shown in Fig. P2.16.

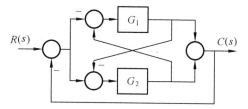

Figure P2.16

P2.17 Draw equivalent signal flow graphs for the block diagrams in Fig. P2.12. Find the transfer functions by use of Mason's formula.

Chapter 3 Time-Domain Analysis of Control Systems

3.1 Introduction

In most control systems, it is an interest topic to evaluate the time response. In the analysis problem, an input signal is applied to a system, and the performance of the system is evaluated by studying the response in the time domain. For instance, if the objective of the control system is to have the output variable follow the input signal as closely as possible, it is necessary to compare the input and the output as function of time.

3.1.1 Transient Response and Steady-State Response

The time response of a control system is usually divided into two parts: transient response and steady-state response. If a time response is denoted by $c(t)$, then the transient response and steady-state response may be denoted by $c_t(t)$ and $c_{ss}(t)$, respectively.

In control system applications, a response can still vary with time when it has reached its steady state. The steady-state response is simply the fixed response when time reaches infinity, i.e.,

$$c_{ss}(t) = \lim_{t \to \infty} c(t) \tag{3.1}$$

Therefore, a sine wave is considered as steady-state response because its behavior is fixed as time approaches infinity. Similarly, if a response is described by $c(t) = t$, it may be defined as a steady-state response.

Transient response is defined as the part of the response that goes to zero as time becomes large. Therefore, $c_t(t)$ has the property of

$$\lim_{t \to \infty} c_t(t) = 0 \tag{3.2}$$

Hence, it can also be stated that the steady-state response is that part of the response, which remains after the transient has died out.

All control systems exhibit transient phenomenon to some extent before a steady state is reached. Since mass, capacitance, inductance, and etc. cannot be avoided in physical systems, the responses cannot follow sudden changes in the input instantaneously, and transients are usually observed.

The transient response of a control system is of importance, since it is a part of the dynamic behavior of the system; and the difference between the response and the input or the desired response, before the steady state is reached, must be closely watched. The steady-state response, when compared with certain reference, gives an indication of the final accuracy of the system. If the steady-state response of the output does not agree with the steady state of the reference exactly, the system is said to have a steady-state error.

3.1.2 Typical Test Signals for Time Responses

Unlike many electrical circuits and communication systems, the input excitations to many practical control systems are not known ahead of time. In many cases, the actual inputs of a control system may vary in random fashions with respect to time. For instance, in a radar tracking system, the position and speed of the target to be tracked may vary in an unpredictable manner, so that they cannot be expressed determinately by a mathematical expression. This causes a problem for the designer, since it is difficult to design the control system so that it will perform satisfactorily to any input signal. For the purposes of analysis and design, it is necessary to assume some basic types of input functions so that the performance of a system can be evaluated with respect to these test signals. By selecting these basic test signals properly, not only the mathematical treatment of the problem is systematized, but also the responses due to these inputs allow the prediction of the system's performance to other more complex inputs. In a design problem, performance specifications may be specified with respect to these test signals so that a system may be designed to meet the specifications.

When the response of a linear time-invariant system is analyzed in the frequency domain, a sinusoid input with variable frequency is used. To facilitate the time-domain analysis, the following test signals are often used.

1. Step function

The step function represents an instantaneous change in the input. For instance, if the input is the angular position of a mechanical shaft, the step input represents a sudden rotation of the shaft. Sometimes, the step signal is also called the position signal. The step function is very useful as a test signal since its initial instantaneous jump in amplitude reveals a great deal about a system's quickness to respond. Also, since the step function has, in principle, a wide band of frequencies in its spectrum, as a result of the jump discontinuity, as a test signal it is equivalent to the application of numerous sinusoidal signals with a wide range of frequencies.

The mathematical representation of a step function is

$$r(t) = \begin{cases} r_0 & t>0 \\ 0 & t<0 \end{cases} \tag{3.3}$$

or

$$r(t) = r_0 \cdot 1(t) \tag{3.4}$$

The step function is not defined at $t=0$. The step function as a function of time is shown in

Fig. 3.1(a). The Laplace transform of step function is obtained as

$$R(s) = \frac{r_0}{s} \tag{3.5}$$

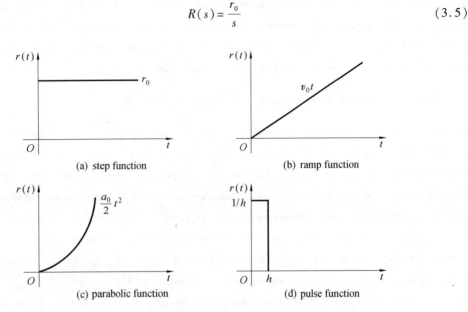

Figure 3.1 Standard test signals

2. Ramp function

In the case of ramp function, the signal increases linearly with respect to time. The ramp function may be regarded as the integral of the step function. If the input variable is of the form of the angular displacement of a shaft, the ramp input represents the constant-speed rotation of the shaft. Sometimes, the ramp signal is also called the velocity signal. A ramp function is one degree faster than a step function and has the ability to test how the system would respond to a signal that changes linearly with time.

Mathematically, a ramp function is represented by

$$r(t) = \begin{cases} v_0 t & t \geq 0 \\ 0 & t < 0 \end{cases} \tag{3.6}$$

or simply

$$r(t) = v_0 t \cdot 1(t) \tag{3.7}$$

The ramp function is shown in Fig. 3.1(b). The Laplace transform of ramp function is obtained as

$$R(s) = \frac{v_0}{s^2} \tag{3.8}$$

3. Parabolic function

The parabolic function may be regarded as the integral of the ramp function. The parabolic function shown in Fig. 3.1(c) is described by

$$r(t) = \begin{cases} \dfrac{a_0 t^2}{2} & t>0 \\ 0 & t<0 \end{cases} \qquad (3.9)$$

or simply

$$r(t) = \dfrac{a_0}{2} t^2 \cdot 1(t) \qquad (3.10)$$

Its Laplace transform is obtained as

$$R(s) = \dfrac{a_0}{s^3} \qquad (3.11)$$

A parabolic function is one degree faster than a ramp function. Sometimes, the parabolic signal is also called the acceleration signal. In practice, we seldom find it necessary to use a test signal faster than a parabolic function. This is because, as we shall show later, to track or follow a faster input, the system is necessary of higher order, which may mean that stability problem will be encountered.

4. Impulse function

A practical pulse function used in engineering is shown as in Fig. 3.1(d), its mathematical expression is given by

$$r(t) = \begin{cases} \dfrac{1}{h} & 0<t\leq h \\ 0 & t<0 \text{ or } t>h \end{cases} \qquad (3.12)$$

where h is the width of the pulse.

In analysis of control systems we often use idea impulse function $\delta(t)$ that is get from a pulse function as $h\to 0$, i.e.

$$\delta(t) = \begin{cases} \infty & t=0 \\ 0 & t\neq 0 \end{cases} \qquad (3.13)$$

and

$$\int_{-\infty}^{\infty} \delta(t)\,\mathrm{d}t = 1 \qquad (3.14)$$

Its Laplace transform is obtained as

$$R(s) = 1 \qquad (3.15)$$

All these test signals have the common features that they are simple to describe mathematically and easy to realize experimentally.

3.1.3 Transient Performance Specifications

The transient portion of the time response is that part which goes to zero as time becomes large. Of cause, the transient response has significance only when a stable system is referred to, since for an unstable system the response does not diminish and is out of control.

The transient performance of a control system is usually characterized by the use of step response. Typical performance specifications that are used to characterize the transient response to a step input include rise time, peak time, maximum overshoot, setting time, and etc. Fig. 3.2 illustrates a typical step response of linear control systems.

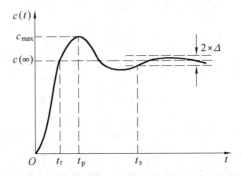

Figure 3.2 Representative step response

1. Rise time t_r

The rise time is a measure of the speed with which the system responds to a sudden change and can be defined in different ways. Usually, it is defined as the time required for the step response to reach the final value first time.

2. Peak time t_p

The peak time is also a measure of the speed of system response and is the time to the maximum of the first overshoot.

3. Maximum (percent) overshoot σ_p

The maximum overshoot is a measure of the oscillatory nature of the response of a stable system and defined as the largest deviation of the output over the step input during the transient state. The amount of maximum overshoot is also used as a measure of the relative stability of the system. The maximum overshoot is often represented as a percentage of the final value of the step response, i. e.

$$\sigma_p = \frac{c_{max} - c(\infty)}{c(\infty)} \times 100\% \tag{3.16}$$

4. Setting time t_s

The setting time is defined as the time required for the step response to settle to within a certain tolerance. The tolerance $\pm \Delta$ is usually taken as 5% or 2% of the final value as shown in Fig. 3.2.

The transient response of a control system may be described in terms of two factors:

① The speed of response, as represented by the rise time, peak time, and setting time.

② The closeness of the response to the desired response, as represented by the overshoot.

As nature would have it, these are contradictory requirements, and a compromise must be obtained.

These four quantities defined above give a direct measure of the transient characteristics of the step response. These quantities are relatively easy to measure when a step response is already plotted. However, analytically these quantities are difficult to determine except for the simple cases.

For many control systems to be encountered the transient response is mainly characterized by the maximum (percent) overshoot and setting time. As for the response with no oscillation, the rise time may be defined as the time required for the step response to rise from 10% to 90% of its final value, the peak time and percent overshoot are not defined.

3.2 Time Response of First-Order System

3.2.1 First-Order Systems

The order of a system is defined as the degree of its characteristic polynomial. One of the simplest dynamic systems studied is that represented by a first-order differential equation. Such a system is known as a first-order system. It may be seen from the previous examples that many different systems may be represented in first-order form. For a first-order system, the generalized transfer function between the input $R(s)$ and the output $C(s)$ may be given by the equation

$$\frac{C(s)}{R(s)} = \frac{1}{Ts+1} \quad (3.17)$$

and shown in block diagram form depicted in Fig. 3.3, where T is called the time constant of the first-order system. This class of systems will be studied in some detail by determining its responses to some common inputs.

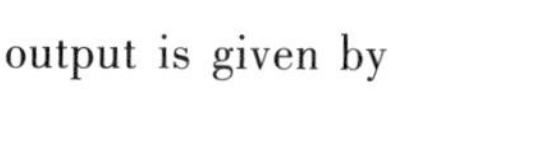

Figure 3.3 First-order system

3.2.2 Step Response

If the input is a unit step function, $R(s) = 1/s$, then the output is given by

$$C(s) = \frac{1}{Ts+1} \cdot \frac{1}{s} = \frac{\frac{1}{T}}{s\left(s+\frac{1}{T}\right)} = \frac{1}{s} - \frac{1}{s+\frac{1}{T}} \quad (3.18)$$

i.e.,

$$c(t) = 1 - e^{-\frac{1}{T}t} \quad (3.19)$$

The magnitude of output may be calculated from

$$c(t)|_{t=0} = 1 - e^0 = 0, \qquad c(t)|_{t=T} = 1 - e^{-1} = 0.632$$
$$c(t)|_{t=2T} = 1 - e^{-2} = 0.865, \qquad c(t)|_{t=3T} = 1 - e^{-3} = 0.950$$
$$c(t)|_{t=4T} = 1 - e^{-4} = 0.982, \qquad c(t)|_{t\to\infty} = 1 - e^{-\infty} = 1$$

First-order systems may be described as reaching 63% of their final value in one time constant and 86% of their final value in $2T$, 95% in $3T$, and 98% in $4T$. The step response curve is plotted in Fig. 3.4. In a practical sense, we consider that a step response will reach to its final value in three or four time constants, i.e., the setting time of the first-order system is

$$t_s = \begin{cases} 3T & \Delta = 5\% \\ 4T & \Delta = 2\% \end{cases} \quad (3.20)$$

Note that the step response curve has an initial slope of $1/T$, i.e.,

$$\left.\frac{dc(t)}{dt}\right|_{t=0} = -\left.\frac{d}{dt}(e^{-t/T})\right|_{t=0} = \frac{1}{T} \quad (3.21)$$

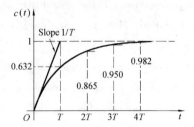

Figure 3.4 Step response of a first-order system

which means that the response would reach its final value in T seconds if it had continued to increase at the initial rate, so that the parameter T is called the time constant of the first-order system and has units of seconds.

3.2.3 Impulse Response

In this case, for a unit impulse, $R(s) = 1$ and the output becomes

$$C(s) = \frac{1}{Ts+1} = \frac{\frac{1}{T}}{s+\frac{1}{T}} \quad (3.22)$$

and

$$c(t) = \frac{1}{T}e^{-\frac{1}{T}t} \quad (3.23)$$

Fig. 3.5 shows the output of the system. Since we have assumed zero initial condition, the output must change instantaneously from 0 at $t = 0_-$ to $1/T$ at $t = 0_+$.

It may be noted, since impulse is the derivative of step function, the impulse response just is the derivative of step response.

3.2.4 Ramp Response

If the input signal is a unit ramp function, i.e. $R(s) = 1/s^2$, then

$$C(s) = \frac{1}{Ts+1} \cdot \frac{1}{s^2} = \frac{1}{s^2} - \frac{T}{s} + \frac{T}{s+\frac{1}{T}} \quad (3.24)$$

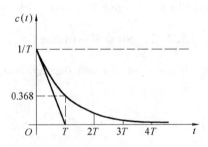

Figure 3.5 Impulse response of first-order system

The time domain response becomes

$$c(t) = t - T + Te^{-\frac{1}{T}t} = t - T(1 - e^{-\frac{1}{T}t}) \quad (3.25)$$

Fig. 3.6 (a) shows the relationship between the input and output for this case. The response consists of a transient part and a steady-state part, and in the steady state the output lags the input by a time equal to the time constant T. It may be shown that if the input were not the unit ramp but $r(t) = At$ then a fixed time, the difference between the input and steady-state output would be AT as shown in Fig. 3.6(b).

(a) $r(t)=t$ (b) $r(t)=At$

Figure 3.6 Ramp response of first-order system

It should be noted that the ramp response is an integral of the step response.

3.2.5 First-Order Feedback Systems

Now that the response of first-order systems has been studied in some detail, one of the principal benefits of feedback control may be illustrated. Suppose it is necessary to control a system known to be of first order. The simplest method will be to implement open-loop control as illustrated in Fig. 3.3. If the time constant is assumed to be $T = 1$ s, then the response to a unit step will be of the form

$$c(t) = 1 - e^{-t} \quad (3.26)$$

as shown in Fig. 3.4, reaching 63% of the steady-state output in 1s. Now, suppose the same first-order system is considered to be the plant in a feedback control system with a variable amplifier gain as controller, as shown in Fig. 3.7.

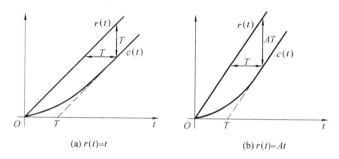

Figure 3.7 First-order feedback system

Now, the relationship between the input and output is given by the closed-loop transfer function

$$\frac{C(s)}{R(s)} = \frac{KG(s)}{1+KG(s)} = \frac{K}{Ts+K+1} \quad (3.27)$$

If the input is a unit step, then

$$C(s) = \frac{K}{Ts+K+1} \cdot \frac{1}{s} = \frac{K/T}{s[s+(K+1)/T]} =$$
$$\frac{K}{K+1} \cdot \frac{1}{s} - \frac{K}{K+1} \cdot \frac{1}{s+(K+1)/T} \quad (3.28)$$

Hence the time response

$$c(t) = \frac{K}{K+1}(1 - e^{-\frac{K+1}{T}t}) \qquad (3.29)$$

Suppose that $T = 1$ and we arbitrarily set $K = 1$, we get

$$c(t) = 0.5(1 - e^{-2t}) \qquad (3.30)$$

This output (step response) is shown in Fig. 3.8, which also includes the step response for the open-loop system (in dashed).

Note the following important observations.

① The steady-state output of the feedback system is 0.5. The feedback system has an error of 0.5 between the input and steady-state output.

② The time constant of the feedback system is 0.5. The feedback system response is faster than that of the open-loop system.

Hence we have improved the system transient response, which is good; however, the error between

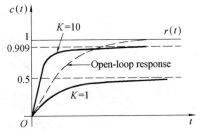

Figure 3.8 Step response of feedback system

the input and steady-state output, which in this case is considerable, is a serious impediment to the system performance.

Notice now, however, that if the gain K is increased to 10, then the output becomes

$$c(t) = 0.909(1 - e^{-11t}) \qquad (3.31)$$

This case, which is also plotted in Fig. 3.8, indicated that the steady-state error between the input and output is now reduced to less than 10% while the closed-loop time constant is 0.09 s, resulting in over 10 times the response speed of the open-loop system.

This example clearly shows that through the use of feedback it is possible to improve the dynamic response of a control system; meanwhile, this process may also have some detrimental effects such as the appearance of an error between the input and steady-state output. However, by using a large value of gain, not only does the system response become very fast, but also the error becomes smaller. In practical case the gain cannot be increased indefinitely without signal saturation occurring, causing the system to become nonlinear.

3.2.6 Poles and Zeros of First-Order Systems

Consider the transfer function of the first-order system developed in detail earlier in the module

$$G(s) = \frac{1}{Ts+1} \qquad (3.32)$$

It is seen that this open-loop system has no zeros and one pole at

$$s = -1/T \qquad (3.33)$$

Recall that the transient response of the system, for any input, was characterized by the time constant T, or more specifically $-1/T$, i.e. the system pole. This feature is not limited to first-order system and may be stated in a wider context as follows:

"A system's transient response is determined by the poles of the transfer function, which in turn are obtained by setting the denominator of the transfer function to zero and solving the resulting expression, known as the characteristic equation."

One may observe that it is the poles that determine the character of the system response, since they will appear in the determination of the output $C(s)$ for any input $R(s)$. The role of the zeros in the determination of the partial-fraction coefficients prior to taking the inverse Laplace transform to obtain the time-domain response. In the general feedback case, i.e.

$$\frac{C(s)}{R(s)} = \frac{G(s)}{1+G(s)H(s)}$$

the characteristic equation is of the form

$$1+G(s)H(s) = 0 \tag{3.34}$$

In the case of that the closed-loop first-order system is described by

$$\frac{C(s)}{R(s)} = \frac{KG(s)}{1+KG(s)} = \frac{K}{Ts+K+1} \tag{3.35}$$

the pole at

$$s = -\frac{K+1}{T} = -\frac{1}{T'} \tag{3.36}$$

is known as a closed-loop pole. The above equation shows that this closed-loop pole is a function of K and the system transient response will therefore be a function of this parameter, confirming our previous observation of the effects of feedback on the first-order system. From Eq. (3.36), larger values of K result in larger values of s, which produce smaller effective time constant T' and faster response.

3.3 Time Response of Second-Order System

3.3.1 Second-Order Systems

The next class of systems to be investigated is called second-order because they will be described by second-order differential equations. Second-order systems are important because their behavior is very different from first-order systems and may exhibit feature such as oscillatory response, or overshoot. Moreover, their analysis generally is helpful to form a basis for the understanding of analysis and design techniques.

1. Normalized second-order system and its transfer function

Consider the block diagram of a second-order system shown in Fig. 3.9, it may be a representation of a position control system, for instance Example 2.5 in chapter 2, or some other systems. The transfer function is obtained as

$$\frac{C(s)}{R(s)} = \frac{K}{Ts^2+s+K} \tag{3.37}$$

However, as in the first-order system, usually the coefficients are written in a manner such that they have some physical meaning. Letting

$$\zeta = \frac{1}{2\sqrt{KT}}, \quad \omega_n = \sqrt{\frac{K}{T}}$$

the transfer function of a normalized second-order system is written as

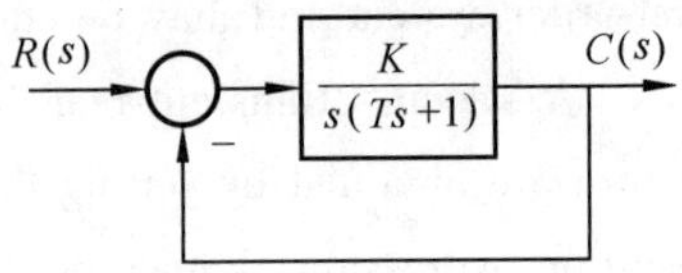

$$\frac{C(s)}{R(s)} = \frac{\omega_n^2}{s^2 + 2\zeta\omega_n s + \omega_n^2} \tag{3.38}$$

Figure 3.9 Normalized second-order system

where ζ is called the dimensionless damping ratio and ω_n is called the natural undamped oscillatory frequency.

2. Poles of second-order system

A second-order system described by Eq. (3.38) has two poles

$$s_{1,2} = -\zeta\omega_n \pm \omega_n \sqrt{\zeta^2 - 1} \tag{3.39}$$

Depending on the value of ζ, these two poles may be either real or a pair of complex as shown on the complex plane in Fig. 3.10. On the s-plane, a closed-loop pole is usually represented

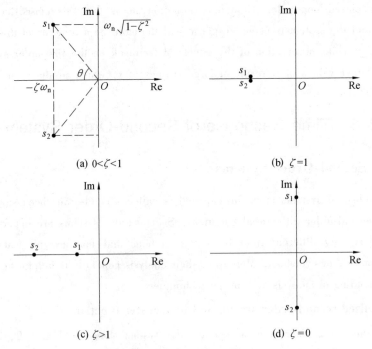

Figure 3.10 Poles of second-order system

with a small solid circle "·".

It is useful to remember the following relationships in the case of $0<\zeta<1$:

① The length of the line joining the pole to the origin is given by

$$l=\sqrt{(\zeta\omega_n)^2+(\omega_n\sqrt{1-\zeta^2})^2}=\omega_n \tag{3.40}$$

② The angle between the line joining the pole to the origin and negative real axis is given by

$$\cos\theta=\frac{\zeta\omega_n}{\omega_n}=\zeta \tag{3.41}$$

3.3.2 Step Response

Considerable information regarding the nature of response of a second-order system may be obtained by studying its behavior when subjected to a unit step input. In such a case, Eq. (3.38) indicates that the output is given by

$$C(s)=\frac{\omega_n^2}{s^2+2\zeta\omega_n s+\omega_n^2}\cdot\frac{1}{s}=\frac{1}{s}-\frac{s+2\zeta\omega_n}{s^2+2\zeta\omega_n s+\omega_n^2} \tag{3.42}$$

Obviously, the response will present different characteristics when the system poles are real or complex due to different values of ζ.

1. Case of $0<\zeta<1$

In this case, two poles of system are complex with negative real part

$$s_{1,2}=-\zeta\omega_n\pm j\omega_n\sqrt{1-\zeta^2} \tag{3.43}$$

The output is given by

$$C(s)=\frac{1}{s}-\frac{s+2\zeta\omega_n}{(s+\zeta\omega_n)^2+\omega_n^2(1-\zeta^2)}$$

Let

$$\omega_d=\omega_n\sqrt{1-\zeta^2} \tag{3.44}$$

then

$$C(s)=\frac{1}{s}-\left[\frac{s+\zeta\omega_n}{(s+\zeta\omega_n)^2+\omega_d^2}+\frac{\zeta\omega_n}{(s+\zeta\omega_n)^2+\omega_d^2}\right] \tag{3.45}$$

Taking the inverse Laplace transform yields

$$c(t)=1-\left[e^{-\zeta\omega_n t}\cos\omega_d t+\frac{\zeta\omega_n}{\omega_d}e^{-\zeta\omega_n t}\sin\omega_d t\right]=1-e^{-\zeta\omega_n t}\left(\cos\omega_d t+\frac{\zeta}{\sqrt{1-\zeta^2}}\sin\omega_d t\right)=$$

$$1-\frac{1}{\sqrt{1-\zeta^2}}e^{-\zeta\omega_n t}\sin(\omega_d t+\theta) \tag{3.46}$$

where

$$\theta=\arccos\zeta \tag{3.47}$$

Note that the transient component of Eq. (3.46) is a decayed oscillatory quantity. Fig. 3.11 shows the step response as a function of the dimensionless time $\omega_n t$ for various values of damping ratio ζ.

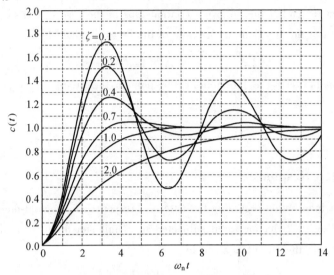

Figure 3.11 Step response of normalized second-order system with various ζ

At this stage, it is important to recognize that when $0<\zeta<1$ the figure shows.

① The system oscillates at frequency ω_d, which is sometimes called the damped oscillatory frequency and just given by the imaginary part of the pole;

② The oscillation is exponentially decayed and the decay rate is controlled by the constant $\zeta\omega_n$, which is called as damping factor and just given by the real part of the pole.

In this case, the system is said to be underdamped due to the underdamped nature of its step response.

2. Case of $\zeta>1$

If $\zeta>1$, then two poles are real and unequal

$$s_{1,2} = -\zeta\omega_n \pm \omega_n\sqrt{\zeta^2-1} \qquad (3.48)$$

and the unit step response is

$$C(s) = \frac{\omega_n^2}{(s-s_1)(s-s_2)} \cdot \frac{1}{s} =$$

$$\frac{1}{s} + \frac{\omega_n^2}{s_1(s_1-s_2)} \cdot \frac{1}{s-s_1} + \frac{\omega_n^2}{s_2(s_2-s_1)} \cdot \frac{1}{s-s_2} \qquad (3.49)$$

Taking inverse Laplace transform

$$c(t) = 1 + \frac{\omega_n^2}{s_1(s_1-s_2)}e^{s_1 t} + \frac{\omega_n^2}{s_2(s_2-s_1)}e^{s_2 t} \qquad (3.50)$$

Eq. (3.50) has a transient component consisting of two decay exponential functions. In the case of $\zeta>1$ the response is of overdamped nature, or the system is said to be overdamped.

3. Case of $\zeta = 1$

For $\zeta = 1$, two poles are real and equal, i.e.,
$$s_{1,2} = -\omega_n \tag{3.51}$$

The unit step response is given by
$$C(s) = \frac{1}{s} - \frac{s+2\omega_n}{(s+\omega_n)^2} = \frac{1}{s} - \left[\frac{\omega_n}{(s+\omega_n)^2} + \frac{1}{s+\omega_n}\right] \tag{3.52}$$

Taking inverse Laplace transform, the result is
$$c(t) = 1 - (\omega_n t + 1)e^{-\omega_n t} \tag{3.53}$$

This equation also has two decay exponential functions terms. In this case, for obvious reason, the system is said to be critical damped.

4. Case of $\zeta = 0$

In this case, two poles are imaginary, i.e.,
$$s_{1,2} = \pm j\omega_n \tag{3.54}$$

the output is
$$C(s) = \frac{1}{s} - \frac{s}{s^2 + \omega_n^2} \tag{3.55}$$

Taking inverse Laplace transform, the result is
$$c(t) = 1 - \cos\omega_n t \tag{3.56}$$

In the case of zero damping, $\zeta = 0$, the response is purely sinusoidal and the system is said to be undamped. As the definition, the natural undamped frequency ω_n corresponds to the frequency of the undamped sinusoid.

3.3.3 Evaluation of Performance Specifications

In practice applications, the evaluation of performance specification for the normalized second-order systems
$$\frac{C(s)}{R(s)} = \frac{\omega_n^2}{s^2 + 2\zeta\omega_n s + \omega_n^2} \tag{3.57}$$

is of interest.

For the case of overdamped ($\zeta > 1$) or critical damped ($\zeta = 1$), there is no overshoot in the step response. As for setting time, it is difficult to determine the exact value, however in case of $\zeta \geq 0.7$ it can be obtained by the empirical formula
$$t_s \approx \frac{1}{\omega_n}(6.45\zeta - 1.7), \quad \Delta = 5\% \tag{3.58}$$

Now we consider the second-order system described by Eq. (3.57), in which, $0 < \zeta < 1$ and the expression of the unit step response is
$$c(t) = 1 - \frac{1}{\sqrt{1-\zeta^2}} e^{-\zeta\omega_n t} \sin(\omega_d t + \theta) \tag{3.59}$$

where $\omega_d = \omega_n\sqrt{1-\zeta^2}$, $\theta = \arccos\zeta$.

(1) Rise time t_r

From Eq. (3.59), for the rise time, we have $c(t_r) = 1$, i.e.,

$$1 - \frac{1}{\sqrt{1-\zeta^2}}e^{-\zeta\omega_n t_r}\sin(\omega_d t_r + \theta) = 1$$

This equation is simplified to

$$\sin(\omega_d t_r + \theta) = 0$$

or

$$t_r = \frac{k\pi - \theta}{\omega_d}$$

where k is a positive integer. Since the rise time is the time for the step response to reach its final value first time, hence

$$t_r = \frac{\pi - \theta}{\omega_d} = \frac{\pi - \theta}{\omega_n\sqrt{1-\zeta^2}} \qquad (3.60)$$

(2) Peak time t_p

The peak time can be obtained by taking the derivative of Eq. (3.59) and setting the result to zero. Thus

$$\left.\frac{dc(t)}{dt}\right|_{t=t_p} = -\frac{\zeta\omega_n e^{-\zeta\omega_n t_p}}{\sqrt{1-\zeta^2}}\sin(\omega_d t_p + \theta) + \frac{\omega_d e^{-\zeta\omega_n t_p}}{\sqrt{1-\zeta^2}}\cos(\omega_d t_p + \theta) =$$

$$\frac{\omega_n e^{-\zeta\omega_n t_p}}{\sqrt{1-\zeta^2}}[-\zeta\sin(\omega_d t_p + \theta) + \sqrt{1-\zeta^2}\cos(\omega_d t_p + \theta)] = 0$$

Since $t_p \neq \infty$, this equation can be simplified to

$$-\zeta\sin(\omega_d t_p + \theta) + \sqrt{1-\zeta^2})\cos(\omega_d t_p + \theta) = 0$$

Since $\cos\theta = \zeta$ and $\sin\theta = \sqrt{1-\zeta^2}$, we get

$$\sin(\omega_d t_p) = 0$$

i.e.,

$$t_p = \frac{k\pi}{\omega_d}$$

where k is a positive integer. Obviously, the peak time occurs at $k=1$. Hence, the peak time is given by

$$t_p = \frac{\pi}{\omega_d} = \frac{\pi}{\omega_n\sqrt{1-\zeta^2}} \qquad (3.61)$$

(3) Maximal (percent) overshoot σ_p

The magnitudes of the maximal overshoot can be obtained by substituting equation (3.61) into Eq. (3.59). Thus

$$c(t_p) = 1 - \frac{1}{\sqrt{1-\zeta^2}}e^{-\zeta\omega_n\frac{\pi}{\omega_n\sqrt{1-\zeta^2}}}\sin\left(\omega_d\frac{\pi}{\omega_d} + \theta\right) =$$

Chapter 3 Time-Domain Analysis of Control Systems

$$1 - \frac{1}{\sqrt{1-\zeta^2}} e^{-\pi\zeta/\sqrt{1-\zeta^2}} \sin\theta = 1 - e^{-\pi\zeta/\sqrt{1-\zeta^2}}$$

By definition, the maximal (percent) overshoot is

$$\sigma_p = e^{-\pi\zeta/\sqrt{1-\zeta^2}} \times 100\% \tag{3.62}$$

Note that the maximal overshoot of the step response is only a function of the damping ratio. The relation between the maximal overshoot and the damping ratio for the second system is shown in Fig. 3.12.

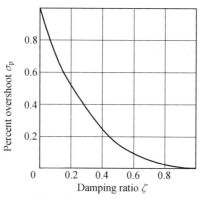

Figure 3.12 σ_p versus ζ

(4) Setting time t_s

From the definition of setting time we have

$$|c(t) - c(\infty)| = \left| \frac{1}{\sqrt{1-\zeta^2}} e^{-\zeta\omega_n t} \sin(\omega_d t + \theta) \right| \leq \Delta, \quad t \geq t_s$$

It is difficult to determine the exact value of the setting time. However, we can see that this occurs approximately when

$$\frac{1}{\sqrt{1-\zeta^2}} e^{-\zeta\omega_n t} \leq \Delta, \quad t \geq t_s$$

or

$$\zeta\omega_n t \geq \ln \frac{1}{\Delta\sqrt{1-\zeta^2}}, \quad t \geq t_s$$

From above equation we take

$$t_s \approx \frac{1}{\zeta\omega_n} \ln \frac{1}{\Delta\sqrt{1-\zeta^2}} \tag{3.63}$$

For small values of ζ, $0 < \zeta < 0.9$, the setting time is given by the approximate formula

$$t_s \approx \frac{3}{\zeta\omega_n}, \quad \Delta = 5\% \tag{3.64}$$

or

$$t_s \approx \frac{4}{\zeta\omega_n}, \quad \Delta = 2\% \tag{3.65}$$

Now reviewing the relationship for the rise time and setting time. It is seen that small value of ζ would yield short rise time, however, a fast setting time requires a large value for ζ. Therefore, a compromise in the value of ζ should be made when all these specifications are to be satisfactorily met in a design problem. Together with the consideration on maximum percent overshoot, a generally accepted range of damping ratio for satisfactory all-around performance is between 0.5 ~ 0.8.

Example 3.1 Fig. 3.13 shows a feedback control system with a fixed plant but a controller with the gain and placement of the pole to be specified by the system designer. Determine the location of the pole and the value of K that satisfy the following step response design requirements:

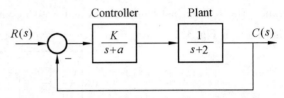

Figure 3.13 Feedback control system of example 3.1

(1) a 5% setting time of no more than 1s and

(2) the overshoot not to exceed 10%. After calculating the appropriate system parameters, determine

(a) the damped oscillatory frequency ω_d,

(b) the peak time t_p.

Solution From the overshoot requirement
$$e^{-\pi\zeta/\sqrt{1-\zeta^2}} \leq 0.1$$
solving this equation yields
$$\zeta = 0.6$$
The 5% setting time is obtained from Eq. (3.64)
$$t_s = \frac{3}{\zeta\omega_n} = 1 \text{ s}$$
hence
$$\omega_n = 5 \text{ rad/s}$$
The closed-loop transfer function of the system takes the form
$$\frac{C(s)}{R(s)} = \frac{K}{s^2+(2+a)s+(K+2a)}$$
Comparing with the normalized form of a second-order system leads to
$$\omega_n = \sqrt{K+2a}, \quad \zeta = \frac{2+a}{2\sqrt{K+2a}}$$
Substituting ζ and ω_n into the equation gives
$$\sqrt{K+2a} = 5, \quad \frac{2+a}{2\sqrt{K+2a}} = 0.6$$
Solving theses equations yields the result
$$a = 4, \quad K = 17$$
The design requirements may therefore be met by setting $K = 17$ and $a = 4$.

The characteristic equation may be obtained directly from the denominator of the closed-loop transfer function derived previously

$$s^2+(2+a)s+(K+2a)=0$$

i. e.,

$$s^2+6s+25=0$$

The roots of this equation are

$$s_{1,2}=-3\pm j4$$

The damped oscillatory frequency is seen to be

$$\omega_d=4 \text{ rad/s}$$

The peak time is given by

$$t_p=\frac{\pi}{\omega_d}=0.785 \text{ s}$$

3.4 Time Response of Higher-Order System

3.4.1 Step Response of Higher-Order System

All the systems which input-output relationships are described by third-order or higher-order differential equations are called the higher-order systems. The transient behavior of a higher-order system is still characterized by its step response

Consider an n-order system

$$\Phi(s)=\frac{C(s)}{R(s)}=\frac{b_0s^m+\ldots b_{m-1}s+b_m}{s^n+a_1s^{n-1}+\ldots+a_{n-1}s+a_n}=\frac{N(s)}{D(s)} \quad (3.66)$$

where $N(s)$ is the numerator polynomial and $D(s)$ is the denominator polynomial. Taking consideration of the requirements on the dynamics, many control systems are designed to present decayed oscillation. Since the poles of $\Phi(s)$ must be real or conjugate complex, without loss of generality, it is assumed that $\Phi(s)$ has p real poles and q pairs of conjugate complex poles, and all these poles are distinct. Then, the system response to a unit step signal is

$$C(s)=\frac{N(s)}{\sum_{i=1}^{p}(s-s_i)\sum_{j=1}^{q}(s^2+2\zeta_j\omega_{nj}s+\omega_{nj}^2)}\cdot\frac{1}{s} \quad (3.67)$$

Expanding this equation into partial-fraction form, the response can be represented in the form

$$C(s)=\frac{A_0}{s}+\sum_{i=1}^{p}\frac{B_i}{s-s_i}+\sum_{j=1}^{q}\frac{C_js+D_j}{s^2+2\zeta_j\omega_{nj}s+\omega_{nj}^2} \quad (3.68)$$

where $A_0, B_i(i=1,2,\ldots,p), C_j$ and $D_j(j=1,2,\ldots,q)$ are coefficients obtained from the partial fraction expansion. Eq. (3.68) shows that the transient part of response contains p decayed exponential components and q decayed sinusoid components if all these poles are negative real or conjugate complex with negative real parts.

Taking Laplace transform, it is not difficult to get the response expression in the time domain, $c(t)$. Thus, the performance specifications may be evaluated by definition from the expression $c(t)$ or the corresponding response curve.

Obviously, the dynamic analysis of higher-order systems is more complex. Moreover, it is difficult to derive the analytic expressions for evaluating the performance specifications. Fortunately, for many higher-order systems, the dominant pole method can be used to simplify the system analysis and evaluation of performance specifications.

3.4.2 Dominant Pole Method

1. Concept of dominant poles

We introduce the concept of dominant pole(s) through an example. Consider the following third-order system

$$\Phi(s) = \frac{C(s)}{R(s)} = \frac{50}{(s+1)(s+5)(s+10)}$$

The unit-step response is

$$C(s) = \frac{50}{(s+1)(s+5)(s+10)} \cdot \frac{1}{s} =$$

$$\frac{1}{s} - \frac{50}{36} \cdot \frac{1}{s+1} + \frac{50}{100} \cdot \frac{1}{s+5} - \frac{50}{450} \cdot \frac{1}{s+10}$$

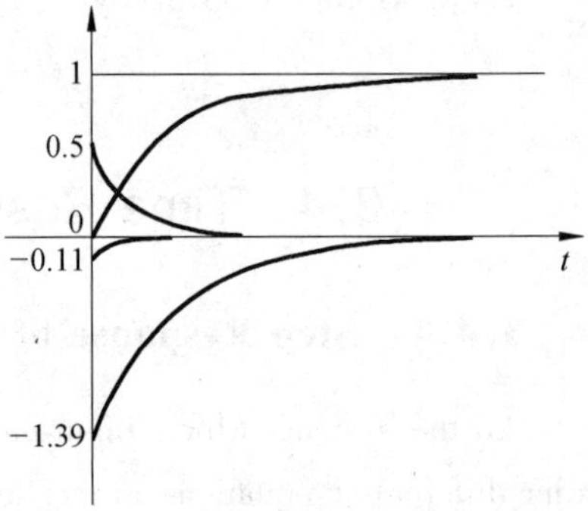

i. e.,

$$c(t) = 1 - 1.39e^{-t} + 0.5e^{-5t} - 0.11e^{-10t}$$

Figure 3.14 Step response and its components of a third-order system

and the response curve and its components are drawn in Fig. 3.14.

Comparing three transient components, it can be seen that both the decay rate and the initial magnitude of each component are closely related to the distance from the corresponding pole to the imaginary axis. The term e^{-t} has considerably slower decay rate and larger initial magnitude than the terms e^{-5t} and e^{-10t}, so that the contributions from the terms e^{-5t} and e^{-10t} may be neglected in the resulting time-domain response for all time but $t \rightarrow 0$, and we can say that the term e^{-t} dominates the response and, correspondingly, $s = -1$ is the dominant pole of the system.

This result may be generalized that, for a transfer function without finite zeros, the following phenomena can be observed:

① The closer the pole to the imaginary axis is, the slower the corresponding response component decays.

② The closer the pole to the imaginary axis is, the greater the initial magnitude of the corresponding component is.

Hence, in many cases, the system response may be approximated by the response of the pole(s) closest to the imaginary axis if all other poles are located sufficient far from the imaginary axis, for example more than 5 times, and such a real pole or a pair of conjugate complex poles is called the dominant pole(s). It should be noted, however, if the transfer function of a system possesses finite zeros and they are located relatively near the pole(s) closest to the imaginary axis, then the zeros will materially affect the transient response of the system and the poles closest to the imaginary axis may be no longer the dominant ones of the system.

2. Response Approximation of Higher-Order System

Consider the system given in Eq. (3.66) and assume that the system has a pair of dominant poles

$$s_{1,2} = -\zeta\omega_n \pm j\omega_n\sqrt{1-\zeta^2} = -\zeta\omega_n \pm j\omega_d \tag{3.69}$$

Then, the step response becomes

$$C(s) = \frac{N(s)}{D(s)} \cdot \frac{1}{s} \approx$$

$$\frac{A_0}{s} + \left[\frac{(s-s_1)N(s)}{D(s)} \cdot \frac{1}{s}\right]_{s=s_1} \cdot \frac{1}{s-s_1} + \left[\frac{(s-s_2)N(s)}{D(s)} \cdot \frac{1}{s}\right]_{s=s_2} \cdot \frac{1}{s-s_2} \tag{3.70}$$

and

$$c(t) \approx A_0 + 2\left.\frac{(s-s_1)N(s)}{sD(s)}\right|_{s=s_1} e^{-\zeta\omega_n t} \cos\left[\left(\omega_d t + \angle \frac{(s-s_1)N(s)}{sD(s)}\bigg|_{s=s_1}\right)\right] \tag{3.71}$$

Note that in Eq. (3.71) the terms of non-dominant poles are neglected, however, the effect of non-dominant poles on the response is reflected in the term of dominant poles.

Applying the dominant pole method, the following equations for calculating performance specifications can be derived from the step response approximation given by Eq. (3.71).

(1) Peak time t_p

According to the definition, the peak time is evaluated by

$$t_p = \frac{1}{\omega_d}\left[\pi - \sum_{i=1}^{m}\angle(s_1-z_i) + \sum_{j=3}^{n}\angle(s_1-s_j)\right] \tag{3.72}$$

where $z_i (i=1,2,\ldots,m)$ are the zeros, s_1 is the dominant pole, and $s_j (j=3,\ldots,n)$ are the non-dominant zeros.

Eq. (3.72) shows that the closed-loop zeros will make t_p shorter; whereas, the non-dominant poles will make it longer. Moreover, in general, the closer to the imaginary axis they are, the stronger their effects are.

(2) Maximal (percent) overshoot σ_p

According to the definition, the maximal (percent) overshoot is evaluated by

$$\sigma_p\% = \frac{\prod_{i=1}^{m}|s_1-z_i|}{\prod_{i=1}^{m}|z_i|} \cdot \frac{\prod_{j=3}^{n}|s_j|}{\prod_{j=3}^{n}|s_1-s_j|} e^{-\zeta\omega_n t_p} \times 100\% \tag{3.73}$$

It can be seen from Eq. (3.73) that the effect of the closed-loop zero z_i on overshoot is to make it larger if $|s_1-z_i|>|z_i|$; whereas, the effect of the non-dominant pole s_j is to make it smaller if $|s_1-s_j|>|s_j|$.

(3) Setting time t_s

The setting time is evaluated by

$$t_s = \frac{1}{\zeta\omega_n}\ln\left[\frac{2}{\Delta}\frac{\prod_{i=1}^{m}|s_1-z_i|}{\prod_{i=1}^{m}|z_i|}\cdot\frac{\prod_{j=2}^{n}|s_j|}{\prod_{j=2}^{n}|s_1-s_j|}\right] \quad (3.74)$$

It can be seen from Eq. (3.74) that the effect of the closed-loop zero z_i on setting time trends to make it longer if $|s_1-z_i|>|z_i|$; whereas, the effect of the non-dominant pole s_j trends to make it smaller if $|s_1-s_j|>|s_j|$.

The performance specifications, for the second-order system with a zero, can be estimated approximately by use of above equations if the zero is far from the imaginary axis.

Example 3.2 The open-loop transfer function of a unity negative feedback system is given by

$$G(s) = \frac{0.5s+1}{s(s+0.5)}$$

Determine the maximal percent overshoot σ_p.

Solution The closed-loop transfer function of this system is

$$\Phi(s) = \frac{0.5s+1}{s^2+s+1}$$

and the poles and zero of the closed-loop system are

$$s_{1,2} = -0.5 \pm j0.866, \quad z_1 = -2$$

respectively.

Since $|z_1| \gg |\text{Re}(s_1)|$, the performance specifications of the system can be estimated by use of the approximate equations for higher-order systems. From Eq. (3.72) ~ (3.74), we have

$$t_p = \frac{1}{\omega_d}[\pi - \angle(s_1-z_i)] = \frac{1}{0.866}(\pi-\arctan\frac{0.866}{2-0.5}) = 3.028 \text{ s}$$

$$\sigma_p\% = \frac{|s_1-z_1|}{|z_1|}e^{-\zeta\omega_n t_p}\times 100\% = \frac{\sqrt{(2-0.5)^2+0.866^2}}{2}\times e^{-(0.5\times 3.02)}\times 100\% = 19.1\%$$

3.4.3 Reducing to Lower-Order Systems

Alternative approach to simply the analysis of higher-order systems possessing dominant poles is to reduce them to lower-order systems. This approach, as another approximation, avoids too much operation in order to obtain the performance specifications.

For example, consider the third-order system

$$\frac{C(s)}{R(s)} = \frac{50}{(s+1)(s+5)(s+10)}$$

In order to obtain the correct transfer function corresponding to the reduced-order system, it is first necessary to express the transfer function in the typical factor form

$$\frac{C(s)}{R(s)} = \frac{1}{(s+1)(0.2s+1)(0.1s+1)}$$

Neglecting the non-dominant poles, the reduced system now becomes

$$\frac{C(s)}{R(s)} = \frac{1}{s+1}$$

The transfer function must be expressed in the standard factor form before the reduction is takes place, otherwise the gain of the transfer function changes as the non-dominant terms are deleted.

The setting time of overdamped second-order systems can be estimated by use of this method if one pole is comparatively farther from the imaginary axis than another one.

Example 3.3 Estimate the setting time t_s of a system with the transfer function

$$\Phi(s) = \frac{2}{s^2+8s+4}$$

Solution From the denominator polynomial of system transfer function, we have

$$\omega_n^2 = 4, \quad 2\zeta\omega_n = 8$$

The damping ratio is

$$\zeta = 2$$

This system is an overdamped second-order system. The poles of the system are

$$s_1 = -0.54, \quad s_2 = -7.46$$

Since $|s_2| \gg |s_1|$, the pole s_1 can be considered as a dominant pole, and the system can be approximated by a first-order system, i.e.,

$$\Phi(s) = \frac{2}{s^2+8s+4} = \frac{2}{(s+0.54)(s+7.46)} = \frac{\frac{2}{0.54 \times 7.46}}{(\frac{1}{0.54}s+1)(\frac{1}{7.46}s+1)} \approx \frac{0.5}{\frac{1}{0.54}s+1}$$

The setting time is

$$t_s \approx 3 \times \frac{1}{0.54} = 5.56 \text{ s}, \quad \Delta = 5\%$$

3.4.4 Third-Order Systems

Consider a third-order system with the transfer function

$$\frac{C(s)}{R(s)} = \frac{s_0\omega_n^2}{(s+s_0)(s^2+2\zeta\omega_n s+\omega_n^2)}$$

where $s_{1,2} = -\zeta\omega_n \pm j\omega_n\sqrt{1-\zeta^2}$ are a pair of conjugate complex poles with negative real part and $s_3 = -s_0$ is a negative real pole. The system response to unit step signal is

$$C(s) = \frac{s_0\omega_n^2}{(s+s_0)(s^2+2\zeta\omega_n s+\omega_n^2)} \cdot \frac{1}{s}$$

Expanding this equation into partial-fraction form and denoting $\omega_d = \omega_n\sqrt{1-\zeta^2}$ yield

$$C(s) = \frac{1}{s} + \frac{A}{s+s_0} + \frac{B}{s+\zeta\omega_n-j\omega_d} + \frac{C}{s+\zeta\omega_n+j\omega_d}$$

where

$$A = \frac{-\omega_n^2}{s_0^2 - 2\zeta\omega_n s_0 + \omega_n^2}$$

$$B = \frac{s_0(2\zeta\omega_n - s_0) - js_0(2\zeta^2\omega_n - \zeta s_0 - \omega_n)/\sqrt{1-\zeta^2}}{2[(2\zeta^2\omega_n - \zeta s_0 - \omega_n)^2 + (2\zeta\omega_n - s_0)^2(1-\zeta^2)]}$$

$$C = \bar{B}$$

and $\bar{B}$ is the conjugate of B.

Taking Laplace transform and letting $\beta = s_0/\zeta\omega_n$ yield

$$c(t) = 1 + Ae^{-s_0 t} + be^{-\zeta\omega_n t}\cos\omega_d t + ce^{-\zeta\omega_n t}\sin\omega_d t, \quad t \geq 0$$

where

$$A = \frac{-\omega_n^2}{s_0^2 - 2\zeta\omega_n s_0 + \omega_n^2} = -\frac{1}{\beta\zeta^2(\beta-2) + 1}$$

$$b = 2\operatorname{Re}B = -\frac{\beta\zeta^2(\beta-2)}{\beta\zeta^2(\beta-2) + 1}$$

$$c = -2\operatorname{Im}B = -\frac{\beta\zeta[\beta\zeta^2(\beta-2) + 1]}{[\beta\zeta^2(\beta-2) + 1]\sqrt{1-\zeta^2}}$$

Substituting these coefficients into the expression of $c(t)$ and rearranging it yield

$$c(t) = 1 - \frac{1}{\beta\zeta^2(\beta-2)+1}e^{-s_0 t} - \frac{1}{\beta\zeta^2(\beta-2)+1}e^{-\zeta\omega_n t}\left[\beta\zeta^2(\beta-2)\cos\omega_d t + \frac{\beta\zeta[\beta\zeta^2(\beta-2)+1]}{\sqrt{1-\zeta^2}}\sin\omega_d t\right], \quad t \geq 0$$

Using a computer simulation, one can determine the response of a system to a unit step input for given ζ, ω_n and various β. The results for $\zeta = 0.45$ and $\omega_n = 1$ are shown in Table 3.1, where the setting time utilizes a 2% criterion.

Table 3.1 Effect of a third pole for $\zeta = 0.45$ and $\omega_n = 1$

s_0	β	Percent overshoot	Setting time
0.444	0.99	0	9.63
0.666	1.47	3.9	6.3
1.111	2.47	12.3	8.81
2.5	5.56	18.6	8.67
20	44.4	20.5	8.37
∞	∞	20.5	8.24

For example, taking $\zeta = 0.45, \omega_n = 1$, and $\beta = 5.56$, results in

$$s_0(=2.5) > 5\zeta\omega_n(=2.25)$$

and we consider that $s_{1,2}$ are a pair of dominant poles.

Neglecting the transient component caused by s_3 and taking consideration of the effect of s_3 on the transient component caused by $s_{1,2}$, from Eq. (3.72) and (3.74), we have

$$t_p = \frac{1}{\omega_n\sqrt{1-\zeta^2}}[\pi+\angle(s_1-s_3)] = \frac{1}{0.893}\left[\pi+\arctan\frac{0.893}{2.5-0.45}\right] = 3.98 \text{ s}$$

$$\sigma_p\% = \frac{|s_3|}{|s_1-s_3|}e^{-\zeta\omega_n t_p} \times 100\% = \frac{2.5}{|2.05+j0.893|}e^{-1.791} = 20.8\%$$

$$t_s \approx \frac{1}{\zeta\omega_n}\ln\left(\frac{2}{\Delta}\cdot\frac{|s_3|}{|s_1-s_3|}\right) = \frac{1}{0.45}\ln\left(\frac{2}{0.02}\times\frac{2.5}{|2.05+j0.893|}\right) = 10.48 \text{ s}, \quad \Delta = 2\%$$

Reducing the third-order system to a second-order system yields

$$\frac{C(s)}{R(s)} = \frac{\omega_n^2}{s^2+2\zeta\omega_n s+\omega_n^2} = \frac{1}{s^2+0.9s+1}$$

Hence, the performance specifications can be approximately evaluated as

$$\sigma_p\% \approx e^{-\pi\zeta/\sqrt{1-\zeta^2}} \times 100\% = e^{-1.791} = 20.5\%$$

$$t_s \approx \frac{4}{\zeta\omega_n} = 8.89 \text{ s}, \quad \Delta = 2\%$$

3.5 Stability of Linear Systems

When the analysis and design of control systems are considered, stability is of utmost importance. From a practical point of view an unstable system is of little value. In practical operation, almost all control systems are subject to extraneous or inherent disturbances, such as fluctuation of load or power source, variation of system parameters or circumstance. If an unstable system subjects to any extraneous or inherent disturbances, its physical variables will deviate and diverge from their normal operating-point as time goes on.

In any case, the concept of stability is used to distinguish two classes of systems: useful and useless. To be useful, a control system must be stable. This type of stable/unstable characterization is referred to as absolute stability. A system possesses absolute stability is called a stable system. Once a system is found to be stable, we can further characterize the degree of stability. Parameters such as overshoot and damping ratio are used to provide indications of the relative stability of a linear time-invariant system in the time domain. In this section we are mainly concerned with the subject of absolute stability of control systems.

Many physical systems are inherently open loop unstable. Introducing feedback is useful to stabilize the unstable plant and then adjust the transient performance with an appropriate controller. For open-loop stable plants feedback is still used to improve the system performance.

3.5.1 Concept and Definition of Stability

The concept of stability can be illustrated by considering a pendulum shown in Fig. 3.15, where O is the pivot. If the pendulum resting at the position A is moved to the position C, it returns to its original equilibrium position A. This equilibrium position and response are said to be stable. On the other hand, if the pendulum is placed upside down at another equilibrium position B and released, it falls and cannot return to its original equilibrium position B. This equilibrium position and response are said to be unstable.

The stability of a control system is defined in a similar manner. A linear time-invariant system is said to be asymptotically stable, or stable for short, if its response to an initial condition caused

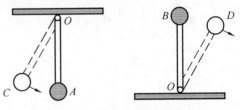

Figure 3.15 Stability of a pendulum

by the extraneous disturbance attenuates gradually as time goes on and approaches zero, i.e., the system approaches the original equilibrium position, as time goes infinite; otherwise, the system is said to be unstable.

Suppose that the linear time-invariant system is described by the differential equation

$$a_0 \frac{d^n c(t)}{dt^n} + a_1 \frac{d^{n-1} c(t)}{dt^{n-1}} + \ldots + a_{n-1} \frac{dc(t)}{dt} + a_n c(t) =$$
$$b_0 \frac{d^m r(t)}{dt^m} + b_1 \frac{d^{m-1} r(t)}{dt^{m-1}} + \ldots + b_{m-1} \frac{dr(t)}{dt} + b_m r(t) \tag{3.75}$$

where $c(t)$ is the output variable and $r(t)$ is the input variable, the coefficients $a_0, a_1, \ldots, a_n$ and $b_0, b_1, \ldots, b_m$ are constants, and $n \geq m$. For zero input,

$$c(t) = \dot{c}(t) = \ldots = c^{(n-1)}(t) = 0$$

satisfies the homogeneous equation and is defined as the equilibrium position of the system. If the system, subjected to nonzero initial condition $c(0), \dot{c}(0), \ldots, c^{(n-1)}(0)$, returns to its equilibrium position as time approaches infinity, the system is said to be asymptotically stable, or stable for short; otherwise the system is unstable.

In a mathematical manner, the definition of stability may be stated: If the zero-input response of the linear time-invariant system described by Eq. (3.75), subjected to nonzero initial condition $c(0), \dot{c}(0), \ldots, c^{(n-1)}(0)$, approaches zero as time approaches infinity, i.e.,

$$\lim_{t \to \infty} c(t) = 0 \tag{3.76}$$

then the system is said to be stable; otherwise the system is unstable.

3.5.2 Necessary condition for Stability

Assume that, for zero-input condition, the system dynamic equation is

$$a_0 \frac{d^n c(t)}{dt^n} + a_1 \frac{d^{n-1} c(t)}{dt^{n-1}} + \ldots + a_{n-1} \frac{dc(t)}{dt} + a_n c(t) = 0 \tag{3.77}$$

Taking Laplace transform under non-zero initial condition and arranging the result yield

$$(a_0 s^n + a_1 s^{n-1} + \ldots + a_{n-1} s + a_n) C(s) = M_0(s) \tag{3.78}$$

or

$$C(s) = \frac{M_0(s)}{a_0 s^n + a_1 s^{n-1} + \ldots + a_{n-1} s + a_n} \tag{3.79}$$

where $M_0(s)$ is a polynomial of s. Since the characteristic equation of the system is

$$a_0 s^n + a_1 s^{n-1} + \ldots + a_{n-1} s + a_n = 0 \tag{3.80}$$

Eq. (3.79) implies that $c(t)$ is governed by the system poles. Thus the condition in Eq. (3.76) requires that the roots of the system must all have negative real parts, i.e., all system poles must locate in the left half of the s-plane.

3.5.3 Algebraic Criterion of Stability

Although the stability of a linear time-invariant system may be checked by means of finding the roots of its characteristic equation, this method sometimes is difficult to implement in practice, because solving higher-order characteristic equation becomes tedious. It should be recognized that it is not necessary to solve the characteristic equation to determine the actual location of all roots of the characteristic equation for checking stability of a system. In fact, we only need to know whether all roots are in the left half of the complex plane. The algebraic criterion can determine the number of roots located in the right half of the complex plane using only the coefficients of characteristic equation without actually solving the characteristic equation for the roots themselves.

1. Necessary condition for stability

Consider that the system characteristic equation is in the polynomial form

$$\Delta(s) = a_0 s^n + a_1 s^{n-1} + \ldots + a_{n-1} s + a_n = 0 \tag{3.81}$$

where all the coefficients are real numbers and $a_0 > 0$.

Assuming that the system characteristic roots are $s_1, s_2, \ldots, s_n$, from the basic knowledge of algebra, the following relations are existed

$$\begin{aligned} \frac{a_1}{a_0} &= (-1) \sum_{i=1}^{n} s_i \\ \frac{a_2}{a_0} &= \sum_{i=1, j=1}^{n} s_i s_j \quad (i \neq j) \\ &\vdots \\ \frac{a_n}{a_0} &= (-1)^n \prod_{i=1}^{n} s_i \end{aligned} \tag{3.82}$$

Obviously, all the ratios must be positive and nonzero unless at least one of the roots has a positive real part.

Therefore, from Eq. (3.82), it is necessary that all coefficients must be positive and nonzero unless at least one roots has a positive real part. This necessary condition can easily be checked by inspection. However, this condition is not sufficient. It is quite often that a polynomial with all its coefficients positive and nonzero still has roots in the right half of the complex plane, and further analysis is required.

2. Hurwits criterion

Consider that the system characteristic equation is in the polynomial form of Eq. (3.81)

$$\Delta(s) = a_0 s^n + a_1 s^{n-1} + \ldots + a_{n-1} s + a_n = 0$$

The necessary and sufficient condition that the system is stable, i. e. all the roots of its characteristic equation lie in the left half of the s-plane, is that all the polynomial's Hurwits determinants $D_i, i = 1, 2, \ldots, n$, must be positive. The polynomial's Hurwits determinants of Eq. (3.81) are given by

$$D_1 = a_1, D_2 = \begin{vmatrix} a_1 & a_3 \\ a_0 & a_2 \end{vmatrix}, D_3 = \begin{vmatrix} a_1 & a_3 & a_5 \\ a_0 & a_2 & a_4 \\ 0 & a_1 & a_3 \end{vmatrix}, \ldots,$$

$$D_n = \begin{vmatrix} a_1 & a_3 & a_5 & a_7 & \cdots & \cdots & a_{2n-1} \\ a_0 & a_2 & a_4 & a_6 & \cdots & \cdots & a_{2n-2} \\ 0 & a_1 & a_3 & a_5 & \cdots & \cdots & a_{2n-3} \\ 0 & a_0 & a_2 & a_4 & \cdots & \cdots & a_{2n-4} \\ 0 & 0 & a_1 & a_3 & \ddots & \cdots & a_{2n-5} \\ \vdots & \vdots & \vdots & \vdots & \vdots & \ddots & \vdots \\ 0 & 0 & 0 & 0 & 0 & 0 & a_n \end{vmatrix} \quad (3.83)$$

where the coefficients with indices larger than n or with negative indices are replaced with zeros.

3. Lienard-Chipart criterion

Lienard-Chipart criterion may be considered as a simplified Hurwits criterion. According to Lienard-Chipart criterion, the necessary and sufficient condition that a system is stable is that (1) the necessary condition for stability must be satisfied, and (2) all the odd-order or all the even-order Hurwits determinants of its characteristic polynomial must be positive.

Example 3.4 Determine the stability of unity negative feedback systems that have the following open-loop transfer functions.

(1) $G_1(s) = \dfrac{100}{(s+1)(s+2)(s+3)}$;

(2) $G_2(s) = \dfrac{10}{s(s-1)(s+5)}$;

(3) $G_3(s) = \dfrac{10(s+1)}{s(s-1)(s+5)}$.

Solution (1) The closed-loop characteristic polynomial is

$$\Delta(s) = (s+1)(s+2)(s+3) + 100 = s^3 + 6s^2 + 11s + 106$$

Since the polynomial has no missing terms and the coefficients are all positive, the necessary condition for stability is satisfied. However, the 2nd-order Hurwits determinations of the polynomial

$$D_2 = \begin{vmatrix} 6 & 106 \\ 1 & 11 \end{vmatrix} = -40$$

is negative. Hence, from the Lienard-Chipart criterion, the closed-loop system is unstable.

(2) The closed-loop characteristic polynomial is
$$\Delta(s) = s(s-1)(s+5) + 10 = s^3 + 4s^2 - 5s + 10$$

Since the polynomial has a negative coefficient of sign, from the necessary condition for stability, we know without applying the Hurwits criterion that the closed-loop system is unstable.

(3) The closed-loop characteristic polynomial is
$$\Delta(s) = s(s-1)(s+5) + 10(s+1) = s^3 = 4s^2 + 5s + 10$$

The polynomial has no missing terms and the coefficients are all positive. Since
$$D_2 = \begin{vmatrix} 4 & 10 \\ 1 & 5 \end{vmatrix} = 10$$

is positive, from Lienard-Chipart criterion, the closed-loop system is stable.

These examples show that there is no certain relation between the stability of open-loop system and closed-loop system.

4. Routh criterion

The application of Hurwits criterion or Lienard-Chipart criterion is not convenient for high-order systems due to evaluation of the polynomial's Hurwits determinants. In this case, the evaluation may be simplified by Routh criterion into a tabulation called Routh array.

This method may be used with following steps:

Step I : Write the characteristic equation in the polynomial form
$$\Delta(s) = a_0 s^n + a_1 s^{n-1} + \ldots + a_{n-1} s + a_n = 0 \tag{3.84}$$
where all the coefficients are real numbers.

Step II : If the necessary condition for stability is satisfied, the Routh array must be formed in the following manner (the example shown is for a sixth-order system):

s^6	a_0	a_2	a_4	a_6
s^5	a_1	a_3	a_5	0
s^4	$B_1 = \dfrac{a_1 a_2 - a_0 a_3}{a_1}$	$B_2 = \dfrac{a_1 a_4 - a_0 a_5}{a_1}$	$B_3 = \dfrac{a_1 a_6 - a_0 \cdot 0}{a_1} = a_6$	0
s^3	$C_1 = \dfrac{B_1 a_3 - a_1 B_2}{B_1}$	$C_2 = \dfrac{B_1 a_5 - a_1 B_3}{B_1}$	$C_3 = \dfrac{B_1 \cdot 0 - a_1 \cdot 0}{a_1} = 0$	0
s^2	$D_1 = \dfrac{C_1 B_2 - B_1 C_2}{C_1}$	$D_2 = \dfrac{C_1 a_6 - B_1 \cdot 0}{C_1} = a_6$	0	0
s^1	$E_1 = \dfrac{D_1 C_2 - C_1 D_2}{D_1}$	0	0	0
s^0	$F_1 = \dfrac{E_1 a_6 - D_1 \cdot 0}{E_1} = a_6$	0	0	0

The array is constructed one row at a time, with each row being labeled on the left-hand side, i.e., with the highest power of s present in the characteristic polynomial for the first row, the next highest power for the second row, and so on. The first two rows are obtained directly from the coefficients of Eq. (3.84), and each of the following rows is derived from its previous two rows as shown. This process is continued until all rows have been completed down to the terms in row s^0. In preparing the Routh array for a given polynomial, some of the elements may not exist. In calculating the entries in the line that follows, these elements are considered to be zeros.

Step Ⅲ: Once the Routh array has been completed, the stability of the system may be determined from Routh criterion, which states that the roots of the characteristic equation (3.84) are all in the left half of the s-plane if all the elements of the first column of the Routh array are of the same sign; moreover, if there are changes of signs in the first column then the number of sign changes indicates the number of roots in the right half of the s-plane. This requirement is both necessary and sufficient.

Example 3.5 Consider if the system described by the characteristic equation
$$\Delta(s) = 2s^4 + s^3 + 3s^2 + 5s + 10 = 0$$
is stable using Routh criterion.

Solution The characteristic equation indicates that the equation has no missing terms and the coefficients are all positive, thus the Routh array is formed as follows:

s^4	2	3	10
s^3	1	5	0
s^2	−7	10	
s^1	6.43	0	
s^0	10		

In the first column there are two changes of sign, from 1 to −7 and from −7 to 6.43, therefore the equation has two roots in the right half s-plane, and the system is unstable.

The coefficients of any row in the Routh array may be multiplied or divided by a positive number without changing the signs of the first column. This may result in simplifying the evaluation of the coefficients and reducing the labor in forming the Routh array, as illustrated in the following example.

Example 3.6 Given a characteristic equation
$$\Delta(s) = s^6 + 3s^5 + 2s^4 + 9s^3 + 5s^2 + 12s + 20 = 0$$
Consider if the system is stable using Routh criterion.

Solution The characteristic equation indicates that the equation has no missing terms and the coefficients are all positive. The Routh array is

s^6	1	2	5	20
s^5	3	9	12	
	1	3	4	(after divided by 3)
s^4	−1	1	20	
s^3	4	24		
	1	6		(after divided by 4)
s^2	7	20		
s^1	22	0		
s^0	20			

Notice that the size of the number has been reduced after dividing the s^5 row by 3 and the s^3 row by 4. The result is unchanged; i. e. , there are two changes of signs in the first column and therefore the equation has two roots in the right half s-plane, and the system is unstable.

The algebraic criterion can be used to determine the range of the open-loop gain or other parameters for which a feedback system may be stable.

Example 3.7 Determine the range of the open-loop gain so that the feedback control system is stable. It is known that the open-loop transfer function of the system is

$$G(s) = \frac{K}{s(0.1s+1)(0.25s+1)}$$

Solution The characteristic equation of the system is

$$D(s) = 0.025s^3 + 0.35s^2 + s + K$$

From the necessary condition of stability we have

$$K > 0$$

Furthermore, by Lienard-Chipart criterion, for the system to be stable the following condition should be satisfied that

$$D_2 = \begin{vmatrix} 0.35 & K \\ 0.025 & 1 \end{vmatrix} > 0$$

i. e. ,

$$K < 14$$

Therefore, the condition for stability is that the open-loop gain must satisfy

$$0 < K < 14$$

In forming the Routh array, there are two special cases for the first column, and each must be treated separately and requires appropriate modifications of the array evaluation procedure.

Special case 1: The first element in a nonzero row is zero.

When the first element in a row is zero but not all the other elements are zeros, the zero in the first column will prevents completion of the array, and it may be replaced with a small positive number ε to overcome this problem.

Example 3.8 Determine the stability of the system described by the characteristic equation
$$\Delta(s) = s^4 + s^3 + 2s^2 + 2s + 5 = 0$$
using Routh criterion.

Solution The characteristic equation has no missing terms and the coefficients are all positive. In this example the Routh array is formed as follows:

s^4	1	2	5
s^3	1	2	0
s^2	ε	5	0
s^1	$2 - \dfrac{5}{\varepsilon} \approx -\dfrac{5}{\varepsilon}$	0	0
s^0	5	0	0

In the first column there are two changes of sign, therefore the system is unstable, and there are two characteristic roots in the right half of the s-plane.

Special case 2: All elements of a row are zero.

When all the elements of one row are zero, the procedure is as follows:

① Form the auxiliary equation from the proceeding row.

② Complete the Routh array by replacing the all-zero row with the coefficients obtained by differentiating the auxiliary equation with respect to s.

In this case, the roots of the auxiliary equation are also the roots of the original equation. Moreover, these roots are located symmetrically about the origin of the complex plane. Notice that the order of the auxiliary equation is always even and indicates the number of symmetrical root pairs.

Example 3.9 A system has the following characteristic equation
$$\Delta(s) = s^3 + 2s^2 + s + 2 = 0$$
Determine the system stability using Routh criterion.

Solution The Routh array is:

s^3	1	1
s^2	2	2
s^1	0	0

It is seen that the s^1 row is a zero row. The auxiliary equation can be obtained from the preceding row, i.e. the s^2 row. Therefore the auxiliary equation is
$$2s^2 + 2 = 0$$
The auxiliary equation is differentiated with respect to s and the new equation is
$$4s = 0$$

The coefficient of this new equation is inserted in the s^1 row, and the Routh array is then completed

s^3	1	1
s^2	2	2
s^1	4	0
s^0	2	

Since there are no changes of sign in the first column, the system has no roots with positive real part. From the auxiliary equation, the system has a pair of pure imaginary roots $s = \pm j1$, and the system is unstable.

3.5.4 Relative Stability

The verification of stability using the algebraic criterion provides only a partial answer to the question of stability. The algebraic criterion ascertains the absolute stability of a system by determining whether any of the roots of the characteristic equation lies on the right half of s-plane. However, if the system satisfies the algebraic criterion and is absolutely stable, it is desirable to determine the relative stability, i.e., how close it is to instability. The relative stability of a system possessing dominant poles with negative real parts, $-\zeta\omega_n$, can also be defined in terms of damping coefficient ζ (limit to the overshoot) and damping factor $\zeta\omega_n$ (limit to the response speed).

Sometimes, if a system is stable, we like to know how far from the imaginary axis is the pole closest to it. The algebraic criterion can be utilized for obtaining this information by shifting the vertical axis in the s-plane to obtain the $\tilde{s}$-plane. Thus, replacing s by $\tilde{s}-a$ in the characteristic polynomial $\Delta(s)$, we get a new polynomial $\Delta(\tilde{s})$. Applying the algebraic criterion to this polynomial will tell us if all the roots of $\Delta(\tilde{s})$ lie in the left half $\tilde{s}$-plane. That is also if all the roots of $\Delta(s)$ lie in the left half of line $s=-a$ in the s-plane.

Example 3.10 Consider the characteristic polynomial in Example 3.7

$$\Delta(s) = 0.025s^3 + 0.35s^2 + s + K$$

Now determine the value of K required so that there are no roots to the right of the line $s=-1$.

Solution Shifting the axis to the left by 1, i.e.,

$$s = \tilde{s} - 1$$

we have

$$\Delta(\tilde{s}) = 0.025(\tilde{s}-1)^3 + 0.35(\tilde{s}-1)^2 + (\tilde{s}-1) + K = \tilde{s}^3 + 11\tilde{s}^2 + 15\tilde{s} + (40K-27)$$

By Lienard-Chipart criterion, in the left half $\tilde{s}$-plane, the following conditions must be satisfied,

① $40K-27>0$, i.e. $K>0.675$;

② $D_2 = \begin{vmatrix} 11 & 40K-27 \\ 1 & 15 \end{vmatrix} > 0$, i. e. $K < 4.8$.

Therefore, if $0.675 < K < 4.8$ then the roots of $\Delta(\tilde{s})$ are all in the left half $\tilde{s}$-plane, i. e., the roots of $\Delta(s)$ are all in the left half of line $s = -1$ in the s-plane.

3.6 Steady-State Error

The time response of a control system may be characterized by the transient response and steady-state response. In a physical system, because of the nature of the particular system, the steady state of the actual response seldom agrees exactly with the desired state. The steady-state error is a criterion used to measure system accuracy in steady state when a specific type of input is applied to a control system. Since the steady-state error may be caused by reference input or disturbance, it also can be considered as the ability of a system to follow the input signal and/or restrain the disturbance signal.

3.6.1 Definitions of Steady-State Error

1. Definitions of error

Theoretically, it is rational to define the difference between the desired output $c_d(t)$ and the actual output $c(t)$ as the error function of a control system, i. e.,

$$E(s) = C_d(s) - C(s) \tag{3.85}$$

In the case of $r(t) = c_d(t)$, the input signal represents the desired output, the error function is simply

$$E(s) = R(s) - C(s) \tag{3.86}$$

For most unity feedback systems, as shown in Fig. 3.16(a), the input is usually can be considered as the desired output; whereas, for the systems shown in Fig. 3.16(b), the command input usually just represents the desired output.

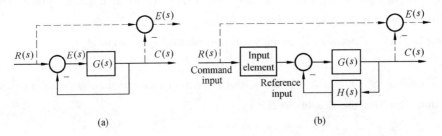

Figure 3.16 Definition of error

Sometimes, however, the input of a given system is only a reference input, and it may be impossible or inconvenient to provide an input that is at the same level or even of the same dimension as the output. For example, it may be necessary to use a low-voltage signal to control the output of high-voltage; in another case for a velocity-control system it is more practical to use a voltage signal or position input to control the velocity of the output shaft. Under these conditions, the error signal cannot be defined simply as the difference between the reference input and the controlled output. Therefore, a nonunity element $H(s)$ is usually incorporated in the feedback path, as shown in Fig. 3.17(a), to form a feedback system with standard structure. For example, if a 10 V reference is used to regulate a 100V voltage supply, then $H(s) = 0.1$. As another example, if a reference voltage input is used to control the output velocity of a velocity control system, then a tachometer is needed in the feedback path, and $H(s) = K_t s$. In these cases it is wised that the error of such a standard nonunity feedback control system is defined as

$$\varepsilon(s) = R(s) - B(s) = R(s) - H(s)C(s) \tag{3.87}$$

where $B(s)$, as shown in Fig. 3.17(a), is the main feedback signal.

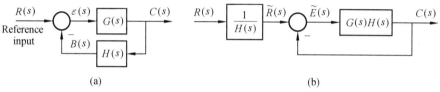

Figure 3.17 Definition of error

Usually, $E(s)$ defined in Eq. (3.85) is called the error at output terminal, and $\varepsilon(s)$ defined in Eq. (3.87) is called the error at input terminal or the deviation. The interrelationship between these two definitions of error can be illustrated by translating Fig. 3.17 (a) into the equivalent block diagram shown in Fig. 3.17(b). Since, functionally, $\tilde{R}(s)$ represents the desired output, we have

$$\tilde{E}(s) = \tilde{R}(s) - C(s) = C_d(s) - C(s) = E(s) \tag{3.88}$$

and

$$\tilde{E}(s) = \frac{1}{1+G(s)H(s)}\tilde{R}(s) = \frac{1}{1+G(s)H(s)} \cdot \frac{1}{H(s)}R(s) = \frac{1}{H(s)}\varepsilon(s) \tag{3.89}$$

Comparing Eq. (3.88) and (3.89) yields a simple relation between two definitions of the error

$$E(s) = \frac{1}{H(s)}\varepsilon(s) \tag{3.90}$$

Especially, for the case of unity negative feedback systems, we have

$$E(s) = \varepsilon(s) \tag{3.91}$$

Eq. (3.90) shows that it is not difficult to obtain $e(t)$ from $\varepsilon(t)$.

2. Steady-state error

As a function of the time, the error signal consists of two parts: the transient component and the steady-state component. Since the steady-state error is meaningful only for the stable systems, the transient component must approach zero as time approaches infinite. Hence, the steady-state error of a feedback control system is defined as the error when time approaches

infinity, i. e. ,
$$e_{ss} = \lim_{t \to \infty} e(t) \qquad (3.92.1)$$
$$\varepsilon_{ss} = \lim_{t \to \infty} \varepsilon(t) \qquad (3.92.2)$$

3.6.2 Evaluation of Steady-State Error Using Final-Value Theorem

The error of a control system may be caused by the input and/or disturbance. At first, in this section, we discuss the error caused by the input. Given a system, the error transfer function $\Phi_e(s)$ or $\Phi_\varepsilon(s)$ can be obtained according to the error definition and the system dynamic description. Thus, the basic expression of error function is

$$E(s) = \Phi_e(s) R(s) \qquad (3.93.1)$$
$$\varepsilon(s) = \Phi_\varepsilon(s) R(s) \qquad (3.93.2)$$

By use of the final-value theorem, if $sE(s)$ or $s\varepsilon(s)$ has no poles that lie on the imaginary axis (except of the original point) and in the right half of s-plane, the steady-state error of the system is given by

$$e_{ss} = \lim_{s \to 0} sE(s) = \lim_{s \to 0} s\Phi_e(s) R(s) \qquad (3.94.1)$$
$$\varepsilon_{ss} = \lim_{s \to 0} s\varepsilon(s) = \lim_{s \to 0} s\Phi_\varepsilon(s) R(s) \qquad (3.94.2)$$

which shows that the steady-state error will depend on both the characteristics of the system and the input demand.

It should be emphasized that steady-state error becomes meaningless if a system is unstable. Since $\Phi_e(s)/\Phi_\varepsilon(s)$ has no poles that lie on the imaginary axis and in the right half of the s-plane if and only if the closed-loop system is stable, so the stability of the closed-loop system should be checked before the steady-state error is calculated. On the other hand, if the input is of step, ramp, and parabolic functions, then all poles of $R(s)$ lie at the origin. Hence if a system is stable and the input is of the typical test signals, then the steady-state error can be calculated by use of the final-value theorem.

Example 3.11 The open-loop transfer function of a unity negative feedback system is
$$G(s) = \frac{K(0.5s+1)}{s(s+1)(2s+1)}$$
Determine the steady-state error for a unit ramp input.

Solution The characteristic equation of the closed-loop system is
$$\Delta(s) = 2s^3 + 3s^2 + (1+0.5K)s + K = 0$$
Since the 2nd-order Hurwits determinant is
$$D_2 = \begin{vmatrix} 3 & K \\ 2 & 1+0.5K \end{vmatrix}$$
the system is stable if and only if $D_2 > 0$, i. e. $0 < K < 6$.

Since $R(s)6 = 1/s^2$, the error function is
$$E(s) = \Phi_e(s) R(s) = \frac{s(s+1)(2s+1)}{s(s+1)(2s+1) + K(0.5s+1)} \cdot \frac{1}{s^2}$$

Therefore, in the case of $0 < K < 6$, the steady-state error is

$$e_{ss} = \lim_{s \to 0} sE(s) = \frac{1}{K}$$

Example 3.12 The speed regulating system is shown in Fig. 3.18. Determine the steady-state error for a unit step input.

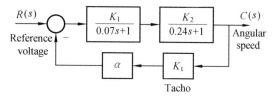

Figure 3.18 Speed regulating system for Example 3.12

Solution The error transfer function is

$$\Phi_\varepsilon(s) = \frac{1}{1 + \dfrac{K_1}{0.07s+1} \cdot \dfrac{K_2}{0.24s+1} \cdot \alpha K_t} = \frac{(0.07s+1)(0.24s+1)}{(0.07s+1)(0.24s+1) + \alpha K_1 K_2 K_t}$$

Since this 2nd-order system must be stable, from the final theorem, we have

$$\varepsilon_{ss} = \lim_{s \to 0} \Phi_\varepsilon(s) R(s) = \lim_{s \to 0} \frac{(0.07s+1)(0.24s+1)}{(0.07s+1)(0.24s+1) + \alpha K_1 K_2 K_t} \cdot \frac{1}{s} = \frac{1}{1 + \alpha K_1 K_2 K_t}$$

Example 3.13 A system is shown in Fig. 3.19, where the error is defined as $E(s) = R(s) - C(s)$. (1) Determine the steady-state error for a unit step input in terms of K_1 and k. (2) Select K_1 so that the steady-state error is zero.

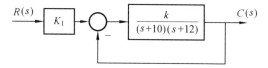

Figure 3.19 System for Example 3.13

Solution (1) The overall transfer function is

$$\Phi(s) = \frac{K_1 k}{(s+10)(s+12) + k}$$

By definition, the error transfer function is

$$E(s) = R(s) - C(s) = R(s) - \Phi(s)R(s) = [1 - \Phi(s)]R(s) = \frac{(s+10)(s+12) + k - K_1 k}{(s+10)(s+12) + k} \cdot \frac{1}{s}$$

Since this system is always stable, from the final theorem, we have

$$\varepsilon_{ss} = \lim_{s \to 0} sE(s) = \lim_{s \to 0} \frac{(s+10)(s+12) + k - K_1 k}{(s+10)(s+12) + k} \cdot \frac{1}{s} = \frac{120 + k - K_1 k}{120 + k}$$

(2) Letting $\varepsilon_{ss} = 0$ yields

$$\frac{120 + k - K_1 k}{120 + k} = 0$$

i. e. , the steady-state error will be zero if

$$K_1 = \frac{120+k}{k} = 1 + \frac{120}{k}$$

3.6.3 Evaluation of ε_{ss} Using Static Error Coefficients

Usually, for many systems, the steady-state performance is characterized with the steady-state error caused by the typical inputs, such as step, ramp, and/or parabolic signals. The evaluation of steady-state error ε_{ss} caused by these typical inputs can be simplified by introducing the concepts of system type and static error constants.

1. System type

In general, the open-loop transfer function of a nonunity negative feedback system may be written as

$$G(s)H(s) = \frac{K \prod_{i=1}^{m} (\tau_i s + 1)}{s^v \prod_{j=v+1}^{n} (T_j s + 1)} \qquad (3.95)$$

where K is the open-loop gain, $\tau_i(i=1,2,\ldots,m)$ and $T_j(j=v+1,v+2,\ldots,n)$ are constants.

The type of a feedback control system refers to the number of pure integral factors in the open-loop transfer function, i.e. the order of the pole of $G(s)H(s)$ at $s=0$.

Note the distinction between the type number and the order of a system. The order of a system is the highest power of s in the characteristic polynomial. For example, a system with open-loop transfer function

$$G(s) = \frac{10(s+1)}{s^2(3s+1)}$$

is of type 2 while the order of the system is 3.

For a feedback system with the standard structure shown in Fig. 3.17(a), if the condition for final-value theorem is satisfied, the expression of steady-state error caused by an input may be rewritten as

$$\varepsilon_{ss} = \lim_{s \to 0} s \cdot \frac{1}{1+G(s)H(s)} \cdot R(s) = \lim_{s \to 0} s \cdot \frac{1}{1+\frac{K}{s^v}} \cdot R(s) = \lim_{s \to 0} \frac{s^{v+1}}{s^v + K} R(s) \qquad (3.96)$$

which shows that the steady-state error is related to the system type, open-loop gain and the reference input $R(s)$.

In what follows the input will be considered to take one of several standard forms (step signal, ramp signal and parabolic signal) in turn, and the steady-state error of feedback system with the standard structure shown in Fig. 3.17(a), will be evaluated as a function of the system type.

2. Static position error constant

If the reference input applied to the system is a step input of magnitude r_0, i.e. $R(s) = r_0/s$, then

$$\varepsilon_{ss} = \lim_{s \to 0} s \cdot \frac{1}{1+G(s)H(s)} \cdot \frac{r_0}{s} = \lim_{s \to 0} \frac{r_0}{1+G(s)H(s)} \quad (3.97)$$

For convenience, we define

$$K_p = \lim_{s \to 0} G(s)H(s) = \lim_{s \to 0} \frac{K}{s^\nu} = \begin{cases} K & \nu = 0 \\ \infty & \nu \geq 1 \end{cases} \quad (3.98)$$

where K_p is called the static position error constant. Then Eq. (3.97) becomes

$$\varepsilon_{ss} = \frac{r_0}{1+K_p} = \begin{cases} \dfrac{r_0}{1+K} & \nu = 0 \, (K_p = K) \\ 0 & \nu \geq 1 \, (K_p \to \infty) \end{cases} \quad (3.99)$$

Eq. (3.99) shows that, for a tape 0 system, ε_{ss} due to step input will always be finite but it will be decreased as the open-loop gain K is increased. On the other hand, for ε_{ss} to be zero, the $G(s)H(s)$ must have at least one pure integral factor.

3. Static velocity error constant

If the reference input applied to the system is $v_0 t \cdot 1(t)$, i.e. $R(s) = v_0/s^2$, then

$$\varepsilon_{ss} = \lim_{s \to 0} s \cdot \frac{1}{1+G(s)H(s)} \cdot \frac{v_0}{s^2} = \lim_{s \to 0} \frac{v_0}{s[1+G(s)H(s)]} =$$

$$\frac{v_0}{\lim_{s \to 0} sG(s)H(s)} \quad (3.100)$$

For convenience, we define

$$K_v = \lim_{s \to 0} sG(s)H(s) = \lim_{s \to 0} \frac{K}{s^{\nu-1}} = \begin{cases} 0 & \nu = 0 \\ K & \nu = 1 \\ \infty & \nu \geq 2 \end{cases} \quad (3.101)$$

where K_v is called the static velocity error constant. Then Eq. (3.100) becomes

$$\varepsilon_{ss} = \frac{v_0}{K_v} = \begin{cases} \infty & \nu = 0 \, (K_v = 0) \\ \dfrac{v_0}{K} & \nu = 1 \, (K_v = K) \\ 0 & \nu \geq 2 \, (K_v \to \infty) \end{cases} \quad (3.102)$$

Eq. (3.102) shows that, in the case of ramp input, $G(s)H(s)$ must have at least two pure integral factors for ε_{ss} to be zero. The type 1 system can follow the ramp input but with a finite error.

4. Static acceleration error constant

If the reference input applied to the system is $\dfrac{a_0}{2}t^2 \cdot 1(t)$, i.e. $R(s) = \dfrac{a_0}{s^3}$, then

$$\varepsilon_{ss} = \lim_{s \to 0} s \cdot \frac{1}{1+G(s)H(s)} \cdot \frac{a_0}{s^3} = \lim_{s \to 0} \frac{a_0}{s^2[1+G(s)H(s)]} = \frac{a_0}{\lim_{s \to 0} s^2 G(s)H(s)} \quad (3.103)$$

For convenience, we define

$$K_a = \lim_{s \to 0} s^2 G(s)H(s) = \lim_{s \to 0} \frac{K}{s^{v-1}} = \begin{cases} 0 & v \leqslant 1 \\ K & v = 2 \\ \infty & v \geqslant 3 \end{cases} \quad (3.104)$$

where K_a is called the static acceleration error constant. Then Eq. (3.103) becomes

$$\varepsilon_{ss} = \frac{a_0}{K_a} = \begin{cases} \infty & v \leqslant 1 (K_a = 0) \\ \frac{a_0}{K} & v = 2 (K_a = K) \\ 0 & v \geqslant 3 (K_a \to \infty) \end{cases} \quad (3.105)$$

which shows that for ε_{ss} to be zero, when the input is a parabolic function, $G(s)H(s)$ must have at least three pure integral factors.

In summary, therefore, the steady-state error presented in the feedback system with standard structure due to the standard inputs can be obtained from the open-loop transfer function if the system is stable. The relation among the error constants, the system types, and the input types are tabulated in Table 3.2.

It should be pointed that if the steady-state error is a function of time, the error constants give only an answer of infinity, and do not provide any information on how the error varies with time. Furthermore, the error constants do not give information on the steady-state error when inputs are other than the three mentioned types.

Table 3.2 Summary of the steady-state error due to step, ramp and parabolic inputs

System type	Error constant			Input		
	K_p	K_v	K_a	Step	Ramp	Parabolic
v				$\varepsilon_{ss} = \frac{r_0}{1+K_p}$	$\varepsilon_{ss} = \frac{v_0}{K_v}$	$\varepsilon_{ss} = \frac{a_0}{K_a}$
0	K	0	0	$\varepsilon_{ss} = \frac{r_0}{1+K}$	$\varepsilon_{ss} \to \infty$	$\varepsilon_{ss} \to \infty$
1	∞	K	0	$\varepsilon_{ss} = 0$	$\varepsilon_{ss} = \frac{v_0}{K}$	$\varepsilon_{ss} \to \infty$
2	∞	∞	K	$\varepsilon_{ss} = 0$	$\varepsilon_{ss} = 0$	$\varepsilon_{ss} = \frac{a_0}{K}$

Example 3.14 The control system shown in Fig. 3.20 is considered.

For the case that $G_c(s) = 10$, the open-loop transfer function is

$$G(s)H(s) = \frac{2.5}{s+0.1}$$

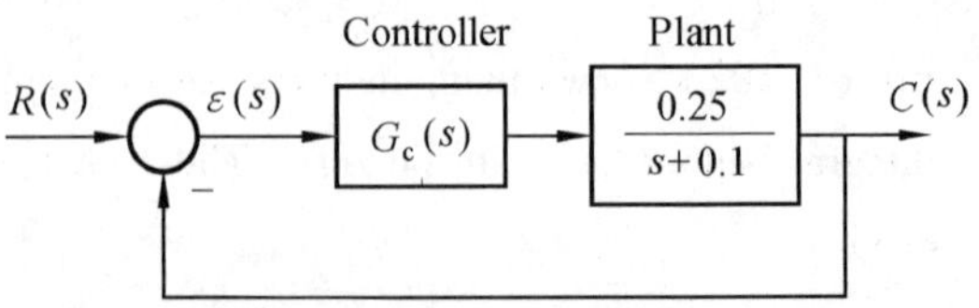

Figure 3.20 Control system for Example 3.13

and the system is stable. We see that the system is type 0, from Eq. (3.98), the static position error constant is

$$K_p = \lim_{s \to 0} G(s) = 25$$

If the input is $r_0 = 5$, the steady-state error is

$$\varepsilon_{ss} = \frac{r_0}{1+K_p} = \frac{5}{26} = 0.192$$

Assume that $G_c(s) = \dfrac{s+0.5}{s}$, the open-loop transfer function becomes

$$G(s) = \frac{2.5(s+0.5)}{s(s+0.1)}$$

and the system is still stable. Since the system is now type 1, the steady-state error is zero for a constant input. In the preceding equation if the parameters of the plant change, the steady-state error remain zero. Note also that if a unit ramp is applied at the input, the steady-state error can be obtained from the velocity error constant. Since

$$K_v = \lim_{s \to 0} sG(s) = 12.5$$

then, the steady-state error is

$$\varepsilon_{ss} = \frac{1}{K_v} = \frac{1}{12.5} = 0.08$$

3.6.4 Degree of Zero-Error

The degree of zero-error is a static accuracy specification of control systems, which is used to measure the ability of a control system to follow the input or to reject the disturbance with standard test signals. A system is said to possess a zero-degree of zero-error with respect to the input if its steady-state error is nonzero even for the step input. A system is said to possess a first-degree of zero-error with respect to the input if its steady-state error for and only for the step input is zero. A system is said to possess a second-degree of zero-error with respect to the input if its steady-state error is zero for the ramp input but not for the parabolic input. A system is said to possess a third-degree of zero-error with respect to the input if its steady-state error is zero even for the parabolic input. Similarly, the degree of a system with respect to the disturbance can be defined.

Consider the system shown in Fig. 3.17(a). It is assumed that the system is stable, the open-loop transfer function is written as

$$G(s)H(s) = \frac{KM(s)}{s^v N(s)} \qquad (3.106)$$

where K is the open-loop gain and v is the system type, and the standard input can be written as

$$R(s) = \frac{R_l}{s^l} \qquad (3.107)$$

where $l = 1, 2, \ldots$ and R_l is a constant. Thus, the error cause by a specified input is

$$\varepsilon(s) = \Phi_\varepsilon(s) R(s) = \frac{1}{1+G(s)H(s)} R(s) = \frac{s^v N(s)}{s^v N(s) + KM(s)} \cdot \frac{R_l}{s^l} \qquad (3.108)$$

and the steady-state error is

$$\varepsilon_{ss} = \lim_{s \to 0} \frac{s^v N(s)}{s^v N(s) + KM(s)} \cdot \frac{R_l}{s^l} = \lim_{s \to 0} \frac{R_l}{s^v + K} s^{v+1-l} \qquad (3.109)$$

It can be seen, from this equation, that if $v=0$ then ε_{ss} will not be zero for any input; if $v=1$ then ε_{ss} will be zero for the step input but not for the ramp or acceleration inputs; if $v=2$ then ε_{ss} will be zero for both the step and ramp inputs but not for the acceleration input. Hence, the system shown in Fig. 3.17(a) possesses v-degree of zero-error.

In more general cases, if the structure of a system is not same as that shown in Fig. 3.17(a), besides of by the definition, the degree of zero-error can be determined by founding the equivalent open-loop transfer function of the overall system; moreover, this system is said to be an equivalent v-type system if the equivalent open-loop transfer function has v integral factors.

3.6.5 Feedforward for Tracking Reference Input

As we know, the steady-state performance of a control system can be improved by increasing the system type and/or the open-loop gain. From the view of dynamic performance, however, both approaches may result in a trouble of stability, making the system unstable or the relative stability decreased. In order to avoid this disadvantage, the compound control is widely used in the system requiring higher accuracy.

The compound system, shown in Fig. 3.21, is used to improve the steady-state performance

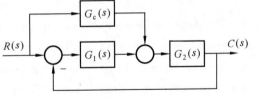

Figure 3.21 Compound control system

of control system by introducing an additional feedforward path based on the feedback control.

From the block diagram in Fig. 3.21, the transfer function of the overall system is

$$\Phi(s) = \frac{C(s)}{R(s)} = \frac{G_1(s) G_2(s) + G_c(s) G_2(s)}{1 + G_1(s) G_2(s)} \qquad (3.110)$$

Taking

$$G_c(s) = \frac{1}{G_2(s)} \qquad (3.111)$$

yields

$$\Phi(s) = 1$$

which means that the system will completely recover the command input in the whole response process.

Physically, since $G_c(s)$ is the reciprocal of $G_2(s)$, it is almost impossible to realize complete recovery in practice. On the other hand, in many cases an approximate recovery is enough for improving the steady-state performance of control systems. As for selection of the position, at which the feedforward signal is applied, attention should be given to reduce both the complexity of $G_c(s)$ and the requirement on the feedforward signal's power. In addition, it should be noted that the feedforward-block transfer function $G_c(s)$ must be stable addition, because it acts in the open loop.

Example 3.15 The block diagram of a position tracking system as shown in Fig. 3.22, where, K_1, K_2, T, and T_c are positive constants. Determine the values of a and b so that the steady-state error is zero for the parabolic input $r(t) = t^2/2$.

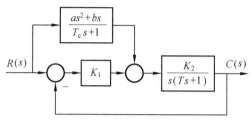

Figure 3.22 Control system for Example 3.15

Solution From the given block diagram, the transfer function of the overall system is

$$\Phi(s) = \frac{C(s)}{R(s)} = \frac{\dfrac{K_1 K_2}{s(Ts+1)} + \dfrac{K_2(as^2+bs)}{s(T_c s+1)(Ts+1)}}{1 + \dfrac{K_1 K_2}{s(Ts+1)}} = \frac{K_1 K_2(T_c s+1) + K_2(as^2+bs)}{(T_c s+1)(Ts^2+s+K_1 K_2)} =$$

$$\frac{K_2[as^2 + (b+K_1 T_c)s + K_1]}{(T_c s+1)(Ts^2+s+K_1 K_2)}$$

Consequently, the error transfer function is

$$E(s) = R(s) - C(s) = [1-\Phi(s)]R(s) = \left[1 - \frac{K_1 K_2(T_c s+1) + K_2(as^2+bs)}{(T_c s+1)(Ts^2+s+K_1 K_2)}\right] \cdot \frac{1}{s^3} =$$

$$\frac{T_c T s^3 + (T_c+T-K_2 a)s^2 + (1-K_2 b)s}{(T_c s+1)(Ts^2+s+K_1 K_2)} \cdot \frac{1}{s^3}$$

By inspection, the system is stable. Letting

$$e_{ss} = \lim_{s \to 0} sE(s) = \lim_{s \to 0} s \cdot \frac{T_c T s^3 + (T_c+T-K_2 a)s^2 + (1-K_2 b)s}{(T_c s+1)(Ts^2+s+K_1 K_2)} \cdot \frac{1}{s^3} = 0$$

results in

$$a = \frac{T+T_c}{K_2}, \quad b = \frac{1}{K_2}$$

3.7 Disturbance Rejection

In any control system there exist some excitations that influence the plant output and that, in general, are off control. We call these excitations disturbances and usually attempt to design the control system such that these disturbances have a minimal effect on the system.

Apart from improving the dynamic response beyond that available with open-loop system, one of other benefits of utilizing feedback control is to make a system less responsive to disturbances. The ability of a system to ignore disturbances is often called its disturbance rejection capability.

3.7.1 Open-Loop and Closed-Loop Disturbance Rejection

Whatever the cause of the disturbance, the resulting effect on the control system will be functionally the same as shown in Fig. 3.23, where $R(s)$ is the reference input, $D(s)$ is the disturbance and $C(s)$ is the output. In our study it is assumed that the system is stable.

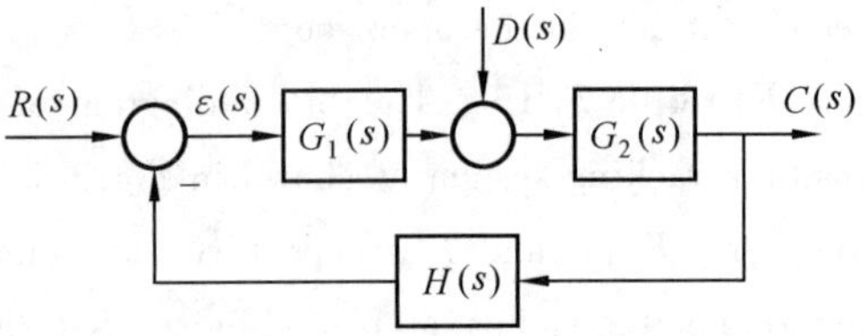

For convenience, let $H(s)=1$ and denote

$$G_1(s) = \frac{k_1 M_1(s)}{s^{v_1} N_1(s)}, \quad G_2(s) = \frac{k_2 M_2(s)}{s^{v_2} N_1(s)}$$

Figure 3.23 Control system with disturbance

where $M_i(s)$ and $N_i(s)$, $i=1,2$, are all polynomials in s, and

$$\lim_{s\to 0} M_i(s) = 1, \quad \lim_{s\to 0} N_i(s) = 1, \quad i=1,2$$

The open-loop gain and system type are given by

$$K = \lim_{s\to 0} s^{v_1+v_2} G_1(s) G_2(s) = k_1 k_2$$

$$v = v_1 + v_2$$

respectively.

By superposition, the system output is

$$C(s) = \frac{G_1(s) G_2(s)}{1+G_1(s) G_2(s)} R(s) + \frac{G_2(s)}{1+G_1(s) G_2(s)} D(s) \tag{3.112}$$

This equation shows that the contribution due to disturbance $D(s)$ may be regarded as a change in the steady-state output established by the input $R(s)$. The magnitude of the increment is considered as a measure of the influence of the disturbance and may be determined from the final-value theorem, if the condition of the theorem is satisfied,

$$\lim_{t\to\infty} \Delta c(t)_{\text{CL}} = \lim_{s\to 0} s \frac{G_2(s)}{1+G_1(s) G_2(s)} D(s) \tag{3.113}$$

For the open-loop system, similarly, the system output is

$$C(s) = G_1(s) G_2(s) R(s) + G_2(s) D(s) \tag{3.114}$$

Again, the effect of disturbance $D(s)$ results in a variation in the steady-state output from that established by the input $R(s)$ and may be determined using the final-value theorem

$$\lim_{t\to\infty} \Delta c(t)_{\text{OL}} = \lim_{s\to 0} s G_2(s) D(s) \tag{3.115}$$

The disturbance rejection ratio for open-and closed-loop system may be found from Eq. (3.115) and (3.113) as

$$\frac{\lim_{t\to\infty}\Delta c(t)_{\text{OL}}}{\lim_{t\to\infty}\Delta c(t)_{\text{CL}}} = \lim_{s\to 0} \frac{s G_2(s) D(s)}{s \dfrac{G_2(s)}{1+G_1(s) G_2(s)} D(s)} =$$

$$\lim_{s\to 0} [1+G_1(s) G_2(s)] = 1 + \lim_{s\to 0} \frac{k_1 M_1(s)}{s^{v_1} N_1(s)} \cdot \frac{k_2 M_2(s)}{s^{v_2} N_2(s)} = 1 + \lim_{s\to 0} \frac{K}{s^v} > 1 \tag{3.116}$$

Eq. (3.116) shows that the ability of the closed-loop system to maintain output better than the open-loop system.

For the system shown in Fig. 3.23, defining $\varepsilon(t) = r(t) - c(t)$, in case of $H(s) = 1$ we have

$$(\varepsilon_{ss,d})_{CL} = \lim_{s \to 0} s \frac{-G_2(s)}{1 + G_1(s)G_2(s)} D(s) \tag{3.117}$$

$$(\varepsilon_{ss,d})_{OL} = \lim_{s \to 0} s [-G_2(s)] D(s) \tag{3.118}$$

$$\frac{(\varepsilon_{ss,d})_{OL}}{(\varepsilon_{ss,d})_{CL}} = \lim_{s \to 0} [1 + G_1(s)G_2(s)] = 1 + \lim_{s \to 0} \frac{K}{s^v} > 1 \tag{3.119}$$

The result of Eq. (3.119) is same as that of Eq. (3.116), hence, the influence of the disturbance also can be reflected from the steady-state error $\varepsilon_{ss,d}$ due to the disturbance.

3.7.2 Disturbance Rejection

1. Effect of disturbance on system

Now consider the effect of step, ramp disturbances on the closed-loop system. Assuming

$$H(s) = \frac{k_3 M_3(s)}{N_3(s)}$$

if the system is stable, we get

$$\varepsilon_{ss,d} = \lim_{s \to 0} s \frac{-G_2(s)H(s)}{1 + G_1(s)G_2(s)H(s)} D(s) = \lim_{s \to 0} s \frac{-\frac{k_2 M_2(s)}{s^{v_2} N_2(s)} \cdot \frac{k_3 M_3(s)}{N_3(s)}}{1 + \frac{k_1 M_1(s)}{s^{v_1} N_1(s)} \cdot \frac{k_2 M_2(s)}{s^{v_2} N_2(s)} \cdot \frac{k_3 M_3(s)}{N_3(s)}} D(s) =$$

$$\lim_{s \to 0} s \frac{-k_2 k_3 s^{v_1}}{s^{v_1 + v_2} + k_1 k_2 k_3} D(s) \tag{3.120}$$

(1) Effect of step disturbance

Assuming the disturbance to be a step function, i.e. $D(s) = d_0/s$, from Eq. (3.120), yields

$$\varepsilon_{ss,d} = \lim_{s \to 0} s \cdot \frac{-k_2 k_3 s^{v_1}}{s^{v_1 + v_2} + k_1 k_2 k_3} \cdot \frac{d_0}{s} = \begin{cases} \dfrac{-k_2 k_3}{1 + k_1 k_2 k_3} d_0 & v_1 = 0, v_2 = 0 \\ \dfrac{-1}{k_1} d_0 & v_1 = 0, v_2 \neq 0 \\ 0 & v_1 = 1, 2, \ldots \end{cases} \tag{3.121}$$

(2) Effect of ramp disturbance

If the disturbance is a ramp function, i.e. $D(s) = d_1/s^2$, we have

$$\varepsilon_{ss,d} = \lim_{s \to 0} s \cdot \frac{-k_2 k_3 s^{v_1}}{s^{v_1 + v_2} + k_1 k_2 k_3} \cdot \frac{d_1}{s^2} = \begin{cases} \infty & v_1 = 0 \\ \dfrac{-1}{k_1} d_1 & v_1 = 1 \\ 0 & v_1 = 2, 3, \ldots \end{cases} \tag{3.122}$$

2. Approaches to disturbances rejection

(1) Serial compensation

It can be seen, from subsection 3.7.1, that the effect of disturbance on a feedback system is closely related to v_1 and k_1, i.e., the number of integral factors and gain in the part of forward path ahead of the disturbance action point, $G_1(s)$. Hence, the disturbance effect may be eliminated by add of a serial compensator $G_c(s)$ containing integral factors into the forward path ahead of the disturbance action point, as shown in Fig. 3.24. Sometimes, It is possible to reject the disturbance by increasing the gain of $G_c(s)$.

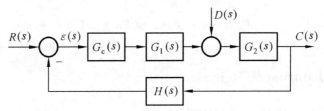

Figure 3.24 Closed-loop system with compensator

It is important to remember that the evaluated error, caused by the input and/or disturbance, is meaningful if and only if the system is stable. Hence, it is necessary to check the system stability again when a compensator $G_c(s)$ is added into the system.

Example 3.15 Consider the control system shown in Fig. 3.25. When the disturbance is a unit step function, evaluate the steady-state error due to the disturbance before the compensator is adopted and design a compensator so that the steady-state error due to the disturbance will be zero.

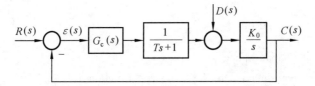

Figure 3.25 Control system for Example 3.15

Solution Before the compensator is adopted, the open-loop transfer function is

$$G_p(s) = \frac{K_0}{s(Ts+1)}$$

the system is a second-order system and it is stable. From Eq. (3.121) we have

$$\varepsilon_{ss} = \varepsilon_{ss,d} = \lim_{s \to 0} s \cdot \frac{-\frac{K_0}{s}}{1 + \frac{K_0}{s(Ts+1)}} \cdot \frac{1}{s} = \lim_{s \to 0} s \cdot \frac{-K_0(Ts+1)}{s(Ts+1) + K_0} \cdot \frac{1}{s} = -1$$

Also from Eq. (3.121), if and only if $G_c(s)$ has at least one integral factor, the steady-state error caused by the step disturbance will be zero. Assuming $G_c(s) = 1/s$ results in a new open-loop transfer function

$$G_c(s)G_p(s) = \frac{K_0}{s^2(Ts+1)}$$

Obviously, the system becomes unstable after the compensator is adopted. Thus, we need redesign another compensator. Now let

$$G_c(s) = \frac{\tau s+1}{s}$$

then the open-loop transfer function becomes

$$G_c(s)G_p(s) = \frac{K_0(\tau s+1)}{s^2(Ts+1)}$$

By algebraic criterion of stability, the system is stable if and only if $\tau > T$. Hence, after the serial compensator

$$G_c(s) = \frac{\tau s+1}{s}, \quad \tau > T$$

is adopted, the steady-state error due to the step disturbance will be zero.

(2) Feedforward compensation

Another method of disturbance rejection is called feedforward compensation and can be applied only if the disturbance is measurable. This method is illustrated as follows with Fig. 3.26.

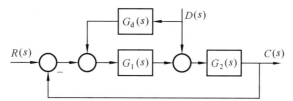

Figure 3.26 Feedforward compensation

Now the transfer function from the disturbance to the output is

$$\frac{C(s)}{D(s)} = \frac{G_2(s) + G_d(s)G_1(s)G_2(s)}{1 + G_1(s)G_2(s)} \qquad (3.123)$$

If the numerator can be made equal to zero, i.e. if

$$G_d(s) = -\frac{1}{G_1(s)} \qquad (3.124)$$

the disturbance will be rejected completely.

Similarly to that of feedforward for reference tracking, this method is often used to improve the steady-state performance of control system. Furthermore, remember that a transfer function is only an approximate model of a physical system. Hence, if Eq. (3.124) is satisfied exactly, the quality of the disturbance rejection in the physical system will depend on the accuracy of the system model.

Problems

P3.1 The unit step response of a certain system is given by
$$c(t) = 1 + e^{-t} - e^{-2t}, \quad t \geq 0$$
(a) Determine the impulse response of the system.

(b) Determine the transfer function $C(s)/R(s)$ of the system.

P3.2 Consider the system described by the block diagram shown in Fig. P3.2(a). Determine the polarities of two feedbacks for each of the following step responses shown in Fig. P3.2(b), where "0" indicates that the feedback is open.

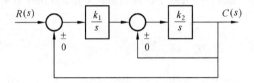

(a) Block diagram

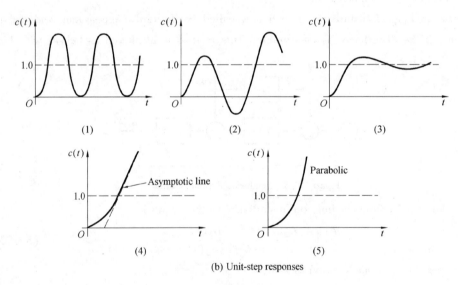

(b) Unit-step responses

Figure P3.2

P3.3 Consider each of the following closed-loop transfer function. By considering the location of the poles on the complex plane, sketch the unit step response, explaining the results obtained.

(a) $\Phi(s) = \dfrac{20}{s^2 + 12s + 20}$

(b) $\Phi(s) = \dfrac{6}{s^3 + 6s^2 + 11s + 6}$

(c) $\Phi(s) = \dfrac{4}{s^2 + 2s + 2}$

(d) $\Phi(s) = \dfrac{12.5}{(s^2 + 2s + 5)(s + 5)}$

P3.4 The open-loop transfer function of a unity negative feedback system is

$$G(s) = \frac{1}{s(s+1)}$$

Determine the rise time, peak time, percent overshoot and setting time (using a 5% setting criterion).

P3.5 A second-order system gives a unit step response shown in Fig. P3.5. Find the open-loop transfer function if the system is a unit negative-feedback system.

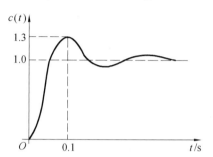

Figure P3.5

P3.6 A feedback system is shown in Fig. P3.6(a), and its unit step response curve is shown in Fig. P3.6(b). Determine the values of k_1, k_2, and a.

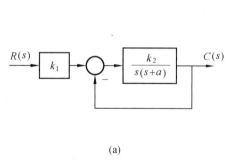

(a)   (b)

Figure P3.6

P3.7 A unity negative feedback system has the open-loop transfer function

$$G(s) = \frac{k}{s(s+\sqrt{2k})}$$

(a) Determine the percent overshoot.

(b) For what range of k the setting time less than 0.75 s (using a 5% setting criterion).

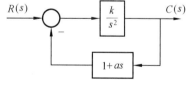

Figure P3.8

P3.8 For the servomechanism system shown in Fig. P3.8, determine the values of k and a that satisfy the following closed-loop system design requirements.

(a) Maximum of 40% overshoot.

(b) Peak time of 4 s.

P3.9 The open-loop transfer function of a unity feedback system is
$$G(s) = \frac{k}{s(s+2)}$$
A step response is specified as: peak time $t_p = 1.1$ s, and percent overshoot $\sigma_p = 5\%$.
(a) Determine whether both specifications can be met simultaneously.
(b) If the specifications cannot be met simultaneously, determine a compromise value for k so that the peak time and percent overshoot are relaxed the same percentage.

P3.10 A control system is represented by the transfer function
$$\frac{C(s)}{R(s)} = \frac{0.33}{(s+2.56)(s^2+0.4s+0.13)}$$
Estimate the peak time, percent overshoot, and setting time ($\Delta = 5\%$), using the dominant pole method, if it is possible.

P3.11 By means of the algebraic criteria, determine the stability of systems that have the following characteristic equations.
(a) $s^3 + 20s^2 + 9s + 20 = 0$
(b) $3s^4 + 10s^3 + 5s^2 + s + 2 = 0$
(c) $s^5 + 2s^4 + 9s^3 + 10s^2 + s + 2 = 0$

P3.12 The characteristic equations for certain systems are given below. In each case, determine the number of characteristic roots in the right-half-plane and the number of pure imaginary roots.
(a) $s^3 - 3s + 2 = 0$
(b) $s^3 + 10s^2 + 16s + 160 = 0$
(c) $s^5 + 3s^4 + 12s^3 + 24s^2 + 32s + 48 = 0$
(d) $s^5 + 2s^4 + 3s^3 + 6s^2 - 4s - 8 = 0$

P3.13 The characteristic equations for certain systems are given below. In each case, determine the value of k and T so that the corresponding system is stable. It is assumed that both k and T are positive numbers.
(a) $s^4 + 2s^3 + 10s^2 + 2s + k = 0$
(b) $s^3 + (T+0.5)s^2 + 4Ts + 50 = 0$

P3.14 The open-loop transfer function of a unity negative feedback system is given by
$$G(s) = \frac{K}{s(0.01s^2 + 0.2\zeta s + 1)}$$
Determine the range of K and ζ in which the closed-loop system is stable.

P3.15 The open-loop transfer function of negative feedback system is given
$$G(s)H(s) = \frac{K(s+1)}{s(Ts+1)(2s+1)}$$
The parameters K and T may be represented in a plane with K as the horizontal axis and T as the vertical axis. Determine the region in which the closed-loop system is stable.

P3.16 A unity negative feedback system has an open-loop transfer function

$$G(s) = \frac{K}{(Ts+1)(nTs+1)(n^2Ts+1)}$$

where $0 \leq n \leq 1$.

(a) Determine the range of K and n so that the system is stable.

(b) Determine the value of K required for stability for $n = 1, 0.5, 0.1, 0.01$, and 0.

(c) Discuss the stability of the closed-loop system as a function of n for a constant K.

P3.17 A unity negative feedback system has an open-loop transfer function

$$G(s) = \frac{K}{s(\frac{s}{3}+1)(\frac{s}{6}+1)}$$

Determine the range of k required so that there are no closed-loop poles to the right of the line $s = -1$.

P3.18 A system has the characteristic equation

$$s^3 + 10s^2 + 29s + k = 0$$

Determine the value of k so that the real part of complex roots is -2, using the algebraic criterion.

P3.19 An automatically guided vehicle is represented by the system in Fig. P3.19.

(a) Determine the value of τ required for stability.

(b) Determine the value of τ when one root of the characteristic equation is $s = -5$, and the values of the remaining roots for the selected τ.

(c) Find the response of the system to a step command for the τ selected in (b).

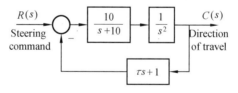

Figure P3.19

P3.20 A thermometer is described by the transfer function $1/(Ts+1)$. It is known that, measuring the water temperature in a container, one minute is required to indicate 98% of the actual water temperature. Evaluate the steady-state indicating error of the thermometer if the container is heated and the water temperature is lineally increased at the rate of 10℃/min.

P3.21 Determine the steady-state error for a unit step input, a unit ramp input, and an acceleration input $t^2/2$ for the following unity negative feedback systems. The open-loop transfer functions are given by

(a) $G(s) = \dfrac{50}{(0.1s+1)(2s+1)}$ (b) $G(s)H(s) = \dfrac{10}{s(s+4)(s+0.5)}$

(c) $G(s) = \dfrac{8(0.5s+1)}{s^2(0.1s+1)}$ (d) $G(s) = \dfrac{10}{s^2(s+1)(s+5)}$

(e) $G(s) = \dfrac{k}{s(s^2+4s+200)}$

P3.22 The open-loop transfer function of a unity negative feedback system is given by
$$G(s) = \dfrac{K}{s(T_1s+1)(T_2s+1)}$$
Determine the values of $K, T_1,$ and T_2 so that the steady-state error for the input, $r(t) = a+bt$, is less than ε_0. It is assumed that $K, T_1,$ and T_2 are positive, a and b are constants.

P3.23 The open-loop transfer function of a unity negative system is given by
$$G(s) = \dfrac{K}{s(Ts+1)}$$
Determine the values of K and T so that the following specifications are satisfied:
(a) The steady-state error for the unit ramp input is less than 0.02.
(b) The percent overshoot is less than 30% and the setting time is less 0.3 s.

P3.24 The block diagram of a control system is shown in Fig. P3.24, where $E(s) = R(s) - C(s)$. Select the values of τ and b so that the steady-state error for a ramp input is zero.

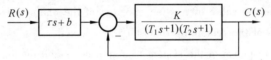

Figure P3.24

P3.25 The block diagram of a compound system is shown in Fig. P3.25. Select the values of a and b so that the steady-state error for a parabolic input is zero.

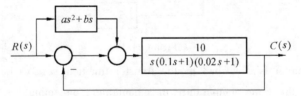

Figure P3.25

P3.26 The block diagram of a system is shown in Fig. P3.26. In each case, determine the steady-state error for a unit step disturbance and a unit ramp disturbance, respectively.

(a) $G_1(s) = K_1, G_2(s) = \dfrac{K_2}{s(T_2s+1)}$

(b) $G_1(s) = \dfrac{K_1(T_1s+1)}{s}, G_2(s) = \dfrac{K_2}{s(T_2s+1)}, T_1 > T_2$

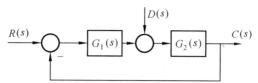

Figure P3.26

P3.27 The block diagram of a compound system is shown in Fig. P3.27, where

$$G_1(s) = \frac{K_1}{T_1 s+1}, \quad G_2(s) = \frac{K_2}{s(T_2 s+1)}, \quad G_3(s) = \frac{K_3}{K_2}$$

Determine the feedforward block transfer function $G_d(s)$ so that the steady-state error due to unit step disturbance is zero.

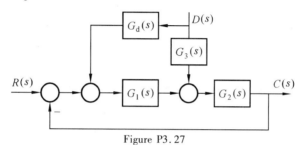

Figure P3.27

Chapter 4 Root Locus Method

4.1 Root Locus Of Feedback System

4.1.1 Introduction

In the previous chapter we have seen that the poles of the closed-loop transfer function, i.e. the roots of the characteristic equation, may be varied by changing the open-loop gain or other parameters. Since the relative stability and the transient performance of the closed-loop system are directly related to the location of the closed-loop roots of the characteristic equation, it is frequently necessary to adjust one or more system parameters in order to obtain suitable root location. Therefore, it is very useful to determine how the roots of the characteristic equation move around the s-plane as a certain parameter of the system is varied.

The root locus method, introduced by W. R. Evans in 1948, is a graphical procedure for sketching the trajectories of the roots of an algebraic equation in the complex plane as a parameter is varied from zero to infinity. This method provides a straightforward method of graphically "solving" the characteristic equation to locate the closed-loop poles on the s-plane. In some cases these locations may be found precisely, while in others may be only approximate. In either case, the information concerning the stability and other performances of the system can be obtained with reasonable accuracy.

The essential principle of the root locus method is that the closed-loop poles (transient response modes) are related to the open-loop zeros and poles, and also to the open-loop gain.

4.1.2 Root Locus of Second-Order System

We introduce the concept of root locus through the following example.

Example 4.1 Consider the second-order system shown in Fig. 4.1. In order to investigate the effect of varying the gain k from zero to infinity, plot the roots of the system characteristic equation in the s-plane.

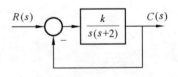

Figure 4.1 Block diagram of a second-order system

Solution In this case, the closed-loop transfer function and characteristic equation of the system are given, respectively, by

$$\Phi(s) = \frac{k}{s^2 + 2s + k} \qquad (4.1)$$

$$\Delta(s) = s^2 + 2s + k = 0 \qquad (4.2)$$

From equation (4.2), these roots are given by

$$s = -1 \pm \sqrt{1-k} \qquad (4.3)$$

First consider the case when $k = 0$. The roots are obtained as $s = 0$ and $s = -2$, which are also the open-loop poles, symboled with "×" in Fig. 4.2. As k is increased from 0 to 1, we have real roots; but for $k>1$, we get a pair of complex conjugate roots with real part equal to -1. The movement of the roots of equation (4.2) is shown in Fig. 4.2.

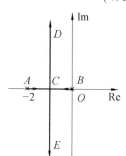

Figure 4.2 Root locus of second-order system

One may visualize the loci starting from the two open-loop poles. The locus starting from point A moves to the right, while the locus starting from point B moves to the left on the real axis until they meet at point C for $k = 1$. For larger values of k, the loci move alone the lines CD and CE, as shown in Fig. 4.2.

Once the root locus has been obtained for a control system, it is possible to determine the variation in system performance when a parameter varies. For example, from the plot of the roots shown in Fig. 4.2, we can conclude that the system is stable for all values of k, and the poles of the closed-loop transfer function are complex only if $k>1$. Furthermore, as k increases with the roots complex, although the system is stable, the value of damping ratio decreases and the overshoot in the transient response increases.

The root locus of each control system can be analyzed in a similar manner to obtain an idea of the variation in its time response resulted from a variation in its parameter.

4.1.3 Root Locus Conditions

The closed-loop system to be considered, in discussing the root locus, may be the nonunity feedback system as shown in Fig. 4.3. Thus, generally, the closed-loop transfer function and characteristic equation are given, respectively, by

$$\frac{C(s)}{R(s)} = \frac{G(s)}{1+G(s)H(s)} \qquad (4.4)$$

and

$$1+G(s)H(s) = 0 \qquad (4.5)$$

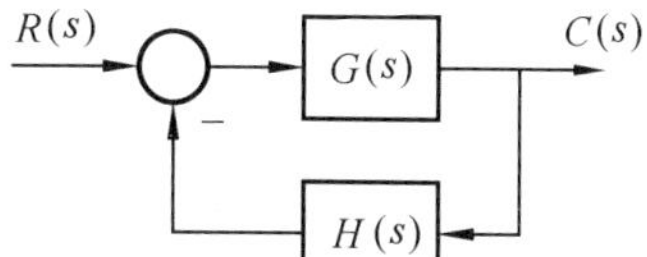

Figure 4.3 Block diagram of closed-loop system

For sketching root locus, the open-loop transfer function is usually expressed in a standard form of poles and zeros, i.e. the Evans form, as follows

$$G(s)H(s) = \frac{K^*(s-z_1)(s-z_2)\ldots(s-z_m)}{(s-p_1)(s-p_2)\ldots(s-p_n)} = \frac{K^* \prod_{i=1}^{m}(s-z_i)}{\prod_{j=1}^{n}(s-p_j)} \qquad (4.6)$$

where $z_i, i = 1, 2, \ldots, m$, and $p_j, j = 1, 2, \ldots, n$, are the zeros and poles of the open-loop transfer function $G(s)H(s)$, respectively. The variable parameter K^* is called the root locus gain to be distinguished from the open-loop gain.

Substituting Eq. (4.6) into (4.5) and rewriting the characteristic equation yield

$$\frac{K^* \prod_{i=1}^{m}(s-z_i)}{\prod_{j=1}^{n}(s-p_j)} = -1 \qquad (4.7)$$

which is sometimes called the root locus equation. To meet the root locus equation, or the characteristic equation, any point in the s-plane to be on the root locus must satisfy the following two conditions simultaneously

$$|G(s)H(s)| = \frac{K^* \prod_{i=1}^{m}|s-z_i|}{\prod_{j=1}^{n}|s-p_j|} = 1 \qquad (4.8)$$

$$\angle G(s)H(s) = \sum_{i=1}^{m}\angle(s-z_i) - \sum_{j=1}^{n}\angle(s-p_j) = (2l+1)\pi, \quad l = 0, \pm 1, \pm 2, \ldots \qquad (4.9)$$

Eq. (4.8) is called the magnitude condition of root locus and Eq. (4.9) is called the angle condition or angle criterion of root locus.

Since we assume that K^* is varied from zero to infinity, Eq. (4.8) can be satisfied for arbitrary point in the s-plane. Hence a point in the s-plane is verified to be on the root locus if and only if the angle condition given by Eq. (4.9) is satisfied.

In principle, the root locus method is performed in two stages:

① Find all points that satisfy the angle condition of root locus in the s-plane.

② Find particular points that satisfy the magnitude condition on the root locus.

4.2 Rules For Plotting Root Locus

It is a very tedious task to obtain the root locus by solving the characteristic equation or searching all points in the s-plane that satisfy the angle condition in a trial-and-error procedure. Fortunately, aided with the properties of the root locus, we will able to obtain an approximate plot, through some "rules" of construction, with much less effort.

The following rules for plotting root locus are developed from the characteristic equation or open-loop transfer function of the closed-loop system. These rules should be regarded only as an aid to the construction of the root locus as they generally do not give the exact plots.

In following discussion it is assumed that the open-loop transfer function is given in the form of poles and zeros

$$G(s)H(s) = \frac{K^* \prod_{i=1}^{m}(s - z_i)}{\prod_{j=1}^{n}(s - p_j)} \qquad (4.10)$$

and the root locus problem with one variable parameter K^* is described by the characteristic equation

$$\prod_{j=1}^{n}(s - p_j) + K^* \prod_{i=1}^{m}(s - z_i) = 0 \qquad (4.11)$$

or by the root locus equation

$$\frac{K^* \prod_{i=1}^{m}(s - z_i)}{\prod_{j=1}^{n}(s - p_j)} = -1 \qquad (4.12)$$

Then, correspondingly, the angle condition of root locus is

$$\sum_{i=1}^{m} \angle(s - z_i) - \sum_{j=1}^{n} \angle(s - p_j) = (2l + 1)\pi, \quad l = 0, \pm 1, \pm 2, \ldots \qquad (4.13)$$

4.2.1 Continuity, Symmetry, and Branch Number of Root Locus

Rule 1 The root locus appears as a family of some continuous branches and the number of the branches is equal to the order of the characteristic equation in Eq. (4.11). Furthermore, these branches are symmetrical with respect to the real axis of the s-plane.

Explanation A branch of the root locus is the trajectory of one root when K^* is varied from 0 to ∞. This rule applies since the number of branches of the root locus must be equal to the number of roots of the characteristic equation in Eq. (4.11). On the other hand, it is assumed that the system models used are rational functions (the ratio of two polynomials) with real coefficients (for physical systems). Hence if the characteristic equation has a complex root, the complex conjugate of that root must also be a root, i.e., the complex roots of real polynomials always appear in complex conjugate pair.

4.2.2 Starting Points and Finishing Points of Root Locus

Rule 2 The root locus starts at the poles of $G(s)H(s)$ and finishes at the zeros of $G(s) \cdot H(s)$ as K^* increases from zero to infinity. If the transfer function has more poles than zeros, i.e. $n > m$, as is the usual case with the models of physical systems, then $n - m$ branches of the root locus will approach infinity as $K^* \to \infty$.

Proof Rewrite Eq. (4.12) as

$$\frac{\prod_{i=1}^{m} |s - z_i|}{\prod_{j=1}^{n} |s - p_j|} = \frac{1}{K^*} \qquad (4.14)$$

As $K^* \to 0$, the value of Eq. (4.14) approaches infinity, correspondingly, s approaches $p_j(j=1,2,\ldots,n)$. As $K^* \to \infty$, the value of Eq. (4.14) approaches zero, correspondingly, s approaches $z_i(i=1,2,\ldots,m)$. If $n>m$, then $G(s)H(s)$ has $n-m$ zeros located at infinity, because for a rational function the total number of poles and zeros must be equal if the poles and zeros at infinity are included.

4.2.3 Asymptotes of Root Locus

Rule 3 If the open-loop transfer function has some zeros at infinity, i.e. if $n>m$, then $n-m$ branches of root locus approach infinity along $n-m$ asymptotes as $K^* \to \infty$. These asymptotes are centered at a point on the real axis given by

$$\sigma_a = \frac{\sum_{j=1}^{n} p_j - \sum_{i=1}^{m} z_i}{n-m} \qquad (4.15)$$

The angles of the asymptotes with respect to the real axis are given by

$$\varphi_a = \frac{(2l+1)\pi}{n-m}, \quad l=0,1,2,\ldots,n-m-1 \qquad (4.16)$$

Proof Eq. (4.11) may be rewritten as

$$\left[s^n - \sum_{j=1}^{n} p_j \cdot s^{n-1} + \ldots + \prod_{j=1}^{m}(-p_j) \right] + K^* \left[s^m - \sum_{i=1}^{m} z_i \cdot s^{n-1} + \ldots + \prod_{i=1}^{m}(-z_i) \right] = 0$$

or

$$\frac{s^n - \sum_{j=1}^{n} p_j \cdot s^{n-1} + \ldots + \prod_{j=1}^{m}(-p_j)}{s^m - \sum_{i=1}^{m} z_i \cdot s^{n-1} + \ldots + \prod_{i=1}^{m}(-z_i)} = -K^*$$

Carrying out the fraction of the left side of this equation by the process of long division, and for large s neglecting all but the first two terms, we get

$$s^{n-m} - \left(\sum_{j=1}^{n} p_j - \sum_{i=1}^{m} z_i \right) s^{n-m-1} = -K^*$$

i.e.,

$$s^{n-m} \left[1 - \frac{1}{s} \left(\sum_{j=1}^{n} p_j - \sum_{i=1}^{m} z_i \right) \right] = -K^*$$

or

$$s \left[1 - \frac{1}{s} \left(\sum_{j=1}^{n} p_j - \sum_{i=1}^{m} z_i \right) \right]^{\frac{1}{n-m}} = (-K^*)^{\frac{1}{n-m}}$$

Making binomial expansion and then retaining only the first two terms in the resulting series, we get

$$s \left[1 - \frac{1}{n-m} \cdot \frac{1}{s} \left(\sum_{j=1}^{n} p_j - \sum_{i=1}^{m} z_i \right) \right] = (-K^*)^{\frac{1}{n-m}}$$

i.e.,

$$s - \frac{1}{n-m}\left(\sum_{j=1}^{n} p_j - \sum_{i=1}^{m} z_i\right) = (-K^*)^{\frac{1}{n-m}}$$

or

$$s = \frac{1}{n-m}\left(\sum_{j=1}^{n} p_j - \sum_{i=1}^{m} z_i\right) + (K^*)^{\frac{1}{n-m}} \cdot e^{\frac{2l+1}{n-m}\pi}, l = 1, 2, \ldots, n-m-1 \quad (4.17)$$

By reference to Fig. 4.4, Eq. (4.17) shows that there are $n-m$ closed-loop poles located at infinity, and the position of each pole is determined by two vectors: a common real vector

$$\sigma_a = \frac{\sum_{j=1}^{n} p_j - \sum_{i=1}^{m} z_i}{n-m}$$

and a distinct complex vector

$$(K^*)^{\frac{1}{n-m}} \cdot e^{\frac{2l+1}{n-m}\pi}, \quad l = 1, 2, \ldots, n-m-1$$

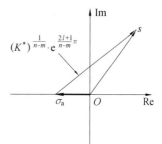

Figure 4.4 Vector diagram of asymptotes

Hence, $n-m$ asymptotes intersect at point σ_a, as given by Eq. (4.15), and the angles φ_a that the asymptotes make with the positive real axis are given by Eq. (4.16).

Example 4.2 Suppose we are required to draw the root locus for the negative feedback system with an open-loop transfer function

$$G(s)H(s) = \frac{K^*}{s(s+1)(s+2)}$$

For this open-loop transfer function there are no zeros, and the three poles are $p_1 = 0$, $p_2 = -1$, and $p_3 = -2$. Locate them in the s-plane with selected symbols. Usually, the open-loop pole is denoted with "×", the open-loop zero is denoted with "∘", and the closed-loop pole is denoted with "·".

From rule 1, the root locus consists of three continuous branches that are symmetrical with respect to the real axis of the s-plane.

From rule 2, since $n = 3$ and $m = 0$, the three branches start from three open-loop poles as $K^* = 0$ and approach infinity as $K^* \to \infty$.

From rule 3, the three branches approach infinity along three asymptotes. The intersection of asymptotes is

$$\sigma_a = \frac{0-1-2}{3-0} = -1$$

and the angles of asymptotes are

$$\varphi_a = \frac{(2l+1)\pi}{3-0} = \pm\frac{\pi}{3}, \pi \quad (l = 0, \pm 1)$$

The result is shown in Fig. 4.5. (To be continued.)

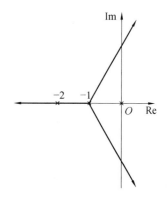

Figure 4.5 Asymptotes of root locus for system in Example 4.2

4.2.4 Root Locus on Real Axis

Rule 4 Any segment of the real axis is a part of the root locus if and only if the total number of open-loop poles and zeros to its right is odd.

Proof The proof of this rule is based on the following observation:

① At any trial point s_t on the real axis, the angles of the vectors drawn from any pair of complex conjugate poles or zeros of $G(s)H(s)$ add up to be zero. For example, In Fig. 4.6 we have

$$\angle(s_t-p_3) + \angle(s_t-p_4) = 0$$

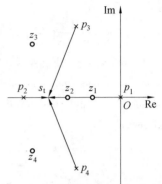

Figure 4.6 Condition for root locus on real axis

Therefore, the only contribution to angle condition in Eq. (4.13) is from the real poles and zeros of $G(s)H(s)$.

② Only the real poles and zeros of $G(s)H(s)$ that lie to the right of the point s_t may contribute to angle condition in Eq. (4.13), since the real poles and zeros that lie to the left of the point contribute zero degrees.

③ Each real pole of $G(s)H(s)$ to the right of the point s_t contributes $-180°$ to Eq. (4.13), each zero of $G(s)H(s)$ to the right of the point s_t contributes $180°$ to Eq. (4.13).

The last observation shows that for s_t to be a point on the root locus, there must be an odd number of open-loop poles and zeros to its right.

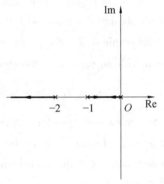

Figure 4.7 Root locus on real axis for Example 4.2

For the system in Example 4.2, the root locus on real axis is shown in Fig. 4.7.

4.2.5 Angles of Departure and Angles of Arrival of Root Locus

The angle of departure (or angle of arrival) is the angle at which the root locus leaves a pole (or approaches a zero), i.e. the angle of the tangent of the root locus at the starting point (or finishing point) with respect to the positive real axis.

Rule 5 The angle of departure of the root locus from the open-loop pole p_k is given by

$$\theta_{p_k} = (2l+1)\pi + \sum_{i=1}^{m}\angle(p_k-z_i) - \sum_{j=1,j\neq k}^{n}\angle(p_k-p_j), \quad l=0,\pm 1,\pm 2\ldots$$

or for brief,

$$\theta_{p_k} = 180° + \sum_{i=1}^{m}\angle(p_k-z_i) - \sum_{j=1,j\neq k}^{n}\angle(p_k-p_j) \qquad (4.18)$$

Similarly the angle of arrival of the root locus at the open-loop zero z_k is given by

$$\theta_{z_k} = 180° + \sum_{j=1}^{n} \angle(z_k - p_j) - \sum_{i=1, j\neq k}^{m} \angle(z_k - z_i) \qquad (4.19)$$

Proof Assume that s_t is a point on a root locus branch leaving the pole p_k and is very near the pole. Then s_t must satisfy Eq. (4.13), that is

$$\sum_{i=1}^{m} \angle(s_t - z_i) - \left[\sum_{j=1}^{k-1} \angle(s_t - p_j) + \angle(s_t - p_k) + \sum_{j=k+1}^{n} \angle(s_t - p_j) \right] = (2l+1)\pi,$$
$$l = 0, \pm 1, \pm 2, \ldots$$

or

$$\angle(s_t - p_k) = -(2l+1)\pi + \sum_{i=1}^{m} \angle(s_t - z_i) - \sum_{j=1, j\neq k}^{n} \angle(s_t - p_j), \quad l = 0, \pm 1, \pm 2, \ldots$$

Letting s_t approach the pole p_k infinitely and taking $l = -1$, with no loss of generality, yields

$$\theta_{p_k} = 180° + \sum_{i=1}^{m} \angle(p_k - z_i) - \sum_{j=1, j\neq k}^{n} \angle(p_k - p_j)$$

Similarly we have

$$\theta_{z_k} = 180° + \sum_{j=1}^{n} \angle(z_k - p_j) - \sum_{i=1, i\neq k}^{m} \angle(z_k - z_i)$$

If a pole or zero is of q-multiple, then the angle of departure is given by

$$\theta_{p_k} = \frac{1}{q}\left[(2l+1)\pi + \sum_{i=1}^{m} \angle(p_k - z_i) - \sum_{\substack{j=1 \\ j\neq k, \ldots, k+q-1}}^{n} \angle(p_k - p_j) \right], \quad l = 0, 1, 2, \ldots, q-1$$

(4.20)

and the angle of arrival is given by

$$\theta_{z_k} = \frac{1}{q}\left[(2l+1)\pi + \sum_{j=1}^{m} \angle(z_k - p_j) - \sum_{\substack{i=1 \\ i\neq k, \ldots, k+q-1}}^{m} \angle(z_k - z_i) \right], \quad l = 0, 1, 2, \ldots, q-1$$

(4.21)

The angle of departure (or arrival) is particularly of interest for complex poles (or zeros) because the information is helpful in completing the root locus. Although this rule is also suitable for the case of real pole or zero. In most case, however, even for the multiple real poles or zeros it is not necessary to determine the angle of departure or arrival by above equations. If q poles or zeros are coincident, then the tangents to the locus at the q-multiple pole or zero are equally spaced over $360°$, i.e., the included angle between any two adjacent branches departing the pole or arriving the zero always keeps

$$\theta = 2\pi/q \qquad (4.22)$$

Example 4.3 A single-loop feedback system has an open-loop transfer function

$$G(s)H(s) = \frac{K^*(s+5)}{s(s^2+4s+8)}$$

Determine the angles of departure of the root locus.

Solution The open-loop transfer function has a zero and three poles as shown in Fig. 4.8,

$$z_1 = -5, \quad p_1 = 0, \quad p_{2,3} = -2 \pm j2$$

The root locus on the real axis can be determined, from Rule 4, as shown in Fig. 4.8. By inspection we get

$$\theta_{p_1} = 180°$$

Then from Eq. (4.18) we have

$$\theta_{p_2} = 180° + \angle(p_2-z_1) - [\angle(p_2-p_1) + \angle(p_2-p_3)] =$$
$$180° + \arctan\frac{2}{3} - \left[\left(180° - \arctan\frac{2}{2}\right) + 90°\right] = -12°$$

Since the root locus is symmetric about the real axis, we get

$$\theta_{p_3} = 12°$$

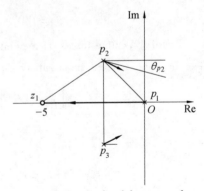

Figure 4.8 Angle of departure of Example 4.3

4.2.6 Breakaway Points and Breakaway Angles of Root Locus

In the case of that the characteristic equation of a closed-loop system has multiple roots as K^* varies, two or more branches of the root locus come together at the breakaway point and then move away from each other, as shown in Fig. 4.9.

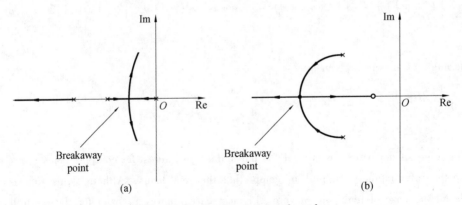

Figure 4.9 Breakaway point of root locus

1. Breakaway points of root locus

Rule 6 The breakaway points on the root locus must satisfy

$$\frac{d}{ds}\left[\frac{\prod_{i=1}^{m}(s-z_i)}{\prod_{j=1}^{n}(s-p_j)}\right] = 0 \qquad (4.23)$$

Proof From Eq. (4.11) the characteristic equation may be rewritten as

$$D(s) = Q(s) + K^* P(s) = 0 \qquad (4.24)$$

where

$$Q(s) = \prod_{j=1}^{n}(s-p_j), \qquad P(s) = \prod_{i=1}^{m}(s-z_i)$$

Suppose that s_b is a breakaway point. Then the characteristic equation has a multiple-order

root at s_b; i.e., we can express the characteristic equation as
$$D(s) = (s-s_b)^\gamma D_1(s)$$
where γ is the order of the root. The derivative of $D(s)$ with respect to s is then
$$D'(s) = \gamma(s-s_b)^{\gamma-1}D_1(s) + (s-s_b)^\gamma D'_1(s) =$$
$$(s-s_b)^{\gamma-1}[\gamma D_1(s) + (s-s_b)D'_1(s)]$$
where the prime indicates the derivative with respect to s. Since $D'(s)$ also has a root of order $(\gamma-1)$ at point s_b, i.e. $D'(s) = 0$, from Eq. (4.24), we get
$$D'(s) = Q'(s) + K^* P'(s) = Q'(s) - \frac{Q(s)}{P(s)}P'(s) = 0$$
which can be expressed as
$$[Q(s)P'(s) - Q'(s)P(s)]_{s=s_b} = 0$$
i.e.,
$$\frac{d}{ds}\left[\frac{\prod_{i=1}^{m}(s-z_i)}{\prod_{j=1}^{n}(s-p_j)}\right] = 0$$

In general, the following conclusions can be made with respect to the breakaway points:

① The condition for the breakaway point given by Eq. (4.23) is necessary, but not sufficient. In other words, any breakaway point must be a solution of Eq. (4.23), and this solution must also be a point on root locus.

② A breakaway point may involve more than two branches of the root locus.

③ A root locus diagram can have more than one breakaway point. Moreover, the breakaway points need not always be on the real axis.

2. Breakaway angles of root locus

Assume that there exists a q-multiple breakaway point when $K^* = K_b^*$, and the corresponding characteristic roots are $s_{b1}, s_{b2}, \ldots, s_{bn}$, among which q roots are coincident. Now, construct a new system with the open-loop poles $s_{b1}, s_{b2}, \ldots, s_{bn}$ and same open-loop zeros as that of the original system. Then the problem of breakaway angles for the original system becomes a problem of angles of departure for the new system. Thus we can make the conclusion:

In general, due to the angle condition, the tangents to the root locus at a breakaway point are equally spaced over 360°. If there is a q-multiple breakaway point, then the included angle between the tangents to any two adjacent branches of the root locus at the breakaway point is
$$\theta = \frac{2\pi}{q} \quad (4.25)$$

Reconsider the system in Example 4.2. Solving the breakaway point equation
$$\frac{d}{ds}\left[\frac{K^*}{s(s+1)(s+2)}\right] = 0$$

results in
$$3s^2+6s+2=0$$
which has roots at
$$s_1 = -0.423, \quad s_2 = -1.577$$

Since $s_2 = -1.577$ does not lie on the root locus, the breakaway point is $s_b = -0.423$.

The included angle between two branches leaving breakaway point is π. Since the segment $[0,-1]$ is a part of the root locus, these two branches leave the real axis vertically. The breakaway point and breakaway angles are shown in Fig. 4.10.

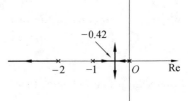

Figure 4.10 Breakaway point and breakaway angle for system in Example 4.2

4.2.7 Intersections of Root Locus with Imaginary Axis

Rule 7. The points where the root locus intersects with imaginary axis of the s-plane, and the corresponding value of K^*, may be given by equation
$$1+G(j\omega)H(j\omega)=0 \qquad (4.26)$$

Explanation If the root locus is intersected with the imaginary axis then the characteristic equation has a pair of pure imaginary roots, i.e. $s=\pm j\omega$, which must satisfy the characteristic equation $1+G(j\omega)H(j\omega)=0$.

In Example 4.2, the characteristic equation
$$1+\frac{K^*}{s(s+1)(s+2)}=0$$
can be rewritten as
$$s^3+3s^2+2s+K^* = 0$$
Substituting $s=j\omega$ into above equation yields
$$\begin{cases} K^* - 3\omega^2 = 0 \\ \omega(2-\omega^2) = 0 \end{cases}$$

The result is $\omega = \pm 1.41$ and $K^* = 6$, which indicates that the root locus intersects the imaginary axis at ± 1.41 while $K^* = 6$. The whole root locus is sketched in Fig. 4.11.

It can be seen from the root locus that the closed-loop system will be stable if $0<K^*<6$.

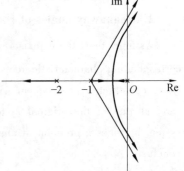

Figure 4.11 Root locus for system in Example 4.2

4.2.8 Other Useful Conclusions

1. Estimation of gain on root locus

Sometimes we need to estimate the root locus gain, or open-loop gain, or static error constants of a system corresponding to a specified point on the root locus.

According to the magnitude condition of root locus, the root locus gain K^* at any point s_1 on the root locus is given by

$$K^* = \frac{\prod_{j=1}^{n} |s_1 - p_j|}{\prod_{i=1}^{m} |s_1 - z_i|} \quad (4.27)$$

On the other hand, if a value of K^* at a specified point on the root locus is given, according to its definition, the open-loop gain K can be obtained by

$$K = \lim_{s \to 0} s^v G(s)H(s) = \lim_{s \to 0} s^v \frac{K^* \prod_{i=1}^{m}(-z_i)}{\prod_{j=1}^{n}(-p_j)} \quad (4.28)$$

where v is the system type.

2. Relation between closed-loop poles and open-loop poles

The closed-loop characteristic equation can be written as

$$\prod_{j=1}^{n}(s - p_j) + K^* \prod_{i=1}^{m}(s - z_i) = 0$$

or

$$\prod_{i=1}^{n}(s - s_i) = 0$$

where $p_j(j=1,2,\ldots,n)$ and $z_i(i=1,2,\ldots,m)$ are open-loop poles and zeros respectively, and $s_j(j=1,2,\ldots,n)$ are the close-loop poles. If $n-m \geq 2$, then we have

$$s^n + \sum_{j=1}^{n}(-p_j)s^{n-1} + \ldots = s^n + \sum_{j=1}^{n}(-s_j)s^{n-1} + \ldots$$

or, in the case of $n-m \geq 2$,

$$\sum_{j=1}^{n} s_i = \sum_{j=1}^{n} p_i \quad (4.29)$$

Sometimes this conclusion is useful to determine the trend of the root locus in the case of $n-m \geq 2$.

3. Closed-loop poles and coefficients of characteristic equation

Assume that the characteristic equation is given by

$$s^n + a_1 s^{n-1} + \ldots + a_{n-1} s + a_n = 0$$

and its roots are $s_1, s_2, \ldots, s_n$, then we have

$$s^n + a_1 s^{n-1} + \ldots + a_{n-1} s + a_n = (s-s_1)(s-s_2)\ldots(s-s_n)$$

According to the relation between the coefficients of an algebraic equation and its roots, we get

$$\sum_{j=1}^{n} s_i = -a_1 \quad (4.30)$$

$$\prod_{j=1}^{n} s_i = (-1)^n a_n \qquad (4.31)$$

4.3 Other Configuration

4.3.1 0° Root Locus

1. 0° root locus and 180° root locus

In previous section, the root locus considered satisfies the characteristic equation

$$1 + G(s)H(s) = 0 \qquad (4.32)$$

where the open-loop transfer function is in the form

$$G(s)H(s) = \frac{K^* \prod_{i=1}^{m}(s-z_i)}{\prod_{j=1}^{n}(s-p_j)} \qquad (4.33)$$

i.e., the root locus is plotted based on the angle condition

$$\sum_{i=1}^{m} \angle(s-z_i) - \sum_{j=1}^{n} \angle(s-p_j) = 180° + 2l\pi, \quad l = \pm 1, \pm 2, \ldots \qquad (4.34)$$

Now, consider a positive feedback system with the open-loop transfer function in Eq. (4.33). In this case, the characteristic equation becomes

$$1 - G(s)H(s) = 0 \qquad (4.35)$$

i.e., the root locus equation becomes

$$\frac{K^* \prod_{i=1}^{m}(s-z_i)}{\prod_{j=1}^{n}(s-p_j)} = 1 \qquad (4.36)$$

and the angle condition of root locus becomes

$$\sum_{i=1}^{m} \angle(s-z_i) - \sum_{j=1}^{n} \angle(s-p_j) = 0° + 2l\pi, \quad l = \pm 1, \pm 2, \ldots \qquad (4.37)$$

Obviously, some of the rules related to the angle condition must be modified for plotting the root locus of Eq. (4.36). For a simple reason, the root locus satisfied Eq. (4.37) is called the 0° root locus, and that satisfied Eq. (4.34) is called the 180° root locus or regular root locus.

2. Basic rules for plotting 0° root locus

It is easy to understand that the rule 3~5 and rule 7 are related to the angle condition of root locus, and they should be modified as follows for plotting 0° root locus.

(1) Modified rule 3: Asymptotes of root locus

The angles of the asymptotes with respect to the real axis are given by

$$\varphi_a = \frac{2l\pi}{n-m}, \quad l = 0, 1, 2, \ldots, n-m-1 \qquad (4.38)$$

(2) Modified rule 4: Root locus on the real axis

Any segment on the root locus is a part of the root locus if and only if the total number of open-loop poles and zeros to its right is not odd.

(3) Modified rule 5: Angle of departure and angle of arrival of the root locus

The angle of departure of the root locus from a complex pole p_k is given by

$$\theta_{p_k} = 0° + \sum_{i=1}^{m} \angle(p_k - z_i) - \sum_{j=1, j \neq k}^{n} \angle(p_k - p_j) \quad (4.39)$$

The angle of arrival of the root locus at a complex zero z_k is given by

$$\theta_{z_k} = 0° + \sum_{j=1}^{n} \angle(z_k - p_j) - \sum_{i=1, i \neq k}^{m} \angle(z_k - z_i) \quad (4.40)$$

(4) Modified rule 7: Intersections of the root locus with the imaginary axis

The points at which the root locus intersects with imaginary axis of the s-plane, and the corresponding values of K^*, can be obtained from the characteristic equation

$$1 - G(j\omega)H(j\omega) = 0 \quad (4.41)$$

Example 4.4 Plot the root locus for the positive feedback control system with an open-loop transfer function

$$G(s)H(s) = \frac{1}{s(s+1)(s+2)}$$

Solution Since the characteristic equation is $1 - G(s)H(s) = 0$ and the root locus satisfies the angle condition in Eq. (4.37), the root locus is a $0°$ root locus. The open-loop poles are $p_1 = 0, p_2 = -1, p_3 = -2$, and there is no zero.

From rule 1, since $n = 3$, the root locus is consisted of three branches.

From rule 2, since $n = 3$ and $m = 0$, the three branches start from three open-loop poles as $K^* = 0$ and approach infinity as $K^* \to \infty$.

From modified rule 3, the asymptotes of root locus intersect at

$$\sigma_a = \frac{0 - 1 - 2}{3 - 0} = -1$$

and the angles of the root locus with respect to the real axis are

$$\varphi_a = \frac{2l\pi}{3-0} = 0, \pm\frac{2\pi}{3} \quad (l = 0, \pm 1)$$

From modified rule 4, the negative real-axis segments $[-2, -1]$ and $[0, \infty]$ form part of the root locus.

From rule 6, by inspection, there is a breakaway point on the real-axis segment $[-2, -1]$. From Example 4.2, the breakaway point equation has two solutions

$$s_1 = -0.423, \quad s_2 = -1.577$$

In this case, however, the breakaway point is $s_b = -1.577$.

The included angle between two branches leaving breakaway point is π, moreover, since the segment $[-2, -1]$ is part of the root loci, these two branches leave the real axis vertically.

From modified rule 7, the characteristic equation for this system is

$$s^3+3s^2+2s-K^*=0$$

Substituting $s=j\omega$ yields the following equation

$$\begin{cases} -3\omega^2-K^*=0 \\ -\omega^3+2\omega=0 \end{cases}$$

Since K^* is varied from 0 to infinity, the only one solution to above equation is $\omega=0$ rad/s, i.e., the root locus does not intersect with the imaginary axis except for the original point.

The final plot is shown in Fig. 4.12.

4.3.2 Parameter Root Locus

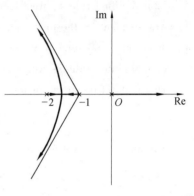

Figure 4.12 0° root locus for system in Example 4.4

In what follows we consider the case in which the varied parameter is not the gain. It is important to remember that a root locus is the locus of characteristic equation roots. We now illustrate the procedure for plotting the root locus with an example.

Example 4.5 Plot the root locus of system shown in Fig. 4.13 as the parameter T_a varies from zero to infinity.

Solution First we write the characteristic equation of the system

$$s(5s+1)+5(T_a s+1)=0$$

Next, grouping terms that are not multiplied by T_a and dividing the equation by these terms yield

$$1+\frac{5T_a s}{5s^2+s+5}=0$$

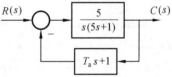

Figure 4.13 Control system in Example 4.5

i.e.,

$$\frac{T_a s}{s^2+0.2s+1}=-1$$

Since this equation is in the form similar to that of Eq. (4.12), we get an equivalent open-loop transfer function

$$G_e(s)H_e(s)=\frac{T_a s}{s^2+0.2s+1}$$

and the equivalent root locus gain is $K_e^*=T_a$.

This equivalent open-loop transfer function has a zero at the origin and two poles at $p_{1,2}=-0.1\pm j0.09$. The root locus is plotted in Fig. 4.14, where the breakaway point is $s_b=-1$, and the angles of departure are $\theta_p=\pm 185.7°$.

In general case, for plotting parameter root locus, the system characteristic equation must be expressed as a polynomial equation. Then, denoting the varied parameter with α, the terms not multiplied by α are grouped into a function $D_e(s)$ and the remaining terms that are multiplied by α are grouped into a function $N_e(s)$, such that the characteristic equation is

expressed as
$$D_e(s) + \alpha N_e(s) = 0 \quad (4.42)$$
Divided by $D_e(s)$, it is resulted in
$$1 + \alpha \frac{N_e(s)}{D_e(s)} = 0 \quad (4.43)$$
This equation can be rewritten as
$$1 + K_e^* \frac{s^m + b_1 s^{m-1} + \ldots + b_m}{s^n + a_1 s^{n-1} + \ldots + a_{n-1} s + a_n} = 0 \quad (4.44)$$
or
$$1 + \frac{K_e^* \prod_{i=1}^{m}(s - z_i)}{\prod_{j=1}^{n}(s - p_j)} = 0 \quad (4.45)$$

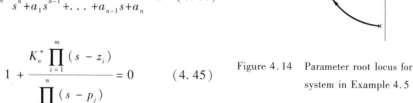

Figure 4.14   Parameter root locus for system in Example 4.5

The root locus is then plotted for the equivalent gain K_e^* with the equivalent open-loop transfer function
$$G_e(s) H_e(s) = \frac{K_e^* \prod_{i=1}^{m}(s - z_i)}{\prod_{j=1}^{n}(s - p_j)}$$

4.4 Application of Root Locus Method

The root locus method provides graphical information, and therefore an approximate sketch can be used to obtain qualitative information concerning the stability and performance of the system including the steady-state performance characterized by the system type and static error constants and the dynamic performance characterized by the percent overshoot and setting time and etc.

In this section the application of root locus method is illustrated with some examples.

Example 4.6   (Parameter design.) Consider a unity negative feedback system with the open-loop transfer function
$$G(s) H(s) = \frac{K}{s(s + 1)(0.5s + 1)}$$
Determine the open-loop gain so that the closed-loop system has a pair of dominant poles with a damping ratio of 0.5.

Solution   Rewriting the open-loop transfer function in the form of poles and zeros
$$G(s) H(s) = \frac{K^*}{s(s+1)(s+2)}$$
the root locus can be plotted as shown in Fig. 4.15 (see Example 4.2).

Plot the constant ζ-lines of $\zeta = 0.5$, as shown in Fig. 4.15. The lines intersect the root locus at
$$s_{1,2} = -0.33 \pm j\,0.58$$
which form a pair of complex conjugate roots of the closed-loop system. Since this is a third-

order system without zeros, from Eq. (4.29), we have
$$s_3 = -2.34$$
Since
$$\frac{|\text{Re}(s_3)|}{|\text{Re}(s_1)|} = \frac{2.34}{0.33} > 7$$
$s_{1,2}$ can be considered as a pair of dominant poles.

Since the closed-loop poles must satisfy the magnitude condition for root locus, letting
$$|G(s)H(s)|_{s=s_3} = \left|\frac{K}{s(s+1)(0.5s+1)}\right|_{s=-2.34} = 1$$
yields
$$K = |s(s+1)(0.5s+1)|_{s=s_3} = 0.525$$

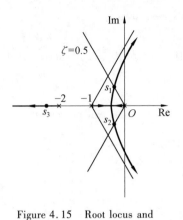

Figure 4.15 Root locus and dominant poles

Example 4.7 (Effect of adding pole.) Comparing the root loci in Fig. 4.16(a) ~ (c).

Fig. 4.16(a) shows the root locus of a system with the open-loop transfer function
$$G(s)H(s) = \frac{k}{s(s+a)}$$

It can be seen that the angles of the asymptotes are ±90°. The system is always stable.

Fig. 4.16(b) shows that the additional pole at $s = -b$ causes the complex part of the root locus to bend toward the right half of the s-plane. The angles of right asymptotes become. The system may become unstable if the value of k exceeds the critical value.

If another additional pole $s = -c$ is added to $G(s)H(s)$, as shown in Fig. 4.16(c), the complex part of root locus is moved farther to the right, and the angles of right asymptotes are changed to ±45°. The stability condition of the system becomes even more restricted.

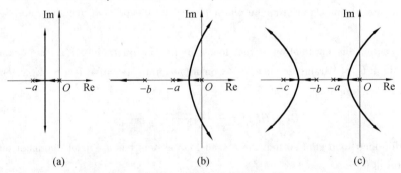

(a) (b) (c)

Figure 4.16 Effect of adding poles

Roughly speaking, adding a pole to the open-loop transfer function in the left half of the s-plane has the effect of pushing the original root locus toward the right half of the s-plane.

Example 4.8 (Effect of adding zero.) Comparing the root loci in Fig. 4.17(a) ~ (d).

Fig. 4.17(b) shows the root locus when a zero at $s = -b$ is added to the open-loop transfer function
$$G(s)H(s) = \frac{k}{s(s+a)}$$

with $b>a$. Comparing with the original root locus shown in Fig. 4.17(a), the resultant root locus is bent toward the left, and the relative stability of the system is improved.

Comparing Fig. 4.17(c) and (d), a similar effect can be observed when a zero at $s=-c$ is added to the open-loop transfer function

$$G(s)H(s) = \frac{k}{s(s+a)(s+b)}$$

with $c>b$.

Roughly speaking, adding a zero to the open-loop transfer function in the left half of the s-plane has the effect of moving the original root locus toward the left half of the s-plane.

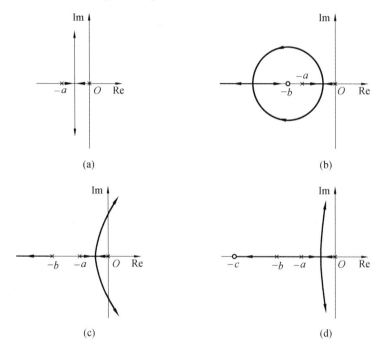

Figure 4.17 Effect of adding zeros

Problems

P4.1 Sketch the root loci for the following open-loop transfer functions when $0<k<\infty$.

(a) $G(s)H(s) = \dfrac{k(s+5)}{s(s+2)(s+3)}$

(b) $G(s)H(s) = \dfrac{k(s+5)}{s(s+1)(s+3)(s+4)}$

(c) $G(s)H(s) = \dfrac{k(s+3)}{s^2+4s+8}$

(d) $G(s)H(s) = \dfrac{k(s+20)}{s(s^2+20s+200)}$

(e) $G(s)H(s) = \dfrac{k}{s(s+3)^2}$

(f) $G(s)H(s) = \dfrac{K}{s(0.05s^2+0.4s+1)}$

P4.2 Consider a unity feedback system with

$$G(s) = \frac{k(s+1)}{s^2+4s+5}$$

(a) Find the angles of departure of the root locus from the complex poles.

(b) Find the entry point for the root locus as it enters the real axis.

P4.3 A unity feedback system has a plant transfer function
$$G(s) = \frac{k(s^2+20)(s+1)}{(s^2-2)(s+10)}$$
Sketch the root locus as k varies.

P4.4 A control system has the open-loop transfer function
$$G(s)H(s) = \frac{k(s^2+4s+8)}{s^2(s+4)}$$
It is desired that the dominant poles have a ζ equal to 0.5. Using the root locus, show that $k=7.35$ is required and the dominant poles are $s=-1.3\pm j2.2$.

P4.5 A unity feedback system has
$$G(s) = \frac{k}{s(s+2)(s+5)}$$
Find (a) the breakaway point on the real axis and the gain for this point, (b) the gain and the roots when two roots lie on the imaginary axis, and (c) the roots when $k=6$. (d) Sketch the root locus.

P4.6 The open-loop transfer function of a unity feedback system is given by
$$G(s) = \frac{k(s+a)}{s^2(s+1)}$$
Determine the values of a so that the root locus will have zero, one, and two breakaway points, respectively, not counting the one at $s=0$. Sketch the root loci for $0<k<\infty$ for all three cases.

P4.7 The transfer functions of a negative feedback system are given by
$$G(s) = \frac{k}{s^2(s+2)(s+5)} \text{ and } H(s) = 1$$
(a) Sketch the root locus for this system. Indicate the crossing points of the locus on the imaginary axis and the corresponding value of k at these points. (b) The transfer function of the feedback loop is now changed to $H(s) = 2s+1$. Determine the stability of the modified system as a function of k. Investigate the effect on the root locus due to this change in $H(s)$.

P4.8 The characteristic equation of a feedback control system is given by
$$\Delta(s) = s^2 + (k+5)s + 2k + 4 = 0$$
Sketch the root locus as a function of k (positive k only) for this system.

P4.9 A unity feedback system has a plant
$$G(s) = \frac{0.25(s+a)}{s^2(s+1)}$$
Sketch the root locus as s function of a (positive a only) for this system.

P4.10 The open-loop transfer function of a control system with positive feedback is given by
$$G(s)H(s) = \frac{k}{s(s^2+4s+4)}$$
Sketch the root locus for this system when $0<k<\infty$.

Chapter 5 Frequency Response Method

5.1 Introduction

It was pointed out that in practice the performance of a feedback control system is more preferably measured by its time-domain response characteristics. However, analytically, it is usually difficult to determine the time response of a control system, especially in the case of higher-order systems. In the design aspects, there are no unified ways of arriving at a designed system directly from the given time-domain specifications, such as the maximum overshoot, rise time, and setting time. The root locus method seems to be more than adequate in predicting the transient behavior of a closed-loop system, given the open-loop transfer function. The weakness of this method is that it relies on the existence of the open-loop transfer function, from which the locus itself is drawn. A very practical and important approach to the analysis and design of a linear feedback control system is the frequency response method. Our ultimate goal will be to use the open-loop frequency response of a system to understand the time-domain behavior of the closed-loop system.

5.1.1 Steady-State Response of Systems to Sinusoidal Inputs

The frequency response of a system is derived from the steady-state response of the system to a sinusoidal input signal. Suppose that the input to a stable linear system with the transfer function $G(s)$ is a sinusoid

$$r(t) = A_r \sin \omega t \tag{5.1}$$

i.e.,

$$R(s) = \frac{A_r \omega}{s^2 + \omega^w} \tag{5.2}$$

then the output is given by

$$C(s) = G(s)R(s) = G(s) \cdot \frac{A_r \omega}{(s-j\omega)(s+j\omega)} =$$

$$\frac{B_1}{s-j\omega} + \frac{B_2}{s+j\omega} + C_g(s) \tag{5.3}$$

where $C_g(s)$ is the collection of all the terms in the partial-fraction expansion that originate in the denominator of $G(s)$. It is assumed that the system is stable, thus the terms in $C_g(s)$ will decay to zero with increasing time. Therefore, only the first two terms in Eq. (5.3) contribute to the steady-state response.

From Eq. (5.2) and (5.3)

$$B_1 = \frac{A_r}{j2} G(j\omega) \tag{5.4}$$

$$B_2 = -\frac{A_r}{j2} G(-j\omega) \tag{5.5}$$

Since, for a given value of ω, $G(j\omega)$ is a complex number, it is convenient to express $G(j\omega)$ as

$$G(j\omega) = |G(j\omega)| e^{j\varphi} \tag{5.6}$$

where

$$\varphi = \angle G(j\omega) \tag{5.7}$$

Then, from Eq. (5.3) through (5.7), the steady-state component of the output is

$$c_s(t) = B_1 e^{j\omega t} + B_2 e^{-j\omega t} = A_r |G(j\omega)| \frac{e^{j\varphi} e^{j\omega t} - e^{-j\varphi} e^{-j\omega t}}{j2} =$$

$$A_r |G(j\omega)| \sin(\omega t + \varphi) = A_c \sin(\omega t + \varphi) \tag{5.8}$$

Now we see that the resulting output in the steady state is a sinusoidal signal at the same angular frequency as the input; it differs from the input waveform only in amplitude and phase angle. Moreover, at a specified frequency, the steady-state amplitude is given by

$$A_c = A_r |G(j\omega)| \tag{5.9}$$

and the phase shift of the output sinusoid relative to the input sinusoid just is

$$\varphi = \angle G(j\omega) \tag{5.10}$$

Example 5.1 The transfer function of a system is

$$G(s) = \frac{5}{s+2}$$

and the input is $2\sin 3t$. Determine the steady-state output of the system.

Solution By inspection, this system is stable. At the frequency $\omega = 3$ we have

$$|G(j\omega)|_{\omega=3} = 1.387, \quad \angle G(j\omega)|_{\omega=3} = -56.3°$$

The steady-state output is given by

$$c_s(t) = A_r |G(j\omega)| \sin(\omega t + \varphi) = 2 \times 1.387 \sin(3t - 56.3°) =$$

$$2.774 \sin(3t - 56.3°)$$

By the way, since the system time constant is 0.5 s, the output would reach steady state approximately 1.5 s ($\Delta = 5\%$) after the application of the input signal.

5.1.2 Frequency Response

We have seen that the response of a stable linear system to a sinusoidal input also is a sinusoidal signal at same frequency as the input. However, the magnitude and phase of the output differ from those of the input sinusoid signal, and the amount of difference is a function of the input frequency. Thus we can investigate the steady-state response of the system to a sinusoidal input as the frequency varies.

For a linear system or element, the ratio of the output steady-state component and the input sinusoid is defined as its frequency characteristic or frequency response. Since both the magnitude and phase of the output steady-state component are functions of the frequency, the magnitude ratio of the output and input sinusoids is defined as the magnitude-frequency characteristic or magnitude characteristic for short, and the phase shift of the output sinusoid relative to the input sinusoid is defined as the phase-frequency characteristic or phase characteristic for short.

One advantage of the frequency response method is the ready availability of sinusoidal test signals for various ranges of frequency and amplitude. Thus the experimental determination of the frequency response of a system is easily accomplished and is the most reliable and uncomplicated method for the experimental analysis of a system. In fact, we can design a control system by frequency-response method without developing a transfer function. Also, one of the most common methods to verify a derived transfer function is to compare the frequency response as calculated from the transfer function with that obtained from measurements on the physical system.

Another advantage of the frequency response method is that the frequency response describing the sinusoidal steady-state behavior of a system can be directly obtained by replacing s with $j\omega$ in the system transfer function $G(s)$, i. e.

$$G(j\omega) = G(s)|_{s=j\omega} = M(\omega) e^{j\varphi(\omega)} \tag{5.11}$$

where $M(\omega)$ is the magnitude-frequency characteristic, and $\varphi(\omega)$ is the phase-frequency characteristic.

Once the analysis and design are carried out in the frequency-domain, the time-domain behavior of the system can be interpreted based on the relationships that exist between the time-domain and the frequency-domain properties. Therefore, we may consider that the main purpose of conducting control systems analysis and design in the frequency-domain merely to use the techniques as a convenient vehicle toward the same objective as with time-domain methods.

The basic disadvantage of the frequency response method for analysis and design is the indirect link between the frequency and time domains. Direct correlations between the frequency response and the corresponding transient response characteristics are somewhat tenuous, and in practice the frequency response characteristic is adjusted by using various criteria that will normally result in a satisfactory transient response.

5.1.3 Frequency Response Plots

There are several analytical techniques for investigating the stability and performance of control systems, which may be classed as frequency response methods. They all involve the study of the relationship among three parameters, magnitude, phase angle and frequency of the frequency response. For analysis and design of control systems, it is useful to draw frequency response $G(j\omega)$ versus frequency ω in some form to characterize the frequency response. The familiar frequency response plots include the polar plot, Bode diagram and Nichols chart.

1. Polar plot

The transfer function of a system can be described in the frequency domain by the relation

$$G(j\omega) = G(s) \vert_{s=j\omega} = \text{Re}[G(j\omega)] + j\text{Im}[G(j\omega)] \tag{5.12}$$

Alternatively it can be represented by a magnitude and a phase as

$$G(j\omega) = \vert G(j\omega) \vert e^{j\angle G(j\omega)} \tag{5.13}$$

One common method of displaying the frequency response is in the form of a polar plot, which is obtained by using either Eq. (5.12) or Eq. (5.13). In such a plot, the magnitude and angle of the frequency response (or its real and imaginary parts) are drawn in the polar coordinates as the frequency ω is varied from zero to infinity.

2. Bode diagram

The limitations of polar plots are readily apparent. The addition of poles or zeros to an existing system requires the recalculation of the frequency response. Furthermore, the calculation of the frequency response in this manner is tedious and does not indicate the effect of the individual pole or zero. The introduction of Bode diagram simplifies the determination of the graphical portrayal of the frequency response.

Bode diagram consists of two separate graphs, one plot of magnitude in decibels versus frequency and another plot of phase versus frequency. These plots require semilog graph paper, as shown in Fig. 5.1, where the logarithmic scale is used for the ω-axis. Moreover, the magnitude-frequency plot does not involve simply magnitude M, but the modulus expressed in decibels

$$L(\omega) = M(\text{db}) = 20\lg M = 20\lg \vert G(j\omega) \vert \tag{5.14}$$

as shown in Fig. 5.1.

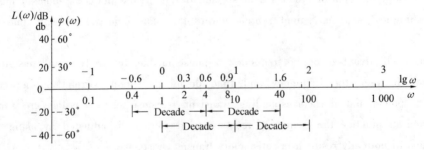

Figure 5.1 Semilog coordinates

In frequency response method the analysis of a system is mainly based on the Bode diagram representation of the open-loop system. The Bode diagram is preferred over the polar plot for several reasons. One, the open-loop frequency response can be quickly drawn by means of straight-line approximations. Two, the use of logarithmic scale for frequency allows equal emphasis to be placed on each decade relating to system performances. Finally, it is very easy to see on the Bode diagram the effect on performance of changing a system parameter.

5.2 Bode Diagrams of Elementary Factors

In root locus method it is particular useful to write the open-loop transfer function in the form of zeros and poles. Instead, in frequency response method the open-loop transfer function is usually written in the form of standard factors, i. e. in the Bode form

$$G(s) = \frac{K \prod (\tau_i s + 1)}{s^v \prod (T_j s + 1) \prod [(s^2/\omega_{nk}^2) + (2\zeta_k/\omega_{nk})s + 1]} \quad (5.15)$$

Usually, a transfer function as shown in Eq. (5.15) has the following essential factors:
① factor K, i. e. a constant, corresponding to a gain element;
② factor s^{-1}, i. e. a pole at the origin, corresponding to an integral element;
③ factor s, i. e. a zero at the origin, corresponding to a derivative element;
④ factor $(Ts+1)^{-1}$, i. e. a real pole, corresponding to an inertial element;
⑤ factor $(\tau s+1)$, i. e. a real zero, corresponding to a first-order derivative element;
⑥ factor $[(s^2+2\zeta\omega_n s+\omega_n^2)/\omega_n^2]^{-1}$, i. e. a pair of complex conjugate poles, corresponding to an oscillatory element.

Each of these terms except K may appear raised to an integral power other than 1.

In such a way, since the magnitude of $G(j\omega)$ in the Bode diagram is expressed in decibels, the product of the factors in $G(j\omega)$ become additions, the phase relations are also added or subtracted from each other in a natural way. The logarithmic magnitude and phase curves can easily be drawn for each factor. Then these curves for each factor can be added together graphically to get the curve for the complete transfer function. The procedure can be further simplified by using asymptotic approximations to these curves, as shown in the following pages.

It is essential for frequency response method to know well the logarithmic magnitude characteristic and phase characteristic of each essential factor, i. e. typical element. The procedure, therefore, is to take the factors one at a time and evaluate the magnitude and phase.

5.2.1 Gain Element

When an open-loop transfer function is written in the Bode form, the constant factor just is the open-loop gain of the system. Hence, this factor is sometimes called the gain element. Since K is frequency-invariant, the logarithmic magnitude

$$L(\omega) = 20 \lg K \quad (5.16)$$

is a horizontal straight line. The phase angle

$$\varphi(\omega) = 0° \quad (5.17)$$

as long as K is positive, is coincided with ω-axis.

The Bode diagram of a constant K is shown in Fig. 5.2.

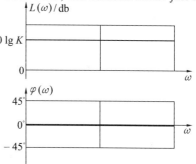

Figure 5.2   Bode diagram for constant K

5.2.2 Integral Element

When the transfer function is written in the Bode form, the factor s^{-1} means a pure integration, so that this factor is sometimes called the integral element. This element has a logarithmic magnitude

$$L(\omega) = 20 \lg \left| \frac{1}{j\omega} \right| = -20 \lg \omega \qquad (5.18)$$

and a phase angle

$$\varphi(\omega) = -90° \qquad (5.19)$$

The magnitude curve is a straight line with a slope of -20 db/dec and intersects the ω-axis at $\omega = 1$.

Similarly, for a multiple pole at the origin, we have

$$L(\omega) = 20 \lg \left| \frac{1}{(j\omega)^v} \right| = -20v \lg \omega \qquad (5.20)$$

and the phase is

$$\varphi(\omega) = -v\, 90° \qquad (5.21)$$

In this case, the slope of log-magnitude curve, due to a multiple pole, is $-20\,v$ db/dec.

The log-magnitude curves and phase curves of s^{-1} and s^{-2} are shown in Fig. 5.3.

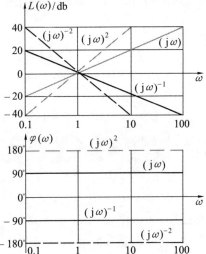

Figure 5.3 Bode diagram for integral and differential elements

5.2.3 Derivative Element

A zero of the transfer function at the origin, i.e. the factor s, is sometimes called the derivative element. It has a logarithmic magnitude

$$L(\omega) = 20 \lg |j\omega| = 20 \lg \omega \qquad (5.22)$$

where the slope is $+20$ db/dec, and a phase angle

$$\varphi(\omega) = 90° \qquad (5.23)$$

The log-magnitude curves and phase curves of derivative element s and s^2 are shown in Fig. 5.3.

5.2.4 Inertial Element

A pole of the transfer function at the negative real axis, i.e. the factor

$$\frac{1}{Ts+1}$$

is sometimes called the inertial element. The magnitude of this element is given by

$$L(\omega) = 20 \lg \frac{1}{\sqrt{T^2\omega^2+1}} = -20 \lg \sqrt{T^2\omega^2+1} \qquad (5.24)$$

The asymptotic curve for $\omega \ll 1/T$ is

$$L(\omega) = -20 \lg 1 = 0 \text{ db} \qquad (5.25)$$

and the asymptotic curve for $\omega \gg 1/T$ is

$$L(\omega) = -20 \lg T\omega = -20 \lg \omega - 20 \lg T \qquad (5.26)$$

which is a straight line of slope $-20\ v$ db/dec and intersects the 0db-line at $\omega = 1/T$. These two asymptotes meet at the frequency $\omega = 1/T$, which is called the break frequency or corner frequency.

The magnitude plot of an inertial element is shown in Fig. 5.4. The exact values of the magnitude-frequency for this element and the values obtained by using the approximation for comparison are given in Table 5.1. We can see that the maximum difference between the actual magnitude and the asymptotic approximation is 3 db. It is reasonable to assume that the higher-frequency asymptote is valid for $\omega > 1/T$ and the lower-frequency asymptote is valid for $\omega > 1/T$.

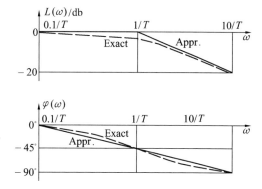

Figure 5.4 Bode diagram for inertial element

The phase angle of this element is

$$\varphi(\omega) = -\arctan T\omega \qquad (5.27)$$

The phase curve and its linear approximation can be obtained as shown in Fig. 5.4. This linear approximation, which passes through the correct phase at the break frequency, is within 6° of the actual phase curve for all frequency. This approximation will provide a useful means for readily determining the form of the phase curve of a transfer function $G(s)$. However, often the accurate phase angle curves are required, and it must be obtained via necessary calculation.

Table 5.1 Exact and approximate values of inertial element

$T\omega$	0.1	0.5	0.76	1	1.31	2	5	10
Exact values of $L(\omega)$/db	-0.04	-1.0	-2.0	-3.0	-4.3	-7.0	-14.2	-20.04
Approximate values of $L(\omega)$/db	0	0	0	0	-2.3	-6.0	-14.0	-20.0
Exact values of $\varphi(\omega)$/degrees	-5.7	-26.6	-37.4	-45.0	-52.7	-63.4	-78.7	-84.3
Linear appr. of $\varphi(\omega)$/degrees	0	-31.5	-39.5	-45.0	-50.3	-58.3	-76.5	-90.0

5.2.5 First-Order Derivative Element

A zero of the transfer function at the negative real axis, i.e. the factor, $G(s) = \tau s + 1$ is sometimes called the first-order derivative element. Its Bode diagram is obtained in the same manner as that of the inertial element. Its asymptotic log-magnitude plot is 0 below the break frequency and is represented by a straight line of slope +20 v db/dec above the break frequency, as shown in Fig.5.5. The true log-magnitude curve lies above the straight line and has a maximum error of +3 db at the break frequency. The phase plot and its linear approximation are shown in Fig. 5.5, where the phase angle is varied from $0°$ to $+90°$.

Note how the Bode diagram for $(\tau s+1)$ factor may be obtained from that of $(Ts+1)^{-1}$ factor simply by reflecting the magnitude and phase plots about the frequency axis, if $\tau = T$.

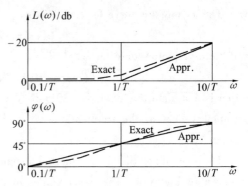

Figure 5.5 Bode diagram for first-order differential element

5.2.6 Oscillatory Element

A pair of complex conjugate poles of the transfer function, i.e. the quadratic factor

$$G(s) = \frac{\omega_n^2}{s^2 + 2\zeta\omega_n s + \omega_n^2}$$

is sometimes called the oscillatory element. It is often occurs in feedback-system transfer function. For $\zeta \geqslant 1$ the quadratic can be factored into two first-order factors with real poles, which can be plotted in the manner shown previously. For $\zeta < 1$, however, this factor contains conjugate poles, and it is plotted without factoring.

The frequency response of the oscillatory element is

$$G(j\omega) = \frac{\omega_n^2}{\omega_n^2 - \omega^2 + j2\zeta\omega_n\omega} \qquad (5.28)$$

or

$$G(j\omega) = \frac{1}{\left(1 - \frac{\omega^2}{\omega_n^2}\right) + j2\zeta\frac{\omega}{\omega_n}} \qquad (5.29)$$

The logarithmic magnitude is

$$L(\omega) = -20 \lg \sqrt{\left(1 - \frac{\omega^2}{\omega_n^2}\right)^2 + \left(2\zeta\frac{\omega}{\omega_n}\right)^2} \qquad (5.30)$$

When $\omega \ll \omega_n$, the asymptotic curve is

$$L(\omega) = 0 \text{ db} \qquad (5.31)$$

which is coincided with the 0db-line. When $\omega \gg \omega_n$, the asymptotic curve is

$$L(\omega) = -20 \lg \frac{\omega^2}{\omega_n^2} = -40 \lg \omega + 40 \lg \omega_n \qquad (5.32)$$

which is a straight line of slope $-40\ v$ db/dec and intersects the 0 db-line at $\omega = \omega_n$. These two asymptotes meet at the break frequency $\omega = \omega_n$, as shown in Fig. 5.6.

At $\omega = \omega_n$ the actual magnitude is

$$|G(j\omega_n)| = \frac{1}{2\zeta} \tag{5.33}$$

or

$$L(\omega_n) = 20 \lg \frac{1}{2\zeta} \tag{5.34}$$

Hence, the exact curve will differ from the asymptotes near the break frequency ω_n, and the deviation is a function of the damping ratio ζ. By taking the derivative of the magnitude in Eq. (5.29) respect to ω and setting it to zero, it shows that the exact magnitude has a peak value for

$$\omega_m = \omega_n \sqrt{1 - 2\zeta^2} \tag{5.35}$$

This indicates that a maximum occurs only if $\zeta < \sqrt{2}/2$. If this condition is satisfied, then the maximum value is obtained by

$$M_m = \frac{1}{2\zeta\sqrt{1-\zeta^2}} \quad \text{for} \quad \zeta < \frac{\sqrt{2}}{2} \tag{5.36}$$

i. e. ,

$$L(\omega_m) = 20 \lg \frac{1}{2\zeta\sqrt{1-\zeta^2}} \tag{5.37}$$

is the maximum magnitude in decibels. The Bode diagram of this element is also shown in Fig. 5.6. Note that the true magnitude may be below or above the straight-line approximate magnitude.

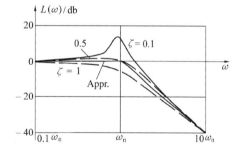

The phase angle is

$$\varphi(\omega) = -\arctan \frac{2\zeta\omega/\omega_n}{1-(\omega/\omega_n)^2} \tag{5.38}$$

At low frequency, the phase is

$$\lim_{\omega \to 0} \varphi(\omega) = -\arctan 0 = 0° \tag{5.39}$$

while at high frequency, taking the appropriate quadrant into consideration, the phase is

$$\lim_{\omega \to \infty} \varphi(\omega) = -180° \tag{5.40}$$

However, the rate at which the phase makes the transition from $0°$ to $-180°$ also depends upon the damping ratio ζ, as shown in Fig. 5.6. Note

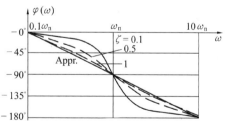

Figure 5.6 Bode diagram for oscillatory factor

also that all phase lines pass through the value $-90°$ at $\omega = \omega_n$. Although not accurate in many cases, a straight-line approximation for the phase curve is sometimes used. This is defined as a linear change from $0°$ to $180°$ beginning one decade below ω_n and finishing one decade above ω_n.

5.2.7 Non-Minimum Phase-Shift Element

In the previous discussions the poles and zeros of a transfer function given by Eq. (5.15) are restricted to the left-half s-plane. However, there are some different cases that may be encountered in constructing a Bode diagram.

1. First-order unstable element

Consider the first-order unstable element

$$G(s) = \frac{1}{Ts-1} \tag{5.41}$$

which has a real pole in the right-half s-plane. The magnitude and phase angle of this element are, respectively, given by

$$L(\omega) = 20 \lg \frac{1}{\sqrt{T^2\omega^2+1}} = -20 \lg \sqrt{T^2\omega^2+1} \tag{5.42}$$

$$\varphi(\omega) = -\arctan \frac{T\omega}{-1} \tag{5.43}$$

The Bode diagram is shown in Fig. 5.7.

Comparing with the inertial element,

$$\frac{1}{Ts+1}$$

which frequency response is shown in Fig. 5.4, there is no change in the magnitude characteristics, and the only difference between these two elements is in the phase-shift characteristics. It can be seen, taking the appropriate quadrant into consideration, that the phase angle of the first-order unstable element is varied from $-180°$ to $-90°$, its phase angle range is greater than that of the inertial element having the same magnitude

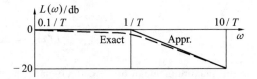
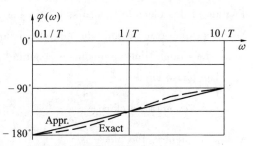

Figure 5.7   Bode diagram for first-order unstable factor

characteristics. Hence, an inertial element is said to be a minimum phase element, and a first-order unstable factor is said to be a non-minimum phase element.

A transfer function is said to be a minimum phase transfer function if it has a minimum phase-shift range among all the transfer functions having the same magnitude characteristics; otherwise, it is said to be a non-minimum phase transfer function. A system or an element having the minimum transfer function is called a minimum phase system or element; otherwise it is called a non-minimum phase system or element. The non-minimum phase-shift will occur when a system contains non-minimum phase components or unstable minor loops.

2. Ideal time delay

Time delay is a non-minimum phase characteristic and usually exists in thermal, hydraulic and pneumatic systems. Suppose that a signal $r(t)$ is applied to an ideal time delay, then the output of this time delay is

$$c(t) = r(t-\tau) \cdot 1(t-\tau) \tag{5.44}$$

where τ is the delay time. Taking Laplace transform, the transfer function of an ideal time delay is

$$G(s) = e^{-\tau s} \tag{5.45}$$

Note that this transfer function is different from all other considered thus far. It is not a ratio of polynomials in s, i.e. is not a rational fraction in s. This fact complicates the analysis of systems that contain ideal time delay. For example, the algebraic criteria cannot be used with systems of this type, because the system characteristic equations are not polynomials in s.

The frequency response of the transfer function in Eq. (5.45) is obtained by replacing s with $j\omega$, i.e.,

$$G(j\omega) = e^{-j\omega\tau} \tag{5.46}$$

Therefore, the logarithmic magnitude is

$$L(\omega) = 0 \text{ db} \tag{5.47}$$

and the phase is

$$\varphi(\omega) = -\tau\omega \text{ rad} = -57.3 \, \tau\omega \text{ deg} \tag{5.48}$$

as shown in Fig. 5.8. The phase lag introduced by the ideal time delay increases without limit as frequency becomes large.

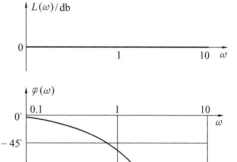

Figure 5.8 Bode diagram for ideal time

5.3 Open-Loop Frequency Response

5.3.1 Compound Bode Diagram

The Bode diagram, especially the Bode diagram of open-loop transfer function, is one of the most powerful graphical tools for analyzing and designing control systems.

As we know, when an open-loop transfer function is written in the form of standard factors, i.e. in the Bode form

$$G(s) = \frac{K \prod (\tau_i s + 1)}{s^v \prod (T_j s + 1) \prod [(s^2 + 2\zeta_k \omega_{nk} s + \omega_{nk}^2)/\omega_{nk}^2]} \tag{5.49}$$

its Bode diagram is obtained by adding the plots due to each individual factor. The simplicity of this method may be illustrated with several examples.

Example 5.2 Plot the Bode diagram of the open-loop transfer function

$$G(s) = \frac{1}{s(Ts+1)}$$

Solution Since the gain is 1, we only need consider two factors. One way to generate the magnitude plot is to begin at low frequency, as we move to the high frequency, we sum graphically the individual magnitude plots. Fig. 5.9 shows the straight-line representation of each factor of the transfer function, together with the resulting magnitude plot. Note how the slope of the straight line changes from −20 db/dec to −40 db/dec at the break frequency. The resultant phase angle may be obtained graphically in a similar way as shown in Fig. 5.9.

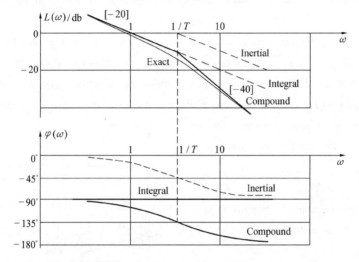

Figure 5.9 Bode diagram of Example 5.2

Since it is known that the error of magnitude between the exact and straight-line approximation is 3db at the break frequency, if it is necessary, a better sketch of the magnitude plot would be obtained.

Example 5.3 Plot the Bode diagram of the open-loop system

$$G(s)H(s) = \frac{100(s+2)}{s(s+1)(s+20)}$$

Solution At first write the transfer function in the Bode form, i.e.,

$$G(s)H(s) = \frac{10(0.5s+1)}{s(s+1)(0.05s+1)}$$

This transfer function have five factors in cascade

$$G_1(s) = 10, \quad G_2(s) = \frac{1}{s}, \quad G_3(s) = \frac{1}{s+1}, \quad G_4(s) = \frac{1}{0.05s+1}, \quad G_5(s) = 0.5s+1$$

Sketch the asymptotic log-magnitude curve $L_i(\omega)$ and phase curve $\varphi_i(\omega) = \angle G_i(j\omega)$ of each factor $G_i(s)$, $i = 1,2,3,4,5$.

Fig. 5.10 shows the straight-line representation and phase angle curve of each factor of the transfer function, together with the resulting magnitude plot and phase plot. Note again how the slop of the straight line changes at each break frequency.

In fact, the composite logarithmic magnitude curve using straight-line approximation can

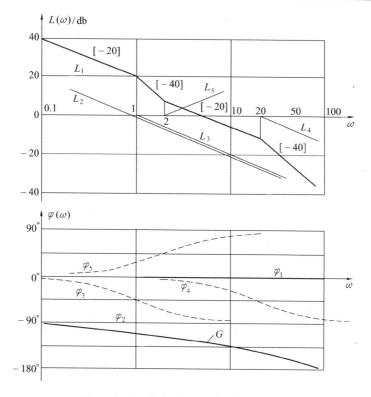

Figure 5.10 Bode diagram for Example 5.3

be drawn directly, as illustrated in the following example.

Example 5.4 Sketch the asymptotic log-magnitude plot of the transfer function

$$G(s) = \frac{5(0.1s+1)}{s(0.5s+1)[(s^2+30s+50^2)/50^2]}$$

Solution This transfer function possesses all the basic factors discussed in the preceding section. The factors, in order of their break frequencies, are as follows:

(a) a constant gain $K=5$;

(b) an integral factor;

(c) an inertial factor with break frequency $\omega_1 = 1/T = 2$;

(d) a first-order differential factor with break frequency $\omega_2 = 1/\tau = 10$;

(e) an oscillatory factor with break frequency $\omega_3 = \omega_n = 50$.

The Bode diagram can be plotted, in order of the break frequencies, according to the following procedures:

Step 1. At the frequencies less than ω_1, the first break frequency, only the constant gain factor (a) and integral factor (b) are effective. All other factors have zero value. Thus, below ω_1 the composite curve has a slope of -20 db/dec and its height at $\omega = 1$ is 20 lg 5 = 14 db.

Step 2. Above ω_1, the inertial factor (c) becomes effective, which has a slope of -20 db/dec and must be added to the terms in step 1. When the slopes are added, the composite curve has a total slope of -40 db/dec in the frequency band from ω_1 to ω_2.

Step 3. Above ω_2, the first-order differential factor (d) becomes effective, which has a slope of +20 db/dec and must be added to obtain the composite curve. The composite curve now has a total slope of -20 db/dec in the frequency band from ω_2 to ω_3.

Step 4. Above ω_3, the last factor (e) must be added. The oscillatory factor has a slope of -40 db/dec; therefore, the total slope of the composite curve is -60 db/dec.

The final composite asymptotic magnitude curve is shown in Fig. 5.11.

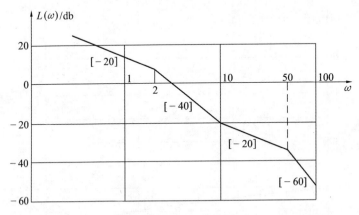

Figure 5.11 Asymptotic log-magnitude plot for Example 5.4

The phase characteristic can be obtained by adding the phases due to each individual factor, or by calculating directly according to the transfer function. Usually the linear approximation of the first-order factor is suitable for the initial analysis or design attempt.

Generally, one may obtain approximate curves for the magnitude and phase angle of a transfer function in order to determine the important frequency ranges. Then, if it is necessary, within the relatively small important frequency ranges the exact magnitude and phase angle can be readily evaluated by using the exact relations.

5.3.2 Estimation of Transfer Function from Bode Diagram

1. Frequency response measurements

A sine signal can be used to measure the open-loop frequency response of a control system. The output magnitudes and phases can be obtained experimentally at a number of frequencies. These data are used to obtain the exact logarithmic magnitude and phase plots. Asymptotes are drawn on the exact magnitude curve, using the fact that their slopes must be multiples of ± 20 db/dec. For a minimum phase transfer function, from the asymptotic magnitude plot, the values of frequency at which the slope changes are identified as the break frequencies of the corresponding factors, then the factors of the transfer function are located. In this way, from these asymptotes, the system type and the approximate time constants are determined, and then the open-loop transfer function can be synthesized.

Care must be exercised in determining whether any poles or zeros of the transfer function are in the right-half s-plane. The phase variation for poles and zeros in the right-half s-plane is different from those in the left-half s-plane. This follows from the fact that, for example, the first-order unstable factor has same magnitude curve as that of inertial factor, but the phase shift is changed.

2. System type and open-loop gain from Bode diagram

Consider the open-loop transfer function written in the Bode form

$$G(s)H(s) = \frac{K\prod(\tau_i s + 1)}{s^v \prod(T_j s + 1) \prod[(s^2 + 2\zeta_k \omega_{nk} s + \omega_{nk}^2)/\omega_{nk}^2]}$$

Obviously, the slope, $-20\ v$ db/dec, of the logarithmic magnitude approximation at the low frequencies below the lowest break frequency is determined by the number of the integral factors in the open-loop transfer function, i.e. the system type. Meanwhile, the height of this segment is depended upon the magnitude of the constant gain, i.e. the open-loop gain K.

(1) Type 0 system

A type 0 system has the value $v = 0$ and will have the asymptotic magnitude plot consisting of a horizontal line with height of 20 lg K = 20 lg K_p at low frequencies, as shown in Fig. 5.12, where K_p is the steady-state position error constant.

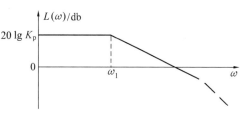

(2) Type 1 system

Figure 5.12 Bode diagram of type 0 system

A type 1 system has the value $v = 1$ and will have the asymptotic magnitude plot with a slope of -20 db/dec at low frequencies, and the height of the low-frequency asymptote or its extension line will be 20 lg K at $\omega = 1$. Moreover, it can be proven that the low-frequency asymptote or its extension line will intersect 0 db-line at $\omega = K = K_v$, where K_v is the static-state velocity error constant. Fig. 5.13 shows some typical magnitude plots of type 1 systems.

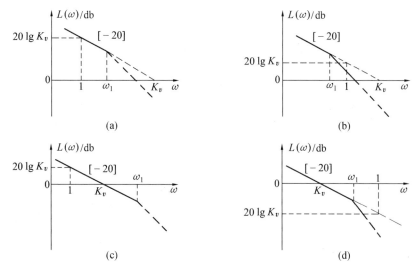

Figure 5.13 Bode diagram of type 1 system

(3) Type 2 system

A type 2 system has the value $v = 2$ and will have the asymptotic magnitude plot with a slope of -40 db/dec at low frequencies, and the height of the low-frequency asymptote or its extension line will be $20 \lg K$ at $\omega = 1$. Moreover, it can be proven that the low-frequency asymptote or its extension line will intersect 0 db-line at $\omega = \sqrt{K} = \sqrt{K_a}$, where K_a is the static-state acceleration error constant. Fig. 5.14 shows the magnitude plot of type 2 systems.

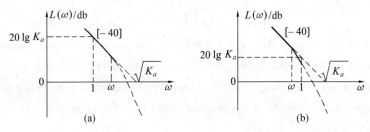

Figure 5.14 Bode diagram of type 2 system

3. Estimation of transfer function

Example 5.5 The asymptotic magnitude plot of a dc servomotor-amplifier combination (the output is the angular velocity of the shaft and the input is the applied voltage) is shown in Fig. 5.15. Determine the transfer function of the dc servomotor-amplifier combination.

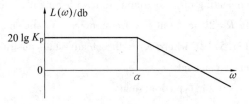

Figure 5.15 Magnitude plot of a dc servomotor-amplifier combination

Solution From the figure it is easily obtained

$$G(s) = \frac{K}{\frac{1}{\alpha}s + 1} = \frac{K\alpha}{s + \alpha}$$

Example 5.6 Estimate the transfer function of a typical type 2 system from the asymptotic log-magnitude plot shown in Fig. 5.16, where the corner frequencies ω_1, ω_2, and the crossover frequency ω_c are known. It is assumed that the transfer function is of minimum phase.

Solution From the asymptotic magnitude plot, it can be seen that the transfer function is of the following form

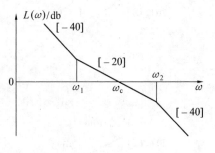

Figure 5.16 Bode diagram of Example 5.6

$$G(s) = \frac{K(\tau s+1)}{s^2(Ts+1)} = \frac{K\left(\dfrac{1}{\omega_1}s+1\right)}{s^2\left(\dfrac{1}{\omega_2}s+1\right)}$$

The plot indicates neither the height of the initial segment or its extension line at $\omega=1$, nor the intersection of the initial segment or its extension line with 0db-line. In this case, gain K can be determined using the expression of asymptotic magnitude, i.e.,

$$L(\omega_c) = 20\lg\left[\frac{K\sqrt{\left(\dfrac{\omega_c}{\omega_1}\right)^2+0}}{\omega_c^2\sqrt{0+1}}\right] = 0 \quad \text{or} \quad \frac{K\dfrac{\omega_c}{\omega_1}}{\omega_c^2} = 1$$

Solving this equation yields

$$K = \omega_1\omega_c$$

Therefore the transfer function is

$$G(s) = \frac{\omega_1\omega_c[(s/\omega_1)+1]}{s^2[(s/\omega_2)+1]}$$

Example 5.7 The magnitude plot of a control element is shown in Fig. 5.17, where the real line is the asymptotic curve and the dashed line is the exact curve. Try to estimate the corresponding transfer function of this element, if it is of minimum phase.

Solution From the magnitude curve, the transfer function has three factors in cascade, and can be written in the form

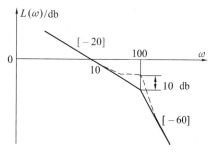

Figure 5.17 Magnitude plot of Example 5.7

$$G(s) = \frac{K\omega_n^2}{s(s^2+2\zeta\omega_n s+\omega_n^2)}$$

The asymptotic magnitude curve intersects ω-axis at $\omega = 10$ rad/s. Since the transfer function has one integral factor, the gain element is

$$K = 10s^{-1}$$

It is easily seen that the corner frequency of the oscillatory element is $\omega_n = 100$ rad/s. Moreover, since the magnitude of the oscillatory factor is $1/2\zeta$ at $\omega = 100$, we have

$$20\lg\frac{1}{2\zeta} = 10$$

Solving this equation yields

$$\zeta = 0.158$$

Therefore the transfer function of this element is

$$G(s) = \frac{10 \cdot 100^2}{s(s^2+2\times 0.158\times 100s+100^2)} = \frac{10^5}{s(s^2+31.6s+10^4)}$$

5.3.3 Polar Plot

One disadvantage of the Bode diagram is that it needs two separate plots showing the variation of the magnitude and the phase shift with frequency. The information contained in these two plots can be combined in polar plot. Hence, frequently, the polar plot of a system is desired.

1. Drawing polar plot

The polar plot can be obtained from Eq. (5.12) or (5.13) and illustrated with the following examples

Example 5.8 Draw the polar plot of a simple RC network in Example 2.1.

Solution The transfer function of this network is

$$G(s) = \frac{1}{RCs+1} = \frac{1}{Ts+1}$$

and the frequency response is

$$G(j\omega) = \frac{1}{jT\omega+1}$$

Then the polar plot is obtained from the relation

$$G(j\omega) = \frac{1}{T^2\omega^2+1} - j\frac{T\omega}{T^2\omega^2+1}$$

or

$$G(j\omega) = \frac{1}{T^2\omega^2+1} e^{-j\arctan\frac{T\omega}{1}}$$

The first step is to consider the cases of $\omega = 0$ and $\omega = \infty$, for $\omega = 0$ we have

$$\text{Re}[G(j\omega)] = 1$$
$$\text{Im}[G(j\omega)] = 0$$

for $\omega = \infty$

$$\text{Re}[G(j\omega)] = \text{Im}[G(j\omega)] = 0$$

These two points are shown in Fig. 5.18.

The locus of the real and imaginary parts is also shown in Fig. 5.18, which is easily shown to be a part of a circle of center at point (0.5, 0). When $\omega = 1/T$, the real and imaginary parts are equal, and the phase angle is $-45°$. The arrow on the polar plot indicates the direction of increasing frequency.

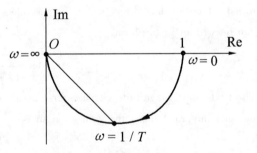

Figure 5.18 Polar plot of Example

Example 5.9 Draw the polar plot of the transfer function

$$G(s) = \frac{10}{s(s+1)}$$

Solution The frequency response is

$$G(j\omega) = \frac{10}{j\omega(j\omega+1)}$$

The magnitude-frequency and phase-frequency characteristics are given by, respectively

$$|G(j\omega)| = \frac{10}{\omega\sqrt{\omega^2+1}}$$

and

$$\angle G(j\omega) = -90° - \arctan \omega$$

The polar plot can be obtained from the relations as given in Table 5.2 and its exact polar plot is shown in Fig. 5.19.

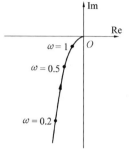

Figure 5.19 Polar plot of Example 5.9

Table 5.2 Frequency response of the transfer function given by Example 5.9.

ω	$\|G(j\omega)\|$	$\angle G(j\omega)$	ω	$\|G(j\omega)\|$	$\angle G(j\omega)$
0	∞	90°	5.0	0.390	-168.7°
0.1	99.5	-95.71°	7	0.202	-171.9°
0.2	49.03	-101.3°	10	0.100	-174.3°
0.5	17.89	-116.6°	20	0.025	-177.1°
1.0	7.07	-135°	50	0.004	-178.9°
2.0	2.24	-153.4°	70	0.002	-179.2°

2. Approximate polar plot

The polar plot is very useful in determining the stability of a closed-loop system from its open-loop frequency response, as will be seen in the next section. In this case, it is often enough for us to draw an approximate curve according to some key parts that can be determined by using following criteria.

Consider the open-loop transfer function described by the general form

$$G(s)H(s) = \frac{K\prod_{i=1}^{m}(\tau_i s + 1)}{s^v \prod_{j=v+1}^{n}(T_j s + 1)} \quad (5.50)$$

for convenience, it is assumed that $n > m$.

(1) Low-frequency beginning of polar plot (as $\omega \to 0$)

Letting $\omega \to 0$ in the frequency response yields

$$\lim_{\omega \to 0} G(j\omega)H(j\omega) = \lim_{\omega \to 0} \frac{K}{(j\omega)^v} \quad (5.51)$$

This equation indicates that the initial position of the open-loop magnitude-phase frequency curve is mainly depended upon the system type, i.e., the value of v. The low-frequency polar plot characteristics (as $\omega \to 0$) of different system types are summarized in Fig. 5.20. The phase angle at $\omega = 0$ is $v(-90°)$.

(2) High-frequency end of polar plot (as $\omega \to \infty$)

The high-frequency end of the polar plot can be determined as follows

$$\lim_{\omega \to \infty} G(j\omega)H(j\omega) = \lim_{\omega \to \infty} \frac{K(j\omega)^m}{(j\omega)^n} = 0 \cdot e^{j(n-m)\cdot(-90°)}$$

(5.52)

Note that since $n>m$ in Eq. (5.49), the polar plot ends at the origin with a phase angle, $(n-m)\cdot(-90°)$, determined by Eq. (5.52).

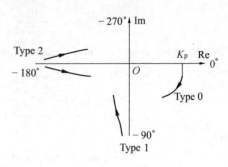

Figure 5.20 Starting position of frequency-phase curve.

(3) Intersection with negative real axis

The frequencies at the intersecting points of the polar plot with the negative real axis are called the phase crossover frequencies, denoted by ω_g, and can be determined by

$$\angle G(j\omega_g)H(j\omega_g) = -180°$$

(5.53)

Then the intersection is given by $G(j\omega_g)H(j\omega_g)$ or $|G(j\omega_g)H(j\omega_g)|$.

Example 5.10 Sketch the approximate polar plot of the transfer function

$$G(s)H(s) = \frac{5}{s(s+1)(s+2)}$$

Solution: This is a system of type 1, according to the criterion (1), the polar plot begins from infinity with a phase angle $-90°$ as $\omega \to 0$. From the criterion (2), the polar plot approaches the origin with a phase of $-270°$ as $\omega \to \infty$. By the definition of phase crossover frequency we have

$$\angle G(j\omega_g)H(j\omega_g) = -90° - \arctan\frac{\omega_g}{1} - \arctan\frac{\omega_g}{2} = -180°$$

Solving this equation yields $\omega_g = \sqrt{2}$ rad/s, and

$$|G(j\omega_g)H(j\omega_g)| = \frac{5}{\sqrt{2}\cdot\sqrt{2+1^2}\cdot\sqrt{2+2^2}} = \frac{5}{6}$$

Thus the polar plot intersects with the real axis at the point $(-5/6, j0)$. The approximate polar plot can be drawn as shown in Fig. 5.21.

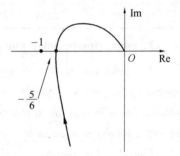

Figure 5.21 Polar plot for Example 5.10

5.4 Nyquist Stability Criterion

5.4.1 Introduction

As stated earlier, in order to be useful, a control system must be stable. The algebraic criterion discussed in Chapter 3 enables us to determine whether a system is stable by examining the characteristic polynomial, corresponding to the denominator of the closed-loop transfer function. The root locus method described in Chapter 4 enables us to determine the relative stability in terms of the damping ratio of the dominant poles. Both of these two methods require the knowledge of the transfer function, which must be a rational function of the complex variable.

In frequency domain the open-loop frequency response of a system will not only be used to determine the stability of the closed-loop system, but also provide sufficient information about the relative stability of the system. Moreover, the frequency response of a system can readily be obtained experimentally by exciting the system with sinusoidal input signals, therefore it can be used even the system parameter values have not been determined.

A frequency domain stability criterion was developed by H. Nyquist in 1932 and remains a fundamental approach to the investigation of the stability of linear control systems. It is based on a theorem in complex variable theory, due to Cauthy, called the principle of the argument.

5.4.2 Principle of the Argument

1. Mapping of function $F(s)$

The principle of the argument is related to the theory of mapping of analytic function of a complex variable. In order to get a thorough appreciation the theory will be reviewed briefly, and then the principle of the argument will be stated without proof.

Let $F(s)$ be a function of the complex variable $s = \sigma + j\omega$. Since in general $F(s)$ will also be complex, we may write

$$F(s) = U(\sigma,\omega) + jV(\sigma,\omega) \tag{5.54}$$

where $U(\sigma,\omega)$ and $V(\sigma,\omega)$ are real functions.

A function $F(s)$, defined in a domain in the s-plane, is said to be analytic in this domain if and only if the derivative dF/ds is continuous in this domain. It can be proved that all rational functions of s are analytic everywhere in the s-plane except for the points of singularities. Hence all transfer functions are analytic in the s-plane except at their poles.

Just as the complex variable s is shown in a plane with real axis σ and the imaginary axis ω, it is possible to illustrate $F(s)$ in a plane with real axis denoted by U and the imaginary axis by V. The former is called s-plane and the latter the F-plane.

Any point in the s-plane will be "mapped" into the F-plane by locating the values of U and V for the given values of σ and ω. For example, consider the function

$$F(s) = \frac{2s+3}{s+5}$$

For $s_t = 1 + j2$, we get

$$F(s_t) = \frac{2(1+j2)+3}{(1+j2)+5} = 0.95 + j0.35$$

This mapping is shown in Fig. 5.22.

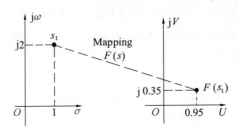

Figure 5.22 Mapping of function $F(s)$

This correspondence between points in the two planes is called a mapping or transformation. For an analytic function, every point in the s-plane maps into a unique point in the F-plane if the function is analytic at this point. The concept can be extended to mapping a line or curve in the s-plane to the F-plane. In particular, a smooth curve in the s-plane will map into a smooth curve in the F-plane if $F(s)$ is analytic at every point on the curve.

2. Principle of the argument

We now consider mapping a closed curve in the s-plane. Such a curve will be called a contour, and denoted as Γ_s. Its map in the F-plane will be denoted by the contour Γ_F. The principle of the argument gives the relationship between the number of poles and zeros of $F(s)$ enclosed by Γ_s and the number of times Γ_F will encircle the origin of the F-plane. It is stated below.

Let $F(s)$ be an analytic function, except for a finite number of poles. If a contour Γ_s in the s-plane encloses Z zeros and P poles of $F(s)$ and does not pass through any pole or zero of $F(s)$, then with the traversal along Γ_s in the clockwise direction the corresponding contour Γ_F in the F-plane encircles the origin of the F-plane N times in the clockwise direction, where $N = Z - P$.

By convention the area within a contour to the right of the traversal of the contour is considered to be the area enclosed by the contour. Therefore we will assume clockwise traversal of a contour to be positive and the area enclosed within the contour to be on the right.

5.4.3 Nyquist Criterion

1. Poles and zeros of $F(s)$

Consider the closed-loop system shown in Fig. 5.23 and denote the open-loop transfer function as

$$G(s)H(s) = \frac{N(s)}{D(s)} \quad (5.55)$$

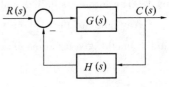

Figure 5.23 Closed-loop system

where $N(s)$ and $D(s)$ are polynomials of s.

Suppose we define the function $F(s)$, previously discussed, to be equal to the characteristic equation of the closed-loop system, i.e.,

$$F(s) = 1+G(s)H(s) = \frac{D(s)+N(s)}{D(s)} \tag{5.56}$$

It is important to note that the open-loop poles are the poles of $F(s)$ and the closed-loop poles are the zeros of $F(s)$.

In order to determine the stability of the closed-loop system, it will be necessary to determine whether there are any closed-loop poles, i.e. zeros of $F(s)$, in the right half s-plane. This may be achieved by

① choosing a contour Γ_s in the s-plane that encloses the entire right half s-plane, and

② plotting Γ_F in the F-plane and determining the number of the encirclements of the origin, N.

2. Nyquist path

Fig. 5.24(a) defines a clockwise path, known as Nyquist path, which encloses the entire right half s-plane. If $F(s)$ has a pole (or a zero) at the origin of the s-plane or at some points on the $j\omega$-axis, the Nyquist path must be modified by making a detour along an infinitesimal semicircle, as shown in Fig. 5.24(b), to meet the condition for the principle of the argument.

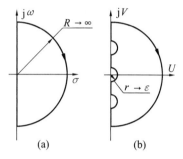

Figure 5.24 Nyquist contour

3. Nyquist stability Criterion

Let $F(s)$ has P poles and Z zeros within the Nyquist path. For a stable system, we must have $Z=0$, i.e., the characteristic equation must not have any root within the Nyquist path. From the principle of the argument, Γ_F in the F-plane will encircle the origin N times in the clockwise direction, i.e.,

$$N = Z - P \tag{5.57}$$

Thus the system is stable if and only if

$$N = -P \tag{5.58}$$

so that Z will be zero. Furthermore, note that the origin of the F-plane is the point $(-1, j0)$ in the GH-plane. Hence, we get the following Nyquist criterion in terms of the open-loop transfer function $G(s)H(s)$.

A feedback control system is stable if and only if the number of counterclockwise encirclements of the point $(-1, j0)$ by the map of the Nyquist contour in the GH-plane Γ_{GH} is equal to the number of poles of the open-loop transfer function $G(s)H(s)$ within the Nyquist contour in the s-plane.

4. Nyquist plot

The map of the Nyquist contour in the GH-plane, Γ_{GH} is called the Nyquist plot of $G(s) \cdot H(s)$. The polar plot of the open-loop transfer function $G(s)H(s)$ is an important part of the

Nyquist plot and can be obtained either experimentally or by computation if the transfer function is known. The procedure for completing the rest of the Nyquist plot will be illustrated by a number of examples.

Example 5.11 Consider a Type 0 system with the open-loop transfer function

$$G(s)H(s) = \frac{6}{(s+1)(0.5s+1)(0.2s+1)}$$

Draw the Nyquist plot and determine the stability of the closed-loop system.

Solution The frequency response is readily calculated, and is given in Table 5.3. We use these values to obtain the polar plot.

Table 5.3 Frequency response of the transfer function given in this example.

ω	Magnitude	Phase angle	ω	Magnitude	Phase angle
0	6	0°	4.123	0.476	−180.0°
0.5	5.18	−46.3°	10.0	0.052	−226.4°
1.0	1.76	−130.2°	20.0	0.007	−247.4°

The Nyquist contour Γ_s in the s-plane is taken as three separate portions: AB, BCD and DA, as shown in Fig. 5.25(a). Thus, we can find their maps in the GH-plane to obtain the Nyquist plot Γ_{GH}, as shown in Fig. 5.25(b).

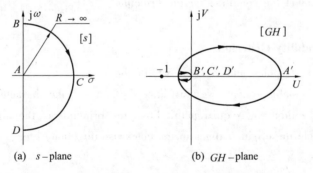

Figure 5.25 Nyquist contour and Nyquist plot for the Example 5.11

(1) The portion AB in the s-plane is the $j\omega$-axis from $\omega = 0$ to $\omega = \infty$. It maps into the polar plot $A'B'$ in the GH-plane because $s = j\omega$, and

$$G(s)H(s)\big|_{s=j\omega} = G(j\omega)H(j\omega)$$

(2) The infinite semicircle BCD maps into the origin of the GH-plane.

(3) The portion DA, which is the negative part of the $j\omega$-axis, maps into the curve $D'A'$ in the GH-plane. It may be noted that $D'A'$ is the mirror image of $A'B'$ about the real axis. This follows from the fact that $G(-j\omega)H(-j\omega)$ is the complex conjugate of $G(j\omega)H(j\omega)$.

The Nyquist plot does not enclose the point $(-1, j0)$ in the GH-plane, hence $N = 0$. Also the transfer function does not have any pole inside the Nyquist contour. This makes the closed-loop system stable since $Z = N + P = 0$.

If we increase the open-loop gain of this system by a factor of 2.5, the resulting Nyquist plot will encircle the point $(-1, j0)$ twice in the GH-plane, as shown in Fig. 5.26. Then, for this case, we get $N = 2$. Since P is still zero, the closed-loop system is unstable, with two poles in the right half s-plane.

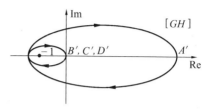

Figure 5.26 Nyquist plot for Example 5.11

Example 5.12 Consider the stability of a type 1 system with the open-loop transfer function

$$G(s)H(s) = \frac{K}{s(T_1 s + 1)(T_2 s + 1)}$$

Solution Since this time the transfer function has a pole at the origin, the Nyquist contour must include a small detour around this pole while still attempting to enclose the entire right-half s-plane. Hence we get the semicircle EFA of infinitesimal radius r. The resulting Nyquist contour Γ_s and the Nyquist plot Γ_{GH} are shown in Fig. 5.27. The procedure for completing Γ_{GH} is given below.

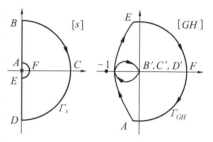

Figure 5.27 Nyquist contour and Nyquist plot for Example 5.12

① The portion AB in the s-plane is the $j\omega$-axis from $\omega = 0_+$ to $\omega \to \infty$, and maps into the polar plot of the frequency response $A'B'$ in the GH-plane.

The phase crossover frequency can be found from Eq. (5.53)

$$\angle G(j\omega_g) H(j\omega_g) = -90° - \arctan T_1 \omega_g - \arctan T_2 \omega_g = -180°$$

hence

$$\omega_g = 1/\sqrt{T_1 T_2}$$

Then the point where the polar plot intersects the real axis can be determined by

$$|G(j\omega_g) H(j\omega_g)| = \frac{K}{\omega_g \sqrt{(T_1 \omega_g)^2 + 1} \cdot \sqrt{(T_2 \omega_g)^2 + 1}} = \frac{K T_1 T_2}{T_1 + T_2}$$

i.e. the polar plot intersects the real axis at point $\left(\dfrac{K T_1 T_2}{T_1 + T_2}, j0 \right)$.

② The infinite semicircle BCD maps into the origin of the GH-plane.

③ The portion DE (negative $j\omega$-axis) maps into the image $D'E'$ of the frequency response.

④ The infinitesimal semicircle EFA can be represented by setting

$$s = \lim_{r \to 0} r\, e^{j\theta}$$

and allowing θ to vary from $-90°$ at $\omega = 0_-$ to $90°$ at $\omega = 0_+$. Because r approaches zero, the mapping for $G(s)H(s)$ is

$$\lim_{r \to 0} G(s)H(s) = \lim_{r \to 0} \frac{K}{T_1 T_2} \cdot \frac{1}{re^{j\theta}} = \infty \cdot e^{-j\theta}$$

It maps into an infinite semicircle $E'F'A'$, as shown, while the angle changes from $90°$ at $\omega = 0°-$ to $-90°$ at $\omega = 0_+$, passing through $0°$ at $\omega = 0$.

By using the Nyquist criterion, the closed-loop system is stable when

$$\frac{KT_1 T_2}{T_1 + T_2} < 1 \quad \text{i.e.} \quad K < \frac{T_1 + T_2}{T_1 T_2}$$

because $Z = N + P = 0$.

Example 5.13 Consider the stability of a type 2 system with the open-loop transfer function

$$G(s)H(s) = \frac{K}{s^2(Ts+1)}$$

Solution The polar plot is obtained when $s = j\omega$. As ω approaches 0_+, we have

$$\lim_{\omega \to 0_+} G(j\omega)H(j\omega) = \lim_{\omega \to 0_+} \frac{K}{(j\omega)^2} = \infty \cdot e^{-j\,180°}$$

As ω approaches $+\infty$, we have

$$\lim_{\omega \to \infty} G(j\omega)H(j\omega) = \lim_{\omega \to +\infty} \frac{K}{T(j\omega)^3} = \infty \cdot e^{-j\,270°}$$

At the small semicircular detour around the origin of the s-plane where $s = \varepsilon e^{j\theta}$, we have

$$\lim_{\varepsilon \to 0} G(s)H(s) = \lim_{\varepsilon \to 0} \frac{K}{(\varepsilon e^{j\theta})^2} = \infty \cdot e^{-j\,2\theta}$$

where $-90° \leqslant \theta \leqslant 90°$. Thus this portion maps into a full circle, while the angle changes from $+180°$ at $\omega = 0_-$ to $-180°$ at $\omega = 0_+$, passing through $0°$ at $\omega = 0$. The complete Nyquist plot Γ_{GH} is shown In Fig. 5.28.

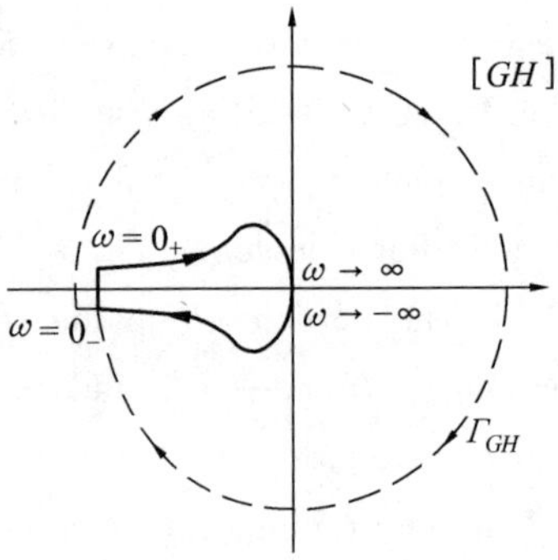

Figure 5.28 Nyquist plot for the transfer function given by Example 5.13

Because of $N = 2$ and $P = 0$, there are two roots of the closed-loop system in the right-half s-plane, and the closed-loop system, irrespective of the gain K, is unstable.

Example 5.14 Consider the stability of two systems with the non-minimum phase open-loop transfer function

(1) $G_1(s) = \dfrac{K}{s(Ts-1)}$

(2) $G_2(s) = \dfrac{K(\tau s+1)}{s(Ts-1)}$

Solution Both transfer functions have an unstable factor, thus we have $P = 1$ for these two systems.

(1) The frequency polar plot is obtained by setting $s = j\omega$. We note that the open-loop phase angle of the first system varies from $-270°$ to $-180°$, and for the polar plot we have

$$\lim_{\omega \to 0_+} |G(j\omega)H(j\omega)| \to \infty \cdot e^{-j\,270°}$$

and
$$\lim_{\omega \to +\infty} G(j\omega)H(j\omega) = 0e^{-j180°}$$
The Nyquist plot is shown in Fig. 5.29(a).

No matter how much the open-loop gain is increased we always have $N=1$, since $P=1$, there are two roots of the closed-loop system in the right-half s-plane, and the closed-loop system is always unstable.

(2) The Nyquist plot of the second system for a specified open-loop gain is drawn as shown in Fig. 5.29(b). In this case we still have $N=1$ and $P=1$, although the Nyquist plot is different from that in case (1). Hence, the closed-loop system is unstable for the specified open-loop gain.

If we increase the open-loop gain, however, the Nyquist plot will be as shown in Fig. 5.29 (c). Now, because $N=-1$ and $P=1$, the closed-loop system becomes stable with a larger open-loop gain.

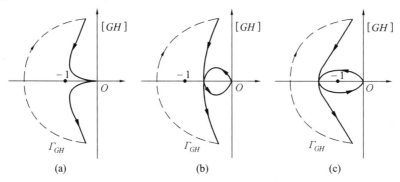

Figure 5.29   Nyquist plots for Example 5.14

5. Simplified Nyquist criterion

(1) Nyquist criterion using half Γ_{GH}

Since the map of the Nyquist path in the GH-plane, i.e. the Nyquist plot Γ_{GH}, is symmetric about the real axis, we can determine the system stability only by utilizing half Γ_{GH}. Hence the Nyquist stability criterion can be stated as follows:

A feedback control system is stable if and only if the number of counterclockwise encirclements of the point $(-1, j0)$ by the Nyquist plot Γ_{GH}, as ω varies from zero to infinity, is equal to $P/2$, where P is the pole number of $G(s)H(s)$ within the Nyquist path in the s-plane.

(2) Nyquist criterion using "crossovers"

It can be seen, from above examples, that the points on segment $(-\infty, -1)$, at which the Nyquist plot intersects with the negative real axis, are important for determining stability of the closed-loop system, because the Nyquist plot must intersect with the negative real axis at this segment if it encircles the $(-1, j0)$ point. It is said that a "crossover" is taken place in the GH-plane if the Nyquist plot intersects with the negative real axis at the segment $(-\infty, -1)$, and it is called the positive crossover if after intersection the phase of Γ_{GH} will be increased, otherwise negative crossover. For example, in Fig. 5.30 Γ_{GH} intersects with negative real axis three times, however, there are only two "crossovers" defined as above. In the plot the positive crossover is denoted by $(+)$, and the negative crossover is denoted by $(-)$.

Given the above definitions, the Nyquist stability criterion can be stated as follows:

A feedback control system is stable if and only if, as ω varies from zero to infinity

$$N_+ - N_- = P/2 \qquad (5.59)$$

where N_+ is the number of positive "crossovers", N_- is the number of negative "crossovers", and P is the number of poles of the open-loop transfer function $G(s)H(s)$ within the right half s-plane.

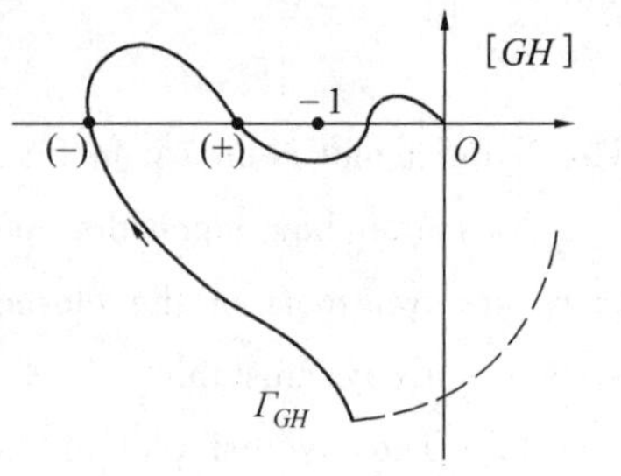

Figure 5.30 Positive "crossover" and negative "crossover"

For example, as ω varies from zero to infinity, in Example 5.11 if the open-loop gain is 6 then we have

$$N_+ - N_- = 0 - 0 = 0$$

the system is stable due to $P = 0$; if the gain is 15 then we have

$$N_+ - N_- = 0 - 2 = -2 \neq P/2$$

the closed-loop system becomes unstable. Similarly, in Example 5.13 we have $N_+ - N_- = 0 - 1 = -1$ and $P = 0$, the system is unstable. As for Example 5.14, we can consider that, as ω varies from zero to infinity, in case of Fig. 5.29(a) we have $N_+ = 0, N_- = 1/2, N_+ - N_- = -1/2 \neq P/2$, the closed-loop system is unstable; in case (b) we still have $N_+ = 0, N_- = 1/2$, $N_+ - N_- = -1/2 \neq P/2$, the closed-loop system is unstable for a smaller open-loop gain; and in case (c), however, we have $N_+ = 1, N_1 = 1/2, N_+ - N_1 = 1/2 = P/2$, the closed-loop system is unstable for a larger open-loop gain.

5.4.4 Stability Using Bode Diagram

As we know, the stability of a system can be determined by using the "crossovers". Note that if a "crossover" is taken place in the GH-plane then the phase-frequency curve will intersect with $-180°$-line while the frequency-magnitude curve is greater than 0 db on the Bode diagram, as shown in Fig. 5.31. Hence the "crossover" can be defined as above, and the Nyquist stability criterion can be defined in

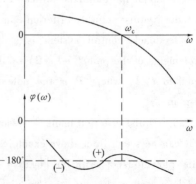

Figure 5.31 "Crossover" in Bode diagram

terms of the 0 db, $-180°$ point on the Bode diagram as above.

Since a "crossover" is related to the Nyquist plot Γ_{GH} containing the map of infinitesimal semicircle detour in the s-plane, the phase curve should be modified when the Nyquist criterion is used on the Bode diagram, i.e. the initial portion of the phase-frequency curve should be moved up by $v\,90°$, where v is the number of integral factors in the open-loop transfer function $G(s)H(s)$.

Example 5.15   Consider the stability of the system with open-loop transfer function

$$G(s)H(s) = \frac{K(\tau s+1)}{s^2(Ts+1)}$$

for two cases: (1) $\tau > T$; (2) $\tau < T$.

Solution   This is a minimum phase transfer function, i.e. $P=0$.

(1) The case of $\tau > T$. The Bode diagram is shown in Fig. 5.32(a). Since $v=2$, make the initial portion of the phase curve move up by $180°$. The phase curve is always above $-180°$-line as ω varies from zero to infinity, and we have

$$N_+ - N_- = 0 - 0 = 0$$

Therefore the closed-loop system is always stable, no matter how much the open-loop gain is increased.

(2) The case of $\tau < T$. The Bode diagram is shown in Fig. 5.32(b). Similarly, the initial portion of the phase curve is moved up by $180°$. In this case, however, there is a negative "crossover" and we have

$$N_+ - N_- = 0 - 1 = -1$$

Therefore the closed-loop system is always unstable, no matter how much the open-loop gain is decreased.

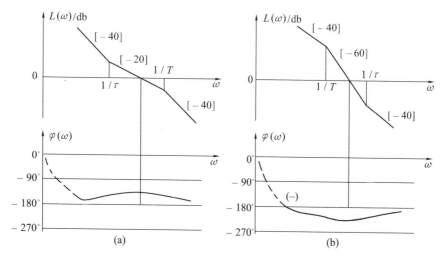

Figure 5.32   Bode diagrams for Example 5.15

5.5 Relative Stability

In most practical situations, in addition to finding out whether a closed-loop system is stable, it is also desired to determine how close it is to instability. This information can be obtained readily from the open-loop frequency response of the system. The proximity of the open-loop frequency response to the point $(-1, j0)$ in the GH-plane is a measure of the relative stability of a closed-loop system. Two commonly used measures of relative stability are the gain margin and phase margin.

5.5.1 Gain Margin

Consider again Example 5.12. As K increases, the polar plot approaches the $(-1, j0)$ point and eventually encircles the $(-1, j0)$ point, and the critical value is

$$K_c = \frac{T_1 + T_2}{T_1 T_2}$$

As K is decreased below this critical value, the stability is increased.

The gain margin is defined to be the gain K_c needed to just make the system frequency response pass through the critical point divided by the actual system gain K, as shown in Fig. 5.33(a), and denoted by K_g, i.e.,

$$K_g = K_c / K \tag{5.60}$$

and the margin between the critical gain and a given gain is a measure of the relative stability. This measure of the relative stability is called the gain margin.

Since the magnitude, for a given frequency, is proportional to the value of K, the gain margin may be stated in terms of the follow lengths in Fig. 5.33(b), i.e.,

$$K_g = \frac{ON}{OA} = \frac{1}{|G(j\omega_g)H(j\omega_g)|} \tag{5.61}$$

where ω_g is the phase crossover frequency and given by

$$\angle G(j\omega_g)H(j\omega_g) = -180° \tag{5.62}$$

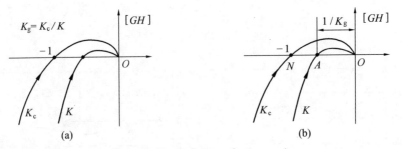

Figure 5.33 Definition of gain margin

5.5.2 Phase Margin

Although gain margin is a measure of the closeness of the polar plot to the critical point, there are occasions when it is ambiguous and misleading. The polar plot of a second-order system is shown in Fig. 5.34(a) that never crosses the negative real axis and is never unstable. For all values of the gains the gain margin is infinity. As the open-loop gain increases, however, the polar plot does get closer to the critical point. The root locus shown in Fig. 5.34(b) confirms that the system becomes more oscillatory even though it is not unstable.

As another example, Fig. 5.35 shows two systems with the same gain margin. However, intuitively, system A is closer to instability than system B. Clearly, another measure of closeness to the critical point is required, and it is called the phase margin.

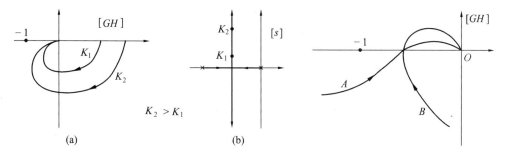

Figure 5.34   Ambiguous gain margin

Figure 5.35   Different systems having the same gain margin

The phase margin is defined as the amount of the minimum phase angle by which the polar plot must be rotated in order to intersect the $(-1, j\,0)$ point for a stable closed-loop system. Again, this is illustrated graphically. The phase margin is indicated by the angle γ in Fig. 5.36. The magnitude of the polar plot $|G(j\omega)H(j\omega)|$ is unity at the gain crossover frequency ω_c.

$$|G(j\omega_c)H(j\omega_c)| = 1 \quad (5.63)$$

Then the phase margin is

$$\gamma = 180° + \angle G(j\omega_c)H(j\omega_c) \quad (5.64)$$

Figure 5.36   Phase margin

Example 5.16   Determine the phase margin of the underdamped second-order system.

Solution   Let the open-loop transfer function of the system be

$$G(s) = \frac{\omega_n^2}{s(s + 2\zeta\omega_n)} \quad (5.65)$$

The magnitude of the frequency response is equal to 1 at the gain crossover frequency ω_c, thus we have

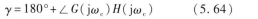

$$\frac{\omega_n^2}{\omega_n\sqrt{\omega_c^2 + 4\zeta^2\omega_n^2}} = 1$$

$$\omega_c^4 + 4\zeta^2\omega_n^2\omega_c^2 - \omega_n^4 = 0$$

Solving for ω_c, we found that

$$\omega_c = \omega_n\sqrt{\sqrt{4\zeta^4+1}-2\zeta^2} \tag{5.66}$$

The phase margin for this system is

$$\gamma = 180° - 90° - \arctan\frac{\omega_c}{2\zeta\omega_n} = 90° - \arctan\frac{\sqrt{\sqrt{4\zeta^4+1}-2\zeta^2}}{2\zeta} =$$

$$\arctan\frac{2\zeta}{\sqrt{\sqrt{4\zeta^4+1}-2\zeta^2}} \tag{5.67}$$

5.5.3 Relative Stability on the BodeDiagram

Although the stability margins may be obtained directly from the polar plot, they are often determined from the Bode diagram as shown in Fig. 5.37.

The gain margin in terms of logarithmic (decibel), K_g(db), is determined from the gain at the phase crossover frequency ω_g

$$K_g(\text{db}) = 20\lg K_g =$$
$$-20\lg|G(j\omega_g)H(j\omega_g)| \tag{5.68}$$

Similarly, the phase margin γ is determined from the phase shift at the gain crossover frequency ω_c

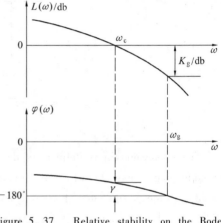

Figure 5.37   Relative stability on the Bode diagram

$$\gamma = 180° + \angle G(j\omega_c)H(j\omega_c) \tag{5.69}$$

5.5.4 Relation of Phase Margin and Gain Margin to Stability

For an underdamping second-order system, from Example 5.16, we have

$$\gamma = \arctan\frac{2\zeta}{\sqrt{\sqrt{4\zeta^4+1}-2\zeta^2}} \tag{5.70}$$

where the phase margin γ is uniquely related to the damping ratio ζ. To those systems which behavior is equivalent to that of a second-order system, it can be shown that the phase margin is related to the effective damping ratio. Satisfactory response is usually obtained with a phase margin 45° to 60°. As an individual gains experience and develops his own particular technique, the value of γ to be used for a particular system becomes more evident.

Usually the gain margin must be positive when expressed in decibels (great than unity as a numeric) for a stable system. A negative gain margin means that the system is unstable.

The damping characteristic of a system is also related to the gain margin. However, the phase margin gives a better estimation of damping characteristic, and therefore of the transient overshoot of the system than the gain margin.

The relationship of stability and gain margin is modified for a conditionally stable system, such as shown in Fig. 5.38. The system for the given gain represented in Fig. 5.38 is stable. Instability can occur both with an increase or a decrease in gain. It is illustrated with the following example.

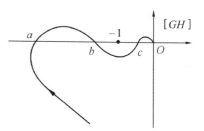

Figure 5.38 Polar plot of a conditionally stable system

Example 5.17 It is assumed that, for the polar plot in Fig. 5.38, if the gain $K = 100$ then $a = -5, b = -2$, and $c = -0.5$. Determine the range in which K must lie for the system to be stable.

Solution Since the magnitude, for a given frequency, is proportional to the value of K, the polar plot will pass through the $(-1, j0)$ point when $K_a = 100/5 = 20$, or $K_b = 100/2 = 50$, or $K_c = 100/0.5 = 200$.

The system is stable if and only if $N_+ - N_- = 0$, i.e. the polar plot intersects with the segment $(-\infty, -1)$ of negative real axis 0, or 2 times. Hence, when $K < 20$ or $50 < K < 200$ the system is stable, otherwise unstable.

5.6 Closed-Loop Frequency-Domain Analysis

In the preceding sections, we were concerned with the frequency response of the open-loop system. From this frequency response we are able to determine the stability and relative stability of the closed-loop system. We stated that the relative stability margins are related, in some undefined way, to the response characteristics of the closed-loop system. However, it is important to keep in mind that ultimately the feedback loop will be closed, and the performance of the closed-loop system needs to be assessed. Hence, in many cases, the closed-loop frequency response will have to be determined.

The closed-loop frequency response may be determined from the closed-loop transfer function, but it can be obtained in a graphical manner directly from the open-loop frequency response if the open-loop frequency response has already been plotted, especially if the open-loop frequency response is get experimentally.

5.6.1 Closed-Loop Frequency Response from Polar Plot

1. Unity-feedback system

At first we consider only unity-feedback systems. The relationship between open-loop and closed-loop frequency responses can be easily established as

$$\Phi(j\omega) = \frac{G(j\omega)}{1+G(j\omega)} = M(\omega) e^{j\theta(\omega)} \tag{5.71}$$

From the polar plot of the open-loop transfer function, Eq. (5.71) can be given an interesting graphical interpretation, as shown in Fig. 5.39. Since for a given ω_t, the vector $\overrightarrow{OP}$ represents the open-loop frequency response $G(j\omega_t)$, the vector $\overrightarrow{AP}$ represents $1+G(j\omega_t)$. Hence, $M(\omega_t)$ is obtained as the ratio of OP to AP, OP/AP. On the other hand, $\theta(\omega_t)$ is obtained as the included angle between $\overrightarrow{OP}$ and $\overrightarrow{AP}$, $\angle(\overrightarrow{OP}-\overrightarrow{AP})$.

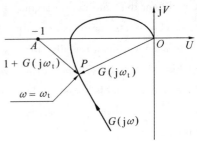

Figure 5.39 Calculation of closed-loop frequency response

The procedure would be, therefore, to pick points on the polar plot at known frequencies and use the above equation to calculate the closed-loop magnitude and phase. If this is repeated for sufficient frequencies, the frequency response may be obtained for the closed-loop system.

Alternatively, this task may be done by considering which points in the G-plane all have the same magnitude. This process generates lines of constant magnitude in the G-plane. Let the coordinates of the G-plane be U and V, we have

$$G(j\omega) = U(\omega) + jV(\omega) \tag{5.72}$$

Therefore, the magnitude of the closed-loop response is

$$M = \frac{|U+jV|}{|1+U+jV|} = \frac{\sqrt{U^2+V^2}}{\sqrt{1+U^2+V^2}} \tag{5.73}$$

Expanding Eq. (5.73) yields

$$(1-M^2)U^2 + (1-M^2)V^2 - 2M^2 U = M^2 \tag{5.74}$$

This equation can be rearranged as

$$\left(U - \frac{M^2}{1-M^2}\right)^2 + V^2 = \left(\frac{M}{1-M^2}\right)^2 \tag{5.75}$$

which, for constant M, is the equation of a circle centered at

$$U = M^2/(1-M^2), \quad V = 0$$

and the radius of the circle is $|M/(1-M^2)|$. A complete family of constant-M circles is shown in Fig. 5.40.

In a similar manner, we can obtain circles of constant closed-loop phase angles. Thus, for Eq. (5.71), the angle relation is

Figure 5.40 Constant-M circles

$$\theta = \angle(U+jV) - \angle(1+U+jV) \quad U = \arctan\frac{V}{U} - \arctan\frac{V}{1+U} \tag{5.76}$$

Taking the tangent of both sides and rearranging, we have

$$U^2 + V^2 + U - \frac{V}{N} = 0 \tag{5.77}$$

where $N = \tan\theta = $ constant. Adding the term

$$\frac{1}{4} + \frac{1}{4N^2}$$

to each side of the equation gives

$$\left(U + \frac{1}{2}\right)^2 + \left(V - \frac{1}{2N}\right)^2 = \frac{1}{4}\left(1 + \frac{1}{N^2}\right) \tag{5.78}$$

which, for constant N, is the equation of a circle centered at

$$U = -0.5, \quad V = 1/(2N)$$

and the radius of the circle is

$$\frac{1}{2}\sqrt{1 + \frac{1}{N^2}}$$

Fig. 5.41 shows a complete family of constant-N circles.

By plotting the polar plot on a complex plane with these constant-M and constant-N circles, the closed-loop frequency response can be determined simply by reading the magnitudes and phase angles for particular frequencies, as shown in Fig. 5.42.

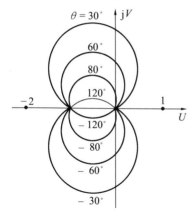
Figure 5.41   Constant-N circles

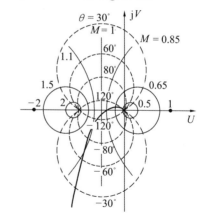
Figure 5.42   Polar plot on constant-M and constant-N circles

2. Non-unity feedback system

If a system does not have unity feedback, some modifications of the system transfer function are required. Consider the closed-loop transfer function as

$$\Phi(s) = \frac{G(s)}{1 + G(s)H(s)} = \frac{G(s)H(s)}{1 + G(s)H(s)} \cdot \frac{1}{H(s)} =$$

$$\frac{\overline{G}(s)}{1 + \overline{G}(s)} \cdot \frac{1}{H(s)} = \overline{\Phi}(s) \cdot \frac{1}{H(s)} \tag{5.79}$$

where $\overline{G}(s) = G(s)H(s), \overline{\Phi}(s) = \overline{G}(s)/[1 + \overline{G}(s)]$. Hence we get

$$M(\omega) = |\overline{\Phi}(j\omega)| \cdot \frac{1}{|H(j\omega)|} = \overline{M}(\omega) \cdot \frac{1}{|H(j\omega)|} \tag{5.80}$$

$$\theta(\omega) = \angle \overline{\Phi}(j\omega) - \angle H(j\omega) = \overline{\theta}(\omega) - \angle H(j\omega) \tag{5.81}$$

where $\overline{M}(\omega)$ and $\overline{\theta}(\omega)$ are the magnitude-frequency and phase-frequency of the equivalent unity-feedback system which can be obtained by means of the constant M and N circles.

5.6.2 Closed-Loop Frequency Response From Bode diagram

Although it is possible to determine the closed-loop frequency response in the manner described, it is more common to derive it from the Bode diagram. Transforming the constant M and N circles to a log-magnitude-phase diagram results in a new diagram called Nichols chart. The constant-M and N circles appear as contours on the Nichols chart shown in Fig. 5.43.

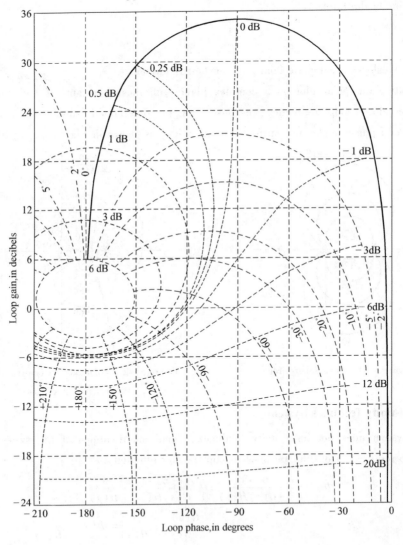

Figure 5.43 Nichols Chart

On Nichols chart the open-loop frequency response is plotted as the logarithmic magnitude in db versus the phase angle in degree, while the constant-M and N lines for the closed-loop frequency response are given in decibels and degrees respectively. An example will illustrate the use of the Nichols chart to determine the closed-loop frequency response.

Example 5.18 Determine the closed-loop frequency response of a unity-feedback system with the open-loop transfer function

$$G(s) = \frac{10(0.5s+1)}{s(s+1)(0.05s+1)}$$

Solution For different values of ω, the open-loop log-magnitudes $L(\omega)$ and phase angles $\varphi(\omega)$ can be found from the Bode diagram shown in Fig. 5.10 (see Example 5.3). Then the closed-loop magnitudes $M(\omega)$ and phase angles $\theta(\omega)$ can be found by plotting the polar plot of $G(j\omega)$ on the Nichols chart. The related data are given in Table 5.4. Finally, from Table 5.4, the closed-loop magnitude-frequency curve can be drawn as shown in Fig. 5.44.

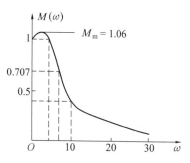

Figure 5.44 Closed-loop magnitude-frequency curve

Table 5.4 The related data of Example 5.18

$\omega/(\text{rad} \cdot \text{s}^{-1})$	0.4	1	2	4.4	6	10	20	30
$L(\omega)/\text{db}$	28	20	8	0	-2.5	-9.5	-14	-22
$\varphi(\omega)/\text{deg}$	-102	-111.4	-114.7	-120.4	-121.5	-129.4	-140.6	-152
M/db	0	0.25	0.5	0	-1.5	-8	-12.5	-22
M	1	1.03	1.06	1	0.84	0.4	0.24	0.08

5.6.3 Performance Specifications in the Frequency Domain

Just as time-domain performance specifications allow different designs of control systems to be compared, certain frequency-domain measures may also be defined and used to assess total system performance. One interesting question is how does the frequency response of a system relate to the expected transient response of the system.

1. Performance of second-order system

Since a large proportion of control systems satisfy the dominant second-order approximation in practice, it is worthwhile looking at a simple second-order system that illustrates why these measures are important and what they tell us about system performance.

Recall the standard second-order system described by the transfer function

$$\Phi(s) = \frac{\omega_n^2}{s^2 + 2\zeta\omega_n s + \omega_n^2} \tag{5.82}$$

The typical frequency response is shown in Fig. 5.45.

As we know, when $\zeta < 1/\sqrt{2}$, at the resonant frequency $\omega_m = \omega_n \sqrt{1-2\zeta^2}$, the magnitude of the frequency response has a peak value

$$M_m = \frac{1}{2\zeta\sqrt{1-\zeta^2}} \quad (5.83)$$

Thus the overshoot to a step input can be related to the resonant peak M_m through the damping ratio ζ. From Eq. (5.83) we have

$$\zeta = \frac{\sqrt{2}}{2}\sqrt{1-\sqrt{1-\frac{1}{M_m^2}}} \quad (5.84)$$

and

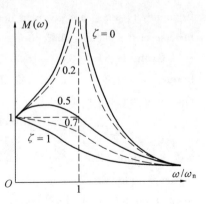

Figure 5.45 Frequency response of second-order system

$$\sigma_p = e^{-\pi\zeta/\sqrt{1-\zeta^2}} \times 100\% = e^{-\pi\sqrt{\frac{M_m - \sqrt{M_m^2-1}}{M_m + \sqrt{M_m^2+1}}}} \times 100\% \quad (5.85)$$

The percent overshoot σ_p is shown as a function of the resonant magnitude M_m in Fig. 5.46. We find that as M_m increases the overshoot to a step input increases. In general, the magnitude M_m indicates the relative stability of a system.

In a closed-loop frequency response, the bandwidth ω_b, at which the frequency response has declined to $\sqrt{2}/2$ of its zero-frequency value is a measure of a system's ability to faithfully reproduce an input signal. According to the definition, we get

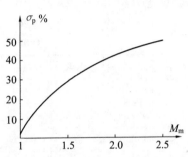

Figure 5.46 σ_p versus M_m

$$|\Phi(j\omega_b)| = \left|\frac{\omega_n^2}{(j\omega_b)^2 + 2\zeta\omega_n(j\omega_b) + \omega_n^2}\right| = \frac{\sqrt{2}}{2}|\Phi(j0)| = \frac{\sqrt{2}}{2} \quad (5.86)$$

Solving this equation yields

$$\omega_b = \omega_n\sqrt{1-2\zeta^2 + \sqrt{2-4\zeta^2+4\zeta^4}} \quad (5.87)$$

Multiplying the rise time t_p and setting time t_s, respectively, on both sides of the equating results in

$$\omega_b t_p = \pi\sqrt{\frac{1-2\zeta^2+\sqrt{2-4\zeta^2+4\zeta^4}}{1-\zeta^2}} \quad (5.88)$$

and

$$\omega_b t_s = \frac{3}{\zeta}\sqrt{1-2\zeta^2+\sqrt{2-4\zeta^2+4\zeta^4}}, \quad \Delta = 5\% \quad (5.89)$$

Thus as ω_b increases, when ζ is constant, both the rise time and setting time of the step response will decrease.

As indicated on the frequency response curve, the bandwidth of a system ω_b can be approximately related to the natural frequency of the system ω_n. As we know, the greater the magnitude of ω_n when ζ is constant, the more rapidly the response approaches the desired steady-state value.

Thus the desirable frequency-domain specifications are relatively small resonant magnitudes and relatively large bandwidths.

2. Performance specifications of higher-order system

Similar to second-order systems, the main frequency-domain specifications of higher-order systems are

(1) resonant magnitude M_m or relative resonant magnitude $M_r = M_m/M(0)$;

(2) bandwidth ω_b.

If a higher-order system does not satisfy the dominant second-order approximation, then its overshoot and setting time may be estimated, from the closed-loop frequency response curve, by following empirical formulas

$$\sigma\% = \left\{ 41 \cdot \ln\left[\frac{M_m \cdot M(\omega_1/4)}{M^2(0)} \cdot \frac{\omega_b}{\omega_{0.5}} \right] + 17 \right\}\% \quad (5.90)$$

$$t_s = \left(13.57 \frac{M_m \omega_b}{M(0)\omega_{0.5}} - 2.51 \right) \cdot \frac{1}{\omega_{0.5}}, \quad \Delta = 5\% \quad (5.91)$$

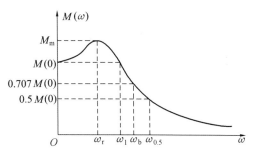

Figure 5.47 Magnitude versus frequency of closed-loop system

where M_m, $M(0)$, ω_r, and ω_b are defined as above, ω_1 is the frequency at which $M(\omega)$ has declined to the value of $M(0)$, and $\omega_{0.5}$ is the frequency at which $M(\omega)$ has declined to the half value of $M(0)$, as shown in Fig. 5.47.

5.7 Open-Loop Frequency-Domain Analysis

For a feedback control system, the closed-loop frequency response is mainly determined by its open-loop frequency response. Especially for a unity-feedback system, the closed-loop frequency response is uniquely determined by its open-loop frequency response. A most interesting problem is what an open-loop frequency response tells us about the closed-loop system performances. In fact, if we divide the entire frequency interested into so-called low, middle and high-frequency portions, each portion of an open-loop frequency response has some characteristics representing the performance of the system quite adequate, this method is very useful for analysis and design of the feedback control systems.

5.7.1 Low-Frequency Portion and Steady-State Performance

As we know, the steady-state error for a specific test input signal is related to the system type (or the number of integral facters) and the open-loop gain. In general, if the open-loop transfer function of a feedback system is written as

$$G(s)H(s) = \frac{K\prod_{i=1}^{m}(\tau_i s + 1)}{s^v \prod_{j=v+1}^{n}(T_j s + 1)} \tag{5.92}$$

then the system type is v, and the open-loop gain is K.

Usually, the low-frequency portion is referred to the range below the first break frequency of the asymptotic log-magnitude-frequency curve. The low frequency portion of the log-magnitude curve, or its extension, will have a slope $-20v\,\text{db/dec}$ and a height $20\lg K$ at $\omega = 1$ or pass through 0 db-line at $\omega = \sqrt[v]{K}$. Hence, the steady-state performance of a system can be estimated from its open-loop log-magnitude plot.

5.7.2 Middle-Frequency Portion and Dynamic Performance

The middle-frequency portion is referred to the segment near the cutoff frequency (i. e. gain crossover frequency) ω_c, which is closely related to the relative stability and response speed of the control systems.

1. Second-order system

Consider a unity feedback second-order system with the open-loop transfer function

$$G(s) = \frac{\omega_n^2}{s(s+2\zeta\omega_n)}, \quad 0 < \zeta < 1 \tag{5.93}$$

From Example 5.16 the crossover frequency is

$$\omega_c = \omega_n \sqrt{\sqrt{4\zeta^2 + 1} - 2\zeta^2} \tag{5.94}$$

and the phase margin for this system is

$$\gamma = \arctan \frac{2\zeta}{\sqrt{\sqrt{4\zeta^2 + 1} - 2\zeta^2}} \tag{5.95}$$

Since the system overshoot to a step input signal

$$\sigma_p = e^{-\pi\zeta/\sqrt{1-\zeta^2}} \times 100\% \tag{5.96}$$

according to the data in Table 5.5, the overshoot σ_p is plotted as a function of the phase margin γ in Fig. 5.48. It can be seen that σ_p to a step input will decreases as γ increases.

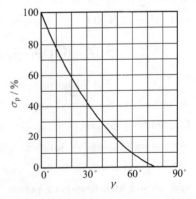

Figure 5.48 σ_p-γ plot

Table 5.5 $\sigma_p - \gamma$

ζ	0	0.2	0.5	0.707	1
$\sigma_p/\%$	100	52.7	16.3	4.3	0
$\gamma/\deg$	0	22.6	51.8	65.5	76.3

Again consider the underdamped second-order system with the open-loop transfer function of Eq. (5.93). Multiplying t_s to both side of Eq. (5.94) yields

$$t_s \omega_c = \frac{3}{\zeta \omega_n} \cdot \omega_n \sqrt{\sqrt{1+4\zeta^4} - 2\zeta^2} = \frac{3\sqrt{\sqrt{1+4\zeta^4} - 2\zeta^2}}{\zeta} \tag{5.97}$$

This equation shows that if ζ is unchanged, i.e. γ is unchanged, the setting time t_s of the step response will decrease as ω_c increases.

2. Higher-order system

It is difficult to find the exact $\sigma_p - \gamma$ and $t_s - \omega_c$ relations for the higher-order systems. However, the tendencies of relative stability and response speed shown by second-order systems are accordant with that of higher-order systems. Usually, in order to ensure a sufficient phase margin to meet the requirement on overshoot, it is suitable for the middle-frequency portion to have a slope of -20db/dec and a sufficient width. Keeping the required phase margin, the response speed can be adjusted by choosing a suitable crossover frequency.

Moreover, for a unity feedback higher-order system, if $34° < \gamma < 90°$, the overshoot and setting time to step input signal can be estimated by following empirical formulas

$$\sigma\% = 0.16 + 0.4\left(\frac{1}{\sin \gamma} - 1\right) \tag{5.98}$$

$$t_s = \frac{\pi}{\omega_c}\left(2 + 1.5\left(\frac{1}{\sin \gamma} - 1\right) + 2.5\left(\frac{1}{\sin \gamma} - 1\right)^2\right), \quad \Delta = 5\% \tag{5.99}$$

5.7.3 High-Frequency Portion and Disturbance Rejection

The high-frequency portion is referred to the frequency range above decade of the crossover frequency ω_c.

Usually, at high frequencies, the magnitude of a system frequency response is small. Hence, for unity-feedback higher-order systems we have

$$|\Phi(j\omega)| = \frac{|G(j\omega)|}{|1 + G(j\omega)|} \approx |G(j\omega)| \tag{5.100}$$

This equation indicates that the ability of a closed-loop system to reject disturbance is almost equivalent to that of the open-loop system.

Since the disturbances applied on a control system are usually with smaller magnitude and higher frequency, the closed-loop system will have better ability to reject disturbance if the open-loop magnitude provides larger amount of attenuation at high-frequency portion.

Problems

P5.1 The open-loop transfer function of a unity feedback system is $G(s) = \dfrac{5}{2s+1}$. Determine the steady-state output of the closed-loop system due to the following input signals:

(a) $r(t) = \sin(t+30°)$

(b) $r(t) = 2\cos(2t-45°)$

P5.2 The unit-step response of a system is
$$c(t) = 1 - 1.8e^{-4t} + 0.8e^{-9t}, \quad t \geq 0$$
Find the frequency response of the system.

P5.3 Plot the asymptotic log-magnitude curves and phase curves for the following transfer functions

(a) $G(s)H(s) = \dfrac{1}{(2s+1)(0.5s+1)}$

(b) $G(s)H(s) = \dfrac{0.5s+1}{s^2}$

(c) $G(s)H(s) = \dfrac{10(s+0.2)}{s^2(s+0.1)}$

(d) $G(s)H(s) = \dfrac{10(s+2)}{s^2(s+5)}$

(e) $G(s)H(s) = \dfrac{8(s+0.1)}{s(s^2+s+1)(s^2+4s+25)}$

P5.4 The asymptotic log-magnitude curves of some systems are given in Fig. P5.4. Determine

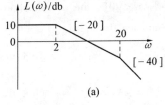
(a)

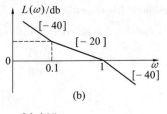
(b)

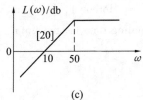
(c)

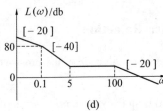
(d)

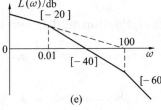
(e)

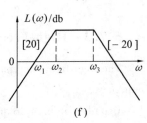
(f)

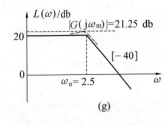
(g)

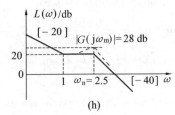
(h)

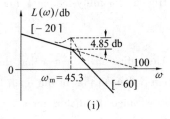
(i)

Figure P5.4

the transfer function and sketch the corresponding asymptotic phase curves for each system. Assume that the systems have minimum phase transfer functions.

P5.5 Fig. P5.5 shows the polar plots of the open-loop transfer functions of some systems. Determine whether the closed-loop systems are stable. In each case, p is the number of the open-loop poles located in the right half s-plane, v is the number of the integral factors in the open-loop transfer function.

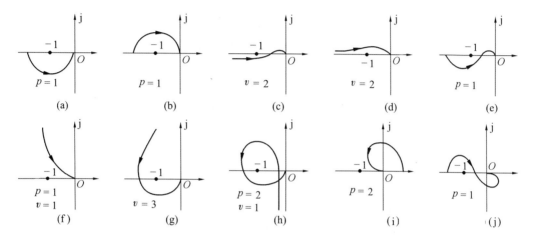

Figure P5.5

P5.6 Sketch the polar plots of the following open-loop transfer functions. Sketch only the portion that is necessary to determine the stability of the closed-loop systems. Determine the stability of the systems by using the Nyquist criterion.

(a) $G(s)H(s) = \dfrac{10}{(0.2s+1)(0.5s+2)(s+1)}$

(b) $G(s)H(s) = \dfrac{100}{s(0.1s+2)(0.4s+1)}$

(c) $G(s)H(s) = \dfrac{100}{s(s+1)(s^2+2s+2)}$

(d) $G(s)H(s) = \dfrac{50}{s(s+2)(s^2+4)}$

(e) $G(s)H(s) = \dfrac{s}{1-0.2s}$

P5.7 Sketch the polar plots of the following open-loop transfer functions, and find the maximum value for the open-loop gain so that the system is stable by using the Nyquist criterion.

(a) $G(s)H(s) = \dfrac{k}{s(s^2+s+4)}$

(b) $G(s)H(s) = \dfrac{k(s+2)}{s^2(s+4)}$

P5.8 A negative feedback system has an open-loop transfer function

$$G(s) = \frac{k(s+0.5)}{s^2(s+2)(s+10)}$$

Plot the Bode diagrams with asymptotic curves and determine whether the system is stable using the Nyquist criterion for $k=10$ and $k=100$, respectively.

P5.9   A unity negative feedback system has the open-loop transfer function

$$G(s) = \frac{11.7}{s(0.05s+1)(0.1s+1)}$$

Determine the crossover frequency and the phase margin.

P5.10   A closed-loop system has the open-loop transfer function

$$G(s) = \frac{Ke^{-\tau s}}{s}$$

(a) Determine the gain K so that the phase margin is $60°$ when $\tau = 0.2s$.

(b) Plot the phase margin versus the time delay τ for K as in part (a).

P5.11   A time-delay system has the open-loop transfer function

$$G(s) = \frac{e^{-\tau s}}{s(s+1)}$$

Determine the time delay τ to maintain stability.

P5.12   The polar plot of a conditionally stable system, for a specific gain $K=50$, is shown in Fig. P5.12. (a) Determine whether the system is stable and find the number of closed-loop poles (if any) in the right half s-plane. Assume that the open-loop characteristic has the minimum phase.
(b) Find the range of K so that the system is stable.

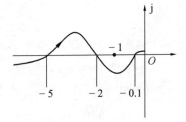

Figure P5.12

P5.13   Consider a unity feedback system with the open-loop transfer function

$$G(s) = \frac{\tau s+1}{s^2}$$

Determine the value of that results in the system with a phase margin of $45°$.

P5.14   Consider a unity feedback system with the open-loop transfer function

$$G(s) = \frac{K}{(0.01s+1)^3}$$

(a) Determine the value of K that results in the system with a phase margin of $45°$.

(b) Determine the gain margin corresponding to the gain obtained in (a).

P5.15   The open-loop transfer function of a unity feedback system is

$$G(s) = \frac{k}{s(s^2+s+100)}$$

(a) Determine the value of open-loop gain so that the system has a gain margin of 20db.

(b) Determine the phase margin corresponding to the gain obtained in (a).

P5.16 The asymptotic logarithmic magnitude curve of a minimum-phase system is shown in Fig. P5.16. Estimate the phase margin and gain margin of the system.

P5.17 Consider a unity feedback system with the open-loop transfer function

$$G(s) = \frac{1\,000}{s+100}$$

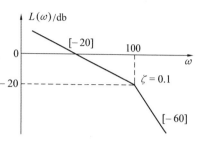

Figure P5.16

Find the bandwidths of the open-loop system and the closed-loop system and compare the results.

P5.18 The open-loop transfer function of a unity feedback system is

$$G(s) = \frac{16}{s(s+2)}$$

(a) Determine the crossover frequency ω_c and the phase margin γ.

(b) Determine the resonant frequency ω_r and the relative resonant peak M_r of the closed-loop system.

P5.19 The asymptotic logarithmic magnitude curve of a unity feedback system is shown in Fig. P5.19. Assume that the system has the minimum-phase.

(a) Determine the open-loop transfer function.

(b) Determine whether the system is stable.

(c) Discuss the effect on the performance specifications $\sigma_p, t_s,$ and e_{ss} if the magnitude curve is translated right by a decade.

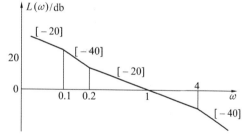

Figure P5.19

Chapter 6 Compensation of Control System

6.1 Introduction

6.1.1 Multiple Constrains in Design

The performance of a feedback control system is of primary importance. We have found that a suitable control system should have some of the following properties.

① It should be stable and present acceptable response to input command, i. e., the controlled variable should follow the changes in the input at a suitable speed without unduly large oscillations or overshoots.

② It should operate with as little error as possible.

③ It should be able to mitigate the effect of undesirable disturbances.

A feedback control system that provided an optimum performance without any necessary adjustments is rare indeed. Usually it is necessary to compromise among the many conflicting and demanding specifications and to adjust the system parameters to provide a suitable and acceptable performance when it is not possible to obtain all the desired optimum specifications.

The preceding chapters have shown that it is often possible to adjust the system parameters in order to provide the desired system response. When the achievement of a simple performance requirement may be met by selecting a particular value of K, the process is called gain compensation. However, we often find that it is not sufficient to adjust a system parameter and thus obtain the desired performance. Rather we are required to reconsider the structure of the system and redesign the system in order to obtain a suitable one. That is, we must examine scheme of the system and obtain a new design that results in a suitable system. Thus the design of a control system is concerned with the arrangement of the system structure and the selection of suitable components and parameters. When we are not able to relax several performance requirements, we must alter the system in some way. The alteration or adjustment of a control system in order to provider a suitable performance is called compensation.

In redesigning a control system to alter the system response, an additional component or device is inserted within the structure of the feedback system to compensate for the performance deficiency. The compensating device may be electric, mechanical, hydraulic, pneumatic, or some other type of devices or networks and is often called a compensator. Commonly an electric network serves as a compensator in many control systems.

Quite often, in practice, the best and simplest way to improve the performance of a control system is to alter, if possible, the plant itself. That is, if the system designer is able to specify and alter the design of the plant, then the performance of the system may be readily improved. For example, to improve the transient behavior of a servomechanism position controller, we often can choose a better motor for the system. Thus a control system designer should recognize that an alteration of the plant maybe result in an improved system. However, often the plant is unalterable or has been altered as much as possible and still results in unsatisfactory performance. Then the addition of compensator becomes useful for improving the performance of the system.

6.1.2 Types of Compensation

The compensator is placed in a suitable location within the system, and can be done in several ways. An additional component may be inserted in the forward path, as shown in Fig. 6.1(a). This is called the cascade or serial compensation. The transfer function of the compensator is designated as $G_c(s)$, whereas that of the original plant (or process) is denoted by $G_p(s)$. Alternatively, the compensator may be placed in the feedback path as shown in Fig. 6.1(b). This is called the feedback compensation. A combination of these two schemes is shown in Fig. 6.1(c). The selection of the compensation scheme depends upon a consideration of the specifications, the power levels at various signal nodes in the system, and the compensators available for us.

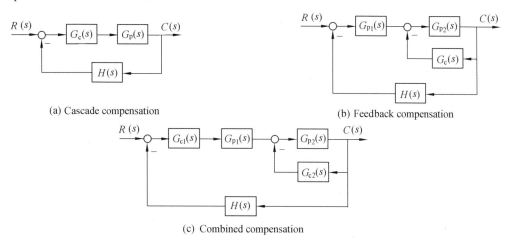

(a) Cascade compensation

(b) Feedback compensation

(c) Combined compensation

Figure 6.1 Types of compensation

6.1.3 Cascade Compensation

Although many different types of compensators can be used, the simplest among them are cascade phase-lead, phase-lag, and phase-lag-lead networks. Each of these can be realized by using an operational amplifier network. The Bode diagram is used to determine a suitable cascade compensator in preference to other frequency response plots. The frequency response of the cascade compensator is added to the frequency response of the uncompensated system.

It is assumed, in below discussion, that the compensator $G_c(s)$, as shown in Fig. 6.1 (a), is used with an uncompensated system so that the overall open-loop gain can be set to satisfy the steady-state error requirement, then $G_c(s)$ is used to adjust the system dynamics favorably without affecting the steady-state error. For convenience, the open-loop transfer function of the uncompensated system, $G_p(s)H(s)$, is denoted by $G_o(s)$.

At first, consider a system described by the open-loop transfer function

$$G_o(s) = \frac{K}{s(0.2s+1)} \tag{6.1}$$

Suppose we wish the closed-loop system to meet the following performance requirements:

① The steady-state error for a unit ramp is to be no more than 0.003 16.

② The phase margin is to be no less than 45°.

For the first requirement, the static velocity error constant can be calculated from equation

$$\varepsilon_{ss} = 1/K_v = 1/K \leqslant 0.003\ 16$$

and thus the required open-loop gain is $K = K_v = 316$. Fig. 6.2 shows the Bode diagram with $K = 316$, which satisfies the steady-state error requirement. It is seen that the gain crossover frequency of the original system is $\omega_{co} = 40$ rad/s, Hence, the phase margin is, by estimating, $\gamma_o \approx 7°$.

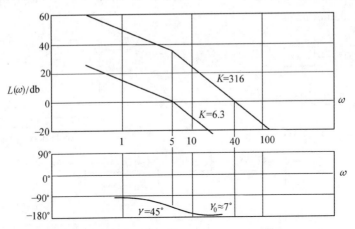

Figure 6.2 Bode diagram for system of Eq. (6.1)

It may also be seen that the phase margin will be about 45° at $\omega = 5$ rad/s; therefore, to meet the second requirement, the magnitude must be zero at this frequency. For this example, it will occur when $K = 6.3$, and the magnitude plot is also shown in Fig. 6.2 for this value of gain. Obviously, it is not possible to satisfy both system performance requirements with a singular value of gain. The system needs to be modified in some way, i.e., the shape of the Bode diagram has to be altered in some way to allow it to achieve both performance requirements.

The system performance requirements stated in the example are typical of those found in many design cases; a steady-state error determines one value of gain while a desired transient response determines another. Note how each requirement relates to a different region of the frequency axis in the Bode diagram.

① The steady-state error relates to the slope and magnitude at low frequency.

② The phase margin relates to the gain crossover frequency, which usually occurs at higher frequency.

In order to meet both requirements, we could do one of two things:

① Keep $K=316$ so as to satisfy the steady-state error requirement but introduce another element that contributes a positive phase angle to the region of $\omega=40$ rad/s.

② Keep the phase plot same, keep $K=316$, but introduce another element that attenuates higher frequencies in such a way that the modified magnitude curve passes through 0 db at $\omega=5$ rad/s.

In each case, the technique involves adding a compensator to the forward path of the being compensated system, as shown in Fig. 6.1 (a), and they are referred to as cascade compensation. In the first case, the technique is known as phase-lead compensation. The second technique is commonly called phase-lag compensation. Lag-lead compensation is a combination of both techniques. Because the open-loop frequency response of the system is altered, the compensator that achieves the required performance is sometimes called a filter.

6.1.4 Approaches to System Design

The performance of a control system can be specified by requirement of certain maximum overshoot and setting time for a step input. Furthermore it is usually necessary to specify the maximum allowable steady-state error for several test signal inputs and disturbances. These performance specifications are related to the location of the poles and zeros of the closed-loop transfer function. Thus the location of the closed-loop poles and zeros can be specified. As we found in chapter 4, the locus of the roots of the closed-loop system can be readily obtained for the variation of one system parameter. However, when the locus of roots does not result in a suitable root configuration, we must add a compensator to alter the locus of the roots as parameter is varied. Therefore we can use the root locus method and determine a suitable compensator transfer function so that the resultant root locus yields the desired closed-loop root configuration.

Alternatively, the performance of a control system can be specified in terms of the relative resonant peak, resonant frequency, and bandwidth of the closed-loop frequency response, or in terms of the phase margin, gain margin and gain crossover frequency of the open-loop frequency response. We can add a suitable compensator, if necessary, in order to satisfy the system performance. The design of the compensator is developed in terms of the frequency response as portrayed on the polar plot, the Bode diagram, or the Nichols chart. Because a cascade transfer function is readily accounted for on a Bode diagram by adding the frequency response of the compensator, we usually prefer to approach the frequency response method by utilizing the Bode diagram.

6.2 Phase-Lead Compensation

6.2.1 Phase-Lead Compensator

The transfer function of the phase-lead compensator takes the form

$$G_c(s) = \frac{\alpha Ts + 1}{Ts + 1} \qquad (6.2)$$

where $\alpha > 1$. The Bode diagram of this transfer function is shown in Fig. 6.3, where

$$\omega_z = \frac{1}{\alpha T}, \quad \omega_p = \frac{1}{T}$$

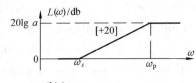

Since the zero occurs first on the frequency axis, the slope of the asymptotic log-magnitude curve between ω_z and ω_p is +20 db/dec.

Figure 6.3 Bode diagram of phase-lead compensator

It is evident that this compensator provides a phase lead, between the output and the input, given by

$$\varphi(\omega) = \angle G_c(j\omega) = \arctan \alpha T\omega - \arctan T\omega \qquad (6.3)$$

Hence it is called a phase-lead compensator.

The phase-lead compensator is a form of high-pass filter, through which the signals at high frequencies are amplified relatively than that at low frequencies. It introduces a gain at high frequencies, which in general is destabilizing. However, its positive phase angle is stabilizing. Hence, we must carefully choose two break frequencies so that the stabilizing effect of the positive phase angle is dominant.

It is easily shown from Eq. (6.2) that the maximum phase lead occurs at a frequency

$$\omega_m = \frac{1}{\sqrt{\alpha} \, T} \qquad (6.4)$$

which is the geometric mean of two break frequencies and may be identified as the halfway point between these two break frequencies on the logarithmic frequency scale. To determine the parameter ensuring maximum value of phase-lead angle, we rewrite Eq. (6.3) as

$$\varphi = \arctan \frac{\alpha T\omega - T\omega}{1 + \alpha T^2 \omega^2} \qquad (6.5)$$

Substituting frequency results in

$$\tan \varphi_m = \frac{\alpha - 1}{2\sqrt{\alpha}} \qquad (6.6)$$

or

$$\sin \varphi_m = \frac{\alpha - 1}{\alpha + 1} \qquad (6.7)$$

The maximum value of the phase angle φ_m is

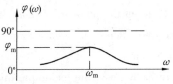

Figure 6.4 Phase lead as a function of α

then a function of α, and the parameter α is given by

$$\alpha = \frac{1+\sin \varphi_m}{1-\sin \varphi_m} \qquad (6.8)$$

Eq. (6.4) and (6.8) are two basic equations of the phase-lead element used for the compensation process. For rapid estimation of α for a required phase lead, Fig. 6.4 may be used.

6.2.2 Phase-Lead Compensation Process

We will first plot the Bode diagram for $G_o(j\omega)$. The uncompensated system is plotted with the desired gain to allow an acceptable steady-state error. Then the phase margin is examined to find whether it satisfies the specification. If the phase margin is not sufficient, phase lead can be added to the phase-frequency curve of the system to increase the phase margin by placing the $G_c(j\omega)$ in a suitable location.

To obtain maximal additional phase lead, we desire to place the compensator so that the frequency ω_m is located at the frequency where the magnitude of the compensated magnitude curve crosses 0 db axis, i.e. at the new gain crossover frequency. The value of the added phase lead required allows us to determine the necessary value for α from Eq. (6.8). The break frequency $\omega_z = 1/(\alpha T)$ is located by noting that the maximum phase lead should occur at ω_m, halfway of two break frequencies. Because the total magnitude for the compensator is $20 \lg \alpha$ (db), we expect a magnitude of $10 \lg \alpha$ (db) at ω_m. The procedure of determining the compensator will be illustrated by some examples.

Example 6.1 Again, consider the system with open-loop transfer function

$$G_o(s) = \frac{K}{s(0.2s+1)}$$

and the two performance specifications are: (a) the steady-state error for a unit ramp is $\varepsilon_{ss} \leq 0.00136$; (b) the phase margin is $\gamma \geq 45°$.

Solution As we evaluated, the steady-state performance requirement will be satisfied if the gain $K = K_v = 316 \text{ s}^{-1}$. The remaining steps are as follows:

① Draw the uncompensated Bode diagram. The Bode diagram of $G_o(j\omega)$ is drawn with $K = 316 s^{-1}$ as shown in Fig. 6.5.

② Evaluate the uncompensated system phase margin γ_o. From the magnitude plot, the uncompensated crossover frequency ω_{co} is 40 rad/s. Thus the phase margin of the uncompensated system is

$$\gamma_o = 180° + \angle G_o(j\omega_{co}) = 180° - 90° - \arctan(0.2 \times 40) = 7°$$

Usually, using the formula to evaluate the phase margin is often easier than drawing the complete phase-angle curve.

③ Determine the necessary additional phase lead φ_m. We need to add a phase-lead compensator so that the phase margin is raised to 45° at the new crossover frequency. Since the new crossover frequency is greater than that of the uncompensated system, the phase lag of the

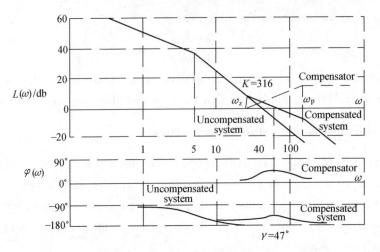

Figure 6.5 Bode diagram of Example 6.1

uncompensated system is greater also. We shall take a maximum phase lead of $45°-7°=38°$ plus a small increment (about $5°$) of phase lead to account for the added phase lag. Thus the necessary additional phase lead is about

$$\varphi_m = 38°+5°=43°$$

④ Evaluate the parameter α. From Eq. (6.8)

$$\alpha = \frac{1+\sin \varphi_m}{1-\sin \varphi_m} = \frac{1+\sin 43°}{1-\sin 43°} = 5.3$$

⑤ Determine the desired crossover frequency ω_c. The maximum phase lead occurs at ω_m, and this frequency will be selected as the new crossover frequency. The magnitude of the phase-lead compensator at ω_m is $10 \lg \alpha = 10 \lg 5.3 = 7.2$ db. The compensated crossover frequency is then evaluated where the magnitude of $G_o(j\omega)$ is

$$-10 \lg \alpha = -7.2 \text{ db}$$

then, from the Bode diagram, we have

$$\omega_m = \omega_c = 62 \text{ rad/s}$$

⑥ Determine the transfer function $G_c(s)$. From Eq. (6.4) we get

$$\omega_p = \frac{1}{T} = 143 \text{ rad/s}, \quad \omega_z = \frac{1}{\alpha T} = 27 \text{ rad/s}$$

Thus the transfer function of the phase-lead compensator is

$$G_c(s) = \frac{0.037s+1}{0.007s+1}$$

⑦ Draw the compensated Bode diagram. The Bode diagrams of $G_c(j\omega)$ and $G_c(j\omega)G_o(j\omega)$ are shown in Fig. 6.5.

⑧ Check the resulting phase margin γ. The final phase margin

$$\gamma = 180°+\angle G(j\omega_c)+\angle G_c(j\omega_c) = 180°-175.4°+42.9° = 47.5°$$

Now both performance requirements are satisfied.

Example 6.2 Consider the system with the open-loop transfer function

$$G_o(s) = \frac{K}{s(0.1s+1)(0.01s+1)}$$

It is desired that the static velocity error constant $K_v \geqslant 100 s^{-1}$, phase margin $\gamma \geqslant 30°$, and crossover frequency $\omega_c \geqslant 45$ rad/s.

Solution Since the system is of type 1, we choose
$$K = K_v = 100 \ s^{-1}$$
The remaining steps are as follows:

① The Bode diagram is plotted with $K = 100 \ s^{-1}$, as shown in Fig. 6.6.

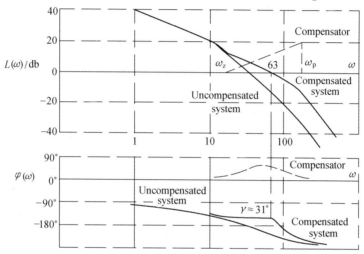

Figure 6.6 Bode diagram of example 6.2

② In this case, the compensated crossover frequency should meet the response speed requirement. If we choose $\omega_c = 50$ rad/s, then the phase angle of $G_o(j\omega)$ at this frequency is
$$\angle G_o(j\omega)|_{\omega=50} = -90° - \arctan(0.1 \times 50) - \arctan(0.01 \times 50) = -195°$$

③ Since it is desired that $\gamma \geqslant 30°$, at least the required phase lead angle should be
$$30° - (180° - 195°) = 45°$$
Taking a mount of safety, we choose $\varphi_m = 55°$, which is provided by the compensator.

④ From Eq. (6.8) we obtain
$$\alpha = \frac{1 + \sin 55°}{1 - \sin 55°} = 10$$
From Eq. (6.4), two break frequencies are
$$\omega_z = \frac{1}{\alpha T} = 15.8 \text{ rad/s}$$
$$\omega_p = \frac{1}{T} = 158 \text{ rad/s}$$
Therefore the transfer function of the compensator is
$$G_c(s) = \frac{\alpha Ts + 1}{Ts + 1} = \frac{0.063s + 1}{0.0063s + 1}$$

⑤ The Bode diagrams of the compensator and the compensated system are shown in Fig.

6.6.

⑥ From Fig. 6.3 we have
$$20 \lg |G_c(\omega_m)| = 10 \lg \alpha = 10 \text{ db}$$
and from Fig. 6.6 we have
$$L_o(\omega_m) = 20 \lg G_o|(j\omega)|_{\omega=\omega_m} = -8 \text{ db}$$
Thus the actual compensated crossover frequency is greater than $\omega_m = 50$ rad/s, and it can be found from Fig. 6.6 that a new compensated crossover frequency is
$$\omega'_c = 63 \text{ rad/s}$$
The final phase margin is then
$$\gamma = 180° + \angle G(j\omega'_c) + \angle G_c(j\omega'_c) = 180° - 203.2° + 54.3° = 31.1°$$
which is greater than the desired value.

6.2.3 Comments on the Applicability and Results

Phase-lead compensation has some distinct advantages over other forms of compensation, whereas it may also be difficult to use. Some observations from the example just analyzed allow a few generalizations to be made regarding phase lead compensation.

① The phase-lead compensation method provides an additional phase lead to limit the system's overshoot to a required value.

② The open-loop (and usually the closed-loop) bandwidths is increased. This is usually beneficial since the inclusion of higher frequencies in the response results in a faster response. It may cause problem, however, if noise exists at the higher frequencies.

③ Problem may occur when the uncompensated phase plot has a steep slope in the vicinity of φ_m. This occurs because, as the new gain crossover point moves to the right, larger and larger phase lead is required from the compensator, demanding very large value of α. This is difficult to achieve when the compensator is realized with physical components. For this reason, value of $\alpha > 15$ should be avoided, and methods to compensate the system using other techniques, such as phase-lag, should be investigated.

6.3   Phase-Lag Compensation

6.3.1 Phase-Lag Compensator

It is often used to add a cascade compensator having phase-lag characteristic. The transfer function of phase-lag compensator is
$$G_c(s) = \frac{\beta T s + 1}{T s + 1} \qquad (6.9)$$
where $\beta < 1$. The Bode diagram of this transfer function is shown in Fig. 6.7, where
$$\omega_p = \frac{1}{T}, \quad \omega_z = \frac{1}{\beta T}$$

Fig. 6.7 indicates that this compensator has a phase lag between the output and input. Hence it is called a phase-lag compensator.

It is important to recognize, however, that the phase lag is not the useful effect of the compensator and the details of the phase part of the Bode diagram are of no consequence at all in the design of the compensator. All that is important is the amount of attenuation provided. When $\omega > \omega_z$, the amount of attenuation is

$$L_c(\omega_z) = 20 \lg \beta \text{ (db)} \qquad (6.10)$$

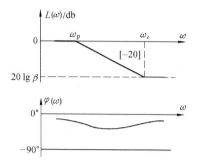

Figure 6.7 Bode diagram of phase-lag compensator

6.3.2 Phase-Lag Compensation Process

In phase-lag compensation, the magnitude part of the uncompensated Bode diagram is attenuated in order to reduce the gain crossover frequency, thereby allowing the uncompensated phase plot to produce the necessary phase margin. The phase-lag compensator is used to provide an attenuation and therefore to lower the crossover frequency of the system. Furthermore, at lower crossover frequency, we usually find that the phase margin of the system is increased, and the specifications can be satisfied.

Of cause, the influence of the phase lag caused by the compensator should be taken into consideration. Usually, the lag phase is about $5° \sim 12°$ if the break frequency corresponding to the zero of the compensator is $\omega_z \approx (0.1 \sim 0.2)\omega_c$.

The compensation procedure will be illustrated with following example.

Example 6.3 Reconsider the system of Example 6.2 and design a phase-lag compensator. The uncompensated transfer function is

$$G_o(s) = \frac{K}{s(0.2s+1)}$$

It is desired that the static velocity error constant $K_v = 316$ while a phase margin of $45°$ is obtained.

Solution ① Draw the uncompensated Bode diagram. The Bode diagram of the uncompensated system with the gain adjusted for the desired error constant $K_v = 316$ is drawn as shown in Fig. 6.8.

② Determine the uncompensated phase margin. From the Bode diagram for Example 6.1, the uncompensated crossover frequency is $\omega_{co} = 40$ rad/s, and the uncompensated phase margin is $\gamma_o \approx 7°$.

③ Determine the new crossover frequency ω_c. Allowing $5°$ for the phase-lag compensator, we locate the frequency where

$$\angle G_o(j\omega) = -180° + 45° + 5° = -130°$$

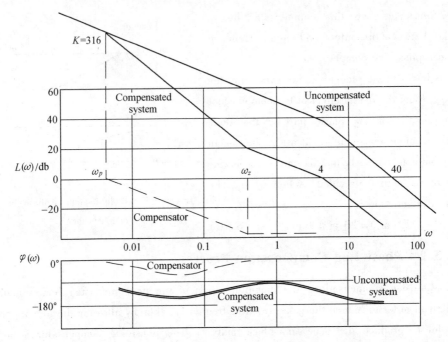

Figure 6.8 Bode diagram with phase-lag compensator added

which is to be the new crossover frequency. From the phase plot, we find that the phase margin requirement would be satisfied if the magnitude curve crossed the 0 dB-line at $\omega_c = 4$ rad/s.

④ Determine the parameter β. From the magnitude plot, the attenuation necessary to cause $\omega = 4$ rad/s to be the new crossover frequency is equal to 38 db. Thus we have

$$20 \lg \beta = -38 \text{ db}$$

i. e.

$$\beta = 0.012\,5$$

⑤ Determine $G_c(s)$. As seen in Figure 6.8, the break frequency ω_z is one decade below the new crossover frequency ω_c and thus ensure only 5° of additional phase lag at ω_c due to the lag compensator. Thus, we have

$$\omega_z = 0.1\omega_c = 0.4 \text{ rad/s}$$
$$\omega_p = \beta\,\omega_z = 0.005 \text{ rad/s}$$

The transfer function of the compensator is then

$$G_c(s) = \frac{2.5\ s+1}{200\ s+1}$$

⑥ Draw the compensated Bode diagram.

⑦ Check the resulting phase margin. The final phase margin is

$$\gamma = 180° + \angle G(j\omega_c) + \angle G_c(j\omega_c) = 180° - 128.7° - 5.7° = 45.6°$$

which is a satisfied result.

Example 6.4 Design a cascade compensator for the system described by the open-loop transfer function is

$$G_o(s) = \frac{K}{s(s+1)(0.5s+1)}$$

such that the static-state error constant $K_v \geq 5s^{-1}$, phase margin of $\gamma \geq 40°$, and gain margin $L(K_g) \geq 10$ db.

Solution In order to meet the requirement of steady-state characteristic, we choose $K = 5$. The remaining steps are as follows:

① Draw the uncompensated Bode diagram. The uncompensated Bode diagram is drawn as shown in Fig. 6.9.

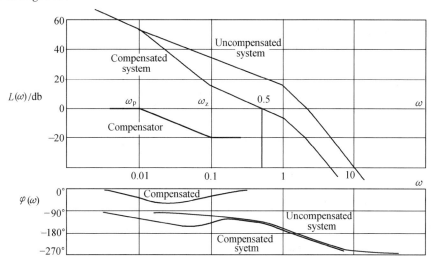

Figure 6.9 Bode diagram of Example 6.4

② Analyze the uncompensated system. The Bode diagram indicates that, the uncompensated crossover frequency $\omega_{co} \approx 2$ rad/s, phase margin $\gamma_o \approx -18°$, and gain margin $L(K_{go}) \approx -7$ db. The uncompensated system is unstable. Since the phase curve has a steeper slope in the vicinity of ω_{co}, phase-lag compensation may be suitable.

③ Determine the new crossover frequency ω_c. Considering the physical realization of the value of βT and allowing 12° for the phase-lag compensator, we locate the frequency where

$$\angle G_o(j\omega) = -180° + 40° + 12° = -128°$$

which is to be the new crossover frequency. From the phase plot, we get

$$\omega_c = 0.5 \text{ rad/s}$$

④ Determine the break frequencies. Since we already take a safe margin of 12°, the break frequency

$$\omega_z = \frac{1}{\beta T} = 0.2\omega_c = 0.1 \text{ rad/s}$$

From the magnitude plot, the attenuation necessary to cause $\omega_c = 0.5$ to be the new crossover frequency is equal to 20 db. Thus we have

$$20 \lg \beta = -20 \text{ db}$$

i. e.
$$\beta = 0.1$$

Thus, another break frequency
$$\omega_p = \frac{1}{T} = 0.01 \text{ rad/s}$$

⑤ Determine $G_c(s)$. The transfer function of the compensator is
$$G_c(s) = \frac{10s+1}{100s+1}$$

⑥ Draw the compensated frequency response.

⑦ Check the characteristic performances of compensated system. From the Bode diagram we have, for the compensated system, $K_v \geq 5s^{-1}$, $\omega_c = 0.5$ rad/s, $\gamma = 40°$, and $L(K_g) = 11$ db. All the design requirements are satisfied.

6.3.3 Comments on the Applicability and Results

① The phase-lag method provides the necessary damping ratio in order to limit the overshoot to the required value.

② The compensation process is somewhat simpler than the phase-lead compensation in that the selection of the break frequencies is not too critical.

③ As can be seen from the compensated system, the phase-lag technique reduces the open- and hence the closed-loop bandwidth, which results in a slower response.

④ Unlike phase-lead compensation, theoretically, phase lag compensation may change the phase margin by more than 90°.

6.4 Phase Lag-Lead Compensation

In some cases, the combined phase lag-lead compensation method may allow more requirements to be met than by using either phase lead or lag compensation along.

6.4.1 Phase Lag-Lead Compensator

The transfer function of phase lag-lead compensator is
$$G_c(s) = \frac{(\beta T_1 s + 1)(\alpha T_2 s + 1)}{(T_1 s + 1)(T_2 s + 1)} \tag{6.11}$$

where $\beta < 1$ and $\alpha > 1$. In the compensator design it is usual to assume that the two break frequencies of the lag portion are lower than the two break frequencies of the lead portion. The Bode diagram of this compensator is shown in Fig. 6.10. Note from the figure how the two portions are separated. Further features of the Bode diagram include the following.

① The magnitude at lower frequencies is 0 db while the magnitude at higher frequencies is $20 \lg(\alpha\beta)$ db. Usually, the compensator provides attenuation only and no gain.

② The phase angle first lags and then leads, but the high- and low-frequency phases are both zeros.

③ The maximum phase-lag and the maximum phase-lead occur between their respective break frequencies.

The phase lag-lead compensator utilizes the best feature of the individual lag and lead portions, usually without their disadvantages. For example, the lag-lead compensation allows the introduction of phase lead to stabilize a system, while providing attenuation at higher frequencies to filter out noise.

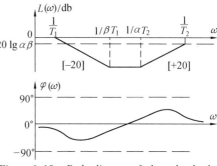

Figure 6.10 Bode diagram of phase lag-lead compensator

6.4.2 Phase Lag-Lead Compensation Process

The process of phase lead-lag compensation is illustrated by following example.

Example 6.5 Design a cascade compensator for the system with open-loop transfer function

$$G_o(s) = \frac{K}{s(0.1s+1)(0.01s+1)}$$

such that the closed-loop system has a static velocity error constant $K_v \geq 100 \text{ s}^{-1}$, a gain crossover frequency $\omega_c = 20$ rad/s, and a phase margin $\gamma \geq 40°$.

Solution The Bode diagram of uncompensated system is drawn with $K = K_v = 100$, as shown in Fig. 6.11, which is the gain needed to satisfy the steady-state error requirement.

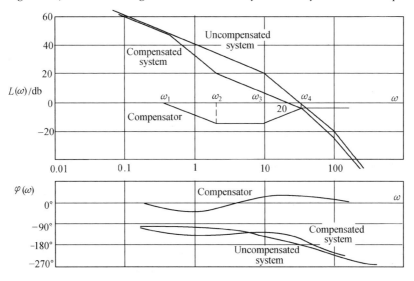

Figure 6.11 Bode diagram for Example 6.5

From the diagram, the uncompensated crossover frequency $\omega_{co} = 30$ rad/s, and the uncompensated phase margin is $\gamma_o \approx 2°$. This tells us that some form of compensation is required. Since the desired crossover frequency is less than the uncompensated crossover frequency, phase-lag compensation is needed. At $\omega = 20$ rad/s we have

$$\angle G_o(j\omega) = -90° - 63.4° - 11.3° \approx -165°$$

Obviously, phase-lead compensation is also needed to ensure the requirement for phase margin.

Taking a safety margin of 5° for the lag portion, we get the maximum additional phase lead is at least

$$\varphi_m = 40° - (180° - 165°) + 5° \approx 30°$$

which occurs at $\omega_m = \omega_c = 20$ rad/s. From Eq. (6.8)

$$\alpha \geq \frac{1 + \sin 30°}{1 - \sin 30°} = 3$$

then the two break frequencies of lead portion should satisfy

$$\omega_3 = \frac{1}{\alpha T_2} \leq 11.5 \text{ rad/s}, \quad \omega_4 = \frac{1}{T_2} \geq 35 \text{ rad/s}$$

Considering the pole at $\omega = 10$ rad/s of uncompensated open-loop system, we take $\omega_3 = 10$ rad/s and $\omega_4 = 35$ rad/s. Measuring the necessary attenuation at $\omega_c = 20$ rad/s, we find that 8db is required, i.e. $20 \lg |G_c(j\omega)|_{\omega=20} = -8$ db. Thus, the Bode diagram for the lead portion can be drawn as shown in Fig. 6.11, where the slope of the magnitude curve between ω_3 and ω_4 is +20 db/dec.

As for the lag portion, again since a safety margin of 5° is taken, we locate

$$\omega_2 = \frac{1}{\beta T_1} = 0.1 \omega_c = 2 \text{ rad/s}$$

By drawing the magnitude plot of the lag portion, we can get

$$\omega_1 = \frac{1}{T_1} = 0.35 \text{ rad/s}$$

Therefore, the transfer function of the lead-lag compensator is

$$G_c(s) = \frac{0.5s + 1}{2.86s + 1} \cdot \frac{0.1s + 1}{0.029s + 1}$$

The Bode diagram of compensated system is shown in Fig. 6.11. The final phase margin is

$$\gamma = 180° - 165° + \angle G_c(j\omega_c) = 15° - 4.7° + 33.3° = 43.6°$$

All the required performance specifications are satisfied.

Since usually the phase lag-lead compensation method allows more performance requirements, it is not known before the compensation is finished whether the system is overconstrained. In some cases, in order to employ a methodical design procedure, it is useful to allocate priorities to the stated requirements and ensure that the design meets the most important ones.

6.5 PID Controller

In the last section the phase lag-lead compensation was developed. This compensation is second order, with a phase-lag compensation followed by a phase-lead compensation. As we saw in that section, the lag-lead compensation offers much more flexibility than does either the phase-lag or the phase-lead compensation separately. In our discussions so far we have considered that the plant to be controlled is completely known to us. In practice this is not always the case. However, it may still be possible to obtain good performances of the closed system by introducing a PID (Proportional-plus-Integral-plus-Derivative) controller, i.e. PID compensator, as shown in the block diagram of Fig. 6.12. In fact, the PID is probably the most commonly used compensator in feedback control systems.

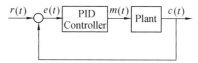

Figure 6.12 System with PID controller

With $e(t)$ the controller input and $m(t)$ the output, the PID controller is defined by the equation

$$m(t) = K_P e(t) + K_I \int_0^t e(\tau) \mathrm{d}\tau + K_D \frac{\mathrm{d}e(t)}{\mathrm{d}t} \tag{6.12}$$

and the transfer function of the controller is

$$G_c(s) = \frac{M(s)}{E(s)} = K_P + \frac{K_I}{s} + K_D s \tag{6.13}$$

The input to plant consists of three components:

① $K_P E$, which is proportional to the error;

② $K_I E/s$, which is proportional to the integral of the error; and

③ $K_D s E$, which is proportional to the derivative of the error.

Quite often it is not necessary to implement all three components to meet the design specifications for a particular control system.

The values of the parameters K_P, K_I, and K_D can often be determined by trial and error, if the plant is not known exactly. If the parameters of the plant are subject to large variation, the parameters of PID controller can be adjusted to improve the performance.

6.5.1 P Controllers

For the proportional controller, in Eq. (6.13), only the proportional gain controller K_P is nonzero. Hence the controller transfer function is

$$G_c(s) = K_P \tag{6.14}$$

and the controller is a pure gain.

This compensator is used in situations in which satisfactory response can be obtained simply by setting a gain in the system, with no dynamic compensation required.

6.5.2 PI Controllers

The transfer function of the PI controller is given by

$$G_c(s) = K_P + \frac{K_I}{s} = \frac{K_P s + K_I}{s} \qquad (6.15)$$

This controller increases the type of the system by 1 and is used to improve the steady-state response.

This controller has a pole at the origin and a zero on the negative real axis. The transfer function can be expressed as

$$G_c(s) = \frac{K_I(\tau s + 1)}{s} \qquad (6.16)$$

where $\tau = K_P/K_I$, it is seen that the zero is located at $s = -K_I/K_P$. The Bode diagram of the PI controller is shown in Fig. 6.13.

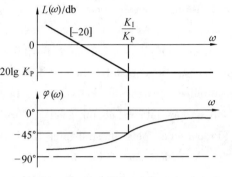

Figure 6.13 Bode diagram of PI controller

Obviously, from the phase plot, the PI controller is phase lag. If the high-frequency gain and the zero of the phase-lag compensator of Fig. 6.7 are held constant and the pole is moved to the origin of the s-plane ($-\infty$ of the $\lg\omega$-axis of the Bode diagram), the PI characteristic of Fig. 6.13 results.

6.5.3 PD Controllers

The transfer function of the PD controller is

$$G_c(s) = K_P + K_D s = K_P\left(\frac{K_D}{K_P}s + 1\right) \qquad (6.17)$$

Hence, $G_c(s)$ has a zero at $s = -K_P/K_D$. The Bode diagram of the PD controller is shown in Fig. 6.14.

It is seen that this controller has a positive phase angle. Therefore, the PD controller is a form of phase-lead compensator and used to improve the system transient response. The design procedure given in section 6.2 can be employed.

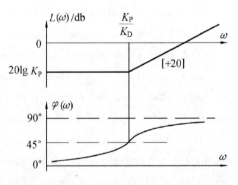

Figure 6.14 Bode diagram of PD controller

In practical implementation, sometimes a problem may be caused by noise, which is invariably present in all physical systems. If a signal is changing rapidly with respect to time, then its derivative is large. Obviously, high-frequency noise through the PD controller will be amplified. Moreover, the higher the frequency is, the more the amplification is. To prevent problems with high-frequency noise, it is usually necessary to add a pole to the PD controller transfer function, such that the transfer function becomes

$$G_c(s) = \frac{K_P + K_D s}{s - p_{add}} = K_P \frac{(K_D/K_P)s + 1}{s - p_{add}} \quad (6.18)$$

The high-frequency gain is now limited to the value of K_D. The pole in this compensator is chosen larger in magnitude than the zero, such that the compensator is still phase-lead. This compensator is now simply a phase-lead compensator of the type in section 6.2.

6.5.4 PID Controllers

The PID controller is employed in control systems in which improvements in both the transient response and the steady-state response are required. The transfer function of the PID controller is given by

$$G_c(s) = K_P + \frac{K_I}{s} + K_D s = \frac{K_D s^2 + K_P s + K_I}{s} \quad (6.19)$$

As described earlier, the integral component is phase-lag and the derivative component is phase-lead. Hence the integral component contributes a low-frequency effect and the derivative component contributes a high-frequency effect. The Bode diagram of the PID controller is shown in Fig.6.15.

Note the similarity between the Bode diagram of the PID controller in Fig. 6.15 and that of the lag-lead compensator in Fig. 6.10. Hence, we can

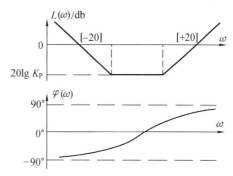

Figure 6.15 Bode diagram of PID controller

see that the PID controller is a form of lag-lead compensator. However, for the PID controller, the three gains are the only parameters to be chosen. The unbounded high-frequency gain of the differentiator can lead to problems, as described before, and a pole is usually added to the derivative path to limit this gain. For this case the transfer function of the controller is

$$G_c(s) = K_P + \frac{K_I}{s} + \frac{K_D s}{s - s_{add}} \quad (6.20)$$

and four parameters must be determined by the design process.

6.6 Feedback Compensation

In order to improve the system performances, besides the cascade compensation, the feedback compensation is often used as another scheme. By using local feedback compensation, almost same effect, as that of cascade compensation, can be obtained. Moreover, additional specific functions for improving system performance are obtained.

6.6.1 Effect of Feedback compensation

The use of local loop of feedback compensation is shown in Fig. 6.16, where $G_{p1}(s)$ is a part of the plant, $G_{p2}(s)$ is another part of the plant, $G_c(s)$ is the feedback compensator. The loop including $G_{p2}(s)$ and $G_c(s)$ is called the local closed loop or minor loop, and the closed loop consisted by $G_{p1}(s)$ in series with the local loop is called the main loop or outer loop.

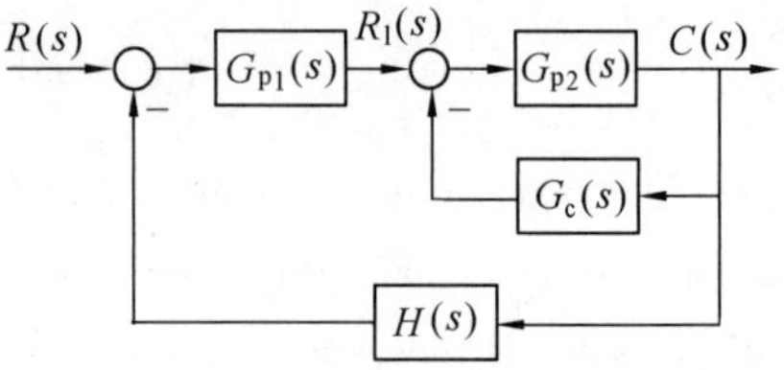

Figure 6.16 Feedback compensation

1. Position feedback

Assume that $G_{p2}(s)$ in Fig. 6.16 is an inertial term with the transfer function

$$G_{p2}(s) = \frac{K_2}{Ts + 1} \tag{6.21}$$

and the local closed loop is formed with a proportional compensator, i.e.,

$$G_c(s) = K_P \tag{6.22}$$

Then the local closed-loop transfer function is

$$\frac{C(s)}{R_1(s)} = \frac{\dfrac{K_2}{Ts + 1}}{1 + \dfrac{K_2 K_P}{Ts + 1}} = \frac{K'_2}{T's + 1} \tag{6.23}$$

where

$$K' = \frac{K_2}{1 + K_2 K_P}, \quad T' = \frac{T}{1 + K_2 K_P}$$

It can be seen that both the time constant and the gain are reduced. Moreover, the larger the proportional gain is, the stronger the negative feedback is. The reduction of the time constant is beneficial to speed the system response. As for the reduction of the gain, which usually is unacceptable, it can be compensated by adding a suitable amplifier in $G_{p1}(s)$.

2. Velocity feedback

Assume that $G_{p2}(s)$ in Fig. 6.16 is consisted of a power amplifier and an actuating motor of a servomechanism, $C(s)$ is an angular signal, then we have

$$G_{p2}(s) = \frac{K_2}{s(T_m s + 1)} \tag{6.24}$$

Consider that $G_c(s)$ denotes a tachometer with the transfer function

$$G_c(s) = K_t s \tag{6.25}$$

Hence the transfer function of local loop is

$$\frac{C(s)}{R_1(s)} = \frac{K_2}{1 + K_2 K_t} \cdot \frac{1}{s\left(\dfrac{T_m}{1 + K_2 K_t} + 1\right)} \tag{6.26}$$

Making $K_2 K_t \gg 1$ results in a simplified equation

$$\frac{C(s)}{R_1(s)} = \frac{1}{K_t s\left(\dfrac{T_m}{1 + K_2 K_t} + 1\right)} \tag{6.27}$$

Comparing with Eq. (6.24), it can be seen that the effect of local feedback is equivalent to add a transfer function

$$G_{eq} = \frac{1}{K_2 K_t} \cdot \frac{T_m s + 1}{\frac{T_m}{K_2 K_t} s + 1} = \frac{1}{\alpha} \cdot \frac{\alpha T s + 1}{T s + 1} \qquad (6.28)$$

ahead of $G_{p2}(s)$.

The effect of velocity feedback is equivalent to that of using a phase-lead compensator, meanwhile the open-loop gain is reduced by a factor $K_2 K_t$.

3. Velocity-derivate feedback

Again consider above example. A velocity-derivate feedback is formed by introducing a derivative network in series with the output of tachometer. Assuming the transfer function of the derivate network be $T_d s/(T_d s+1)$, we have

$$G_c(s) = \frac{K_t T_d s^2}{T_d s + 1} \qquad (6.29)$$

Through a simple derivation, the transfer function of the local loop can be obtained and is given by

$$\frac{C(s)}{R_1(s)} = \frac{K_2(T_d s + 1)}{s[(T_m T_d s^2 + (T_m + K_2 K_t T_d + T_d) s + 1)]} \qquad (6.30)$$

Comparing with Eq. (6.24), it can be seen that the effect of local feedback is equivalent to add a transfer function

$$G_{eq} = \frac{(T_m s + 1)(T_d s + 1)}{T_m T_d s^2 + (T_m + K_2 K_t T_d + T_d) s + 1} \qquad (6.31)$$

ahead of $G_{p2}(s)$. Making $K_2 K_t \gg 1$ and $K_2 K_t T_d \gg T_m$, the denominator of right side in Eq. (6.31) can be approximated as

$$T_m T_d s^2 + \left(\frac{T_m}{K_2 K_t} + K_2 K_t T_d\right) s + 1$$

or

$$\left(\frac{T_m}{K_2 K_t} s + 1\right)(K_2 K_t T_d s + 1)$$

Therefore Eq. (6.31) can be approximated as

$$G_{eq} = \frac{(T_m s + 1)(T_d s + 1)}{\left(\frac{T_m}{K_2 K_t} s + 1\right)(K_2 K_t T_d s + 1)} \qquad (6.32)$$

which is a lag-lead compensator.

Comparing Eq. (6.30) with (6.24) again, it is easily seen that, the loop gain of the inner loop is same as that before the velocity-derivative feedback is adapted. It can be proven that the open-loop gain of original system will not be changed when a feedback compensation is introduced, if the number of pure derivative factors contained in the feedback compensator is greater than that in the enclosed part $G_{p2}(s)$.

4. Effect on sensitivity of systems to parameter variation

One of the important advantages of a feedback system is that the effect of the variation of the parameters of the plant is reduced.

Consider the system shown in Fig. 6.16, and assume that

$$G_{p2}(s) = \frac{K_2}{Ts + 1}, \quad G_c(s) = K_p$$

In the case of no position feedback, if the transfer coefficient is changed from K_2 to $K_2 + \Delta K_2$, the relative increment is $\Delta K_2 / K_2$.

For the case of position feedback, the transfer coefficient is

$$K'_2 = \frac{K_2}{1 + K_2 K_p}$$

and the increment caused by parameter variation is

$$\Delta K'_2 = \frac{\partial K'_2}{\partial K_2} \Delta K_2 = \frac{\Delta K_2}{(1 + K_2 K_p)^2}$$

Hence, the relative increment becomes

$$\frac{\Delta K'_2}{K'_2} = \frac{1}{1 + K_2 K_p} \cdot \frac{\Delta K_2}{K_2} \tag{6.33}$$

This equation shows that, by adding a position feedback, the relative increment of the transfer coefficient is reduced by the factor $(1 + K_2 K_p)$.

6.6.2 Feedback Compensation

When feedback compensation is adopted as shown in Fig. 6.16, the block diagram can be redrawn as shown in Fig. 6.17, where

$$G_d(s) = \frac{G_{p2}(s)}{1 + G_{p2}(s) G_c(s)} \tag{6.34}$$

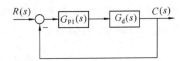

Figure 6.17 Equivalent representation for block diagram in Figure 6.16

In order to apply feedback compensation new technique must be developed. First some approximations with the straight-line log-magnitude curves are needed, and then an exact procedure is developed. When, the transfer function of Eq. (6.34) can be approximated by

$$G_d(j\omega) = \frac{G_{p2}(j\omega)}{1 + G_{p2}(j\omega) G_c(j\omega)} \approx G_{p2}(j\omega) \quad \text{for} \quad |G_{p2}(j\omega) G_c(j\omega)| \ll 1 \tag{6.35}$$

and

$$G_d(j\omega) = \frac{G_{p2}(j\omega)}{1 + G_{p2}(j\omega) G_c(j\omega)} \approx \frac{1}{G_c(j\omega)} \quad \text{for} \quad |G_{p2}(j\omega) G_c(j\omega)| \gg 1 \tag{6.36}$$

Still undefined is the condition when $|G_{p2}(j\omega) G_c(j\omega)| \approx 1$, in which case neither Eq. (6.35) nor Eq. (6.36) is applicable. In the approximate procedure this condition is neglected, and Eq. (6.35) and (6.36) are used when $|G_{p2}(j\omega) G_c(j\omega)| < 1$ and $|G_{p2}(j\omega) G_c(j\omega)| > 1$, respectively.

These approximations can be illustrated with a simple example. Assume that $G_{p2}(s)$ represents a motor described by

$$G_{p2}(s) = \frac{K_m}{T_m s + 1}$$

Let the feedback $G_c(s) = 1$. This problem is sufficiently simple to be solved exactly algebraically. However, use is made of the logarithmic magnitude curve and the approximate condition. The approximate log-magnitude curve is sketched in Fig. 6.18.

From Fig. 6.18 it is seen that $|G_{p2}(j\omega)G_c(j\omega)| > 1$ for all frequencies below ω_1. With the approximation of Eq. (6.38), $G_d(j\omega)$ can be represented by $1/G_c(j\omega)$ for frequencies below. Also, it is seen that $|G_{p2}(j\omega)G_c(j\omega)| < 1$ for all frequencies above ω_1. Therefore $G_d(j\omega)$ can be represented, as shown in Fig. 6.18, by the line of zero slope and amplitude for frequencies up to ω_1 and the line of -40 db/dec slope above ω_1. The equation of $G_d(s)$ therefore has a quadratic in the denominator with $\omega_n = \omega_1$

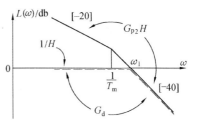

Figure 6.18 Log-magnitude plot of a motor

$$G_d(s) = \frac{\omega_1^2}{s^2 + 2\zeta\omega_1 s + \omega_1^2}$$

Of course, $G_d(j\omega)$ can be obtained algebraically for this simple case from Eq. (6.36), with the result given as

$$G_d(s) = \frac{K_m}{T_m s^2 + s + K_m} = \frac{\dfrac{K_m}{T_m}}{s^2 + \dfrac{1}{T_m}s + \dfrac{K_m}{T_m}}$$

Thus we have $\omega_1 = \omega_n = \sqrt{K_m/T_m}$ and $\zeta = 1/(2\sqrt{K_m T_m})$. Note that the approximate result is basically correct but a detail information, in this case the value of ζ is missing. The approximate phase angle curve can be drawn to correspond to the approximate log-magnitude curve of $G_d(j\omega)$.

Problems

P6.1 A unity feedback system has

$$G(s) = \frac{1350}{s(s+2)(s+30)}$$

(a) Plot the asymptotic log-magnitude plot, and determine the gain crossover frequency and phase margin.

(b) A phase-lead compensator has the transfer function

$$G_c(s) = \frac{0.25s + 1}{0.025s + 1}$$

Repeat (a) for the compensated system, and estimate the maximum overshoot and setting time using the empirical formulas.

P6.2 A unity feedback control system has the open-loop transfer function

$$G(s) = \frac{K}{s(s+1)(0.25s+1)}$$

(a) Design a cascade compensator so that the velocity error constant $K_v \geq 5$ rad/s and the phase margin $\gamma \geq 45°$.

(b) Redesign a cascade compensator if it also is desired that the gain crossover frequency $\omega_c \geq 2$ rad/s.

P6.3 A unity feedback control system has the open-loop transfer function

$$G(s) = \frac{K}{s(0.2s+1)(0.05s+1)}$$

Design a cascade compensator so that the velocity error constant $K_v \geq 12$ rad/s, the percent overshoot $\sigma_p \leq 25\%$, and the setting time $t_s \leq 1(s)$.

P6.4 A unity feedback control system has the open-loop transfer function

$$G(s) = \frac{K}{s(0.25s+1)(0.05s+1)}$$

where K is set equal to 10 in order to achieve a specified $K_v = 3.33$ rad/s. Design a cascade compensator so that the relative resonant peak $M_r \leq 1.4$, and the resonance frequency $\omega_r \geq 10$ rad/s.

P6.5 A unity feedback control system has a plant

$$G(s) = \frac{20}{s(0.1s+1)(0.05s+1)}$$

Select a compensator $G_c(s)$ so that the phase margin is at least $75°$. Use a two-stage lead compensator

$$G_c(s) = \frac{k(\alpha_1 T_1 s + 1)(\alpha_2 T_2 s + 1)}{(T_1 s + 1)(T_2 s + 1)}$$

It is required that the error for a ramp input is 0.5% of the magnitude of the ramp input.

P6.6 A unity feedback control system has the open-loop transfer function

$$G(s) = \frac{40}{s(0.2s+1)(0.0625s+1)}$$

(a) Design a phase-lead compensator so that the phase margin γ is equal to $30°$ and the gain margin is 10 ~ 20 db.

(b) Design a phase-lag compensator if it is desired that the phase margin γ is equal to $50°$ and the gain margin is 30 ~ 40 db.

P6.7 A position control system has the open-loop transfer function

$$G(s) = \frac{3}{s(s+1)(0.5s+1)}$$

keeping the open-loop gain unchanged, determine a phase-lag compensator that will provider a phase margin of 45°.

P6.8 A unity feedback control system has a plant transfer function

$$G(s) = \frac{40}{s(s+2)}$$

It is desired to attain a steady-state error to a ramp input, $r(t) = A\,t$, of less than $0.05A$ and a phase margin of 30°. It is also desired to have a crossover frequency, ω_c, of 10 rad/s. Determine whether a lead or lag compensator is required.

P6.9 A control system is shown in Fig. P6.9. Determine the gain of preamplifier and design a phase-lag compensator so that the velocity error constant $K_v \geqslant 4s^{-1}$ and the relative resonant peak $M_r = 1.4$.

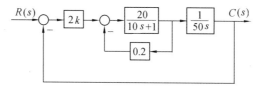

Figure P6.9

P6.10 A unity feedback control system has a plant

$$G(s) = \frac{1}{s(s+1)(0.5s+1)}$$

The compensator is selected as a PI controller so that the steady-state error for a step input is equal to zero. We then have

$$G_c(s) = k_1 + \frac{k_2}{s} = \frac{k_1 s + k_2}{s}$$

Determine a suitable $G_c(s)$ so that (a) the percent overshoot is 5% or less; (b) the setting time (2% criterion) is less than 6 seconds; (c) the velocity error constant K_v is greater than 0.9; and (d) the peak time for a step input is minimized.

P6.11 Consider a unity feedback control system with

$$G(s) = \frac{20}{s(s+2)(s+3)}$$

We wish to add a lead-lag compensator

$$G_c(s) = \frac{(s+0.15)(s+0.7)}{(s+0.015)(s+7)}$$

Show that the phase margin of the compensated system is 75° and the gain margin is 24 db.

P6.12 A unity feedback control system has a plant

$$G(s) = \frac{160}{s^2}$$

Select a lead-lag compensator so that the percent overshoot is less than 5% and the setting time (2% criterion) is less than second. It also is desired that the acceleration constant is greater than 7500.

P6.13 A feedback system has the open-loop transfer function

$$G(s) = \frac{10(0.316s+1)}{(0.1s+1)(0.01s+1)}$$

determine the cascade compensator $G_c(s)$ required to make the system meet the desired open-loop frequency response characteristics shown in Fig. P6.13

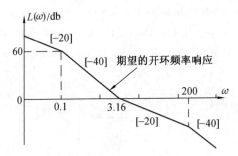

Figure P6.13

P6.14 Consider a control system with a feedback-compensation as shown in Fig. P6.14, where

$$G(s) = \frac{100}{s(1.1s+1)(0.025s+1)}, G_c(s) = 0.25s$$

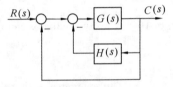

Figure P6.14

(a) Plot the asymptotic log-magnitude curves for the uncompensated system and the compensated system.
(b) Find the equivalent open-loop transfer function. (c) Estimate the phase margin of the compensated system.

P6.15 A control system with minimum phase has the open-loop log-magnitude plot as shown in Fig. P6.15. It is desired to eliminate the resonant peak at the corner frequency 20 rad/s by a feedback-compensation as shown in Fig. P6.15. Determine the transfer function of the feedback compensator.

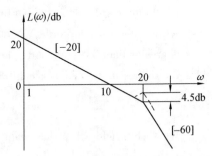

Figure P6.15

Chapter 7 Nonlinear System Analysis

7.1 Introduction

It should be stated again that all physical systems are inherently nonlinear. An essentially linear system is linear only over certain ranges of operation. As has been stated previously, we cannot model a physical system exactly. Usually, by increasing the order of the linear model, the accuracy of the model is increased. However, the point is reached at which adding order to the model will not significantly improve the model. For those cases in which the model accuracy is still not sufficient, it will be necessary to add nonlinearities.

Given a linear system, we can always determine system stability, for example, by using the Routh-Hurwits criterion, the Nyquist criterion, or other techniques. No such statement can be made concerning nonlinear systems. There is no general nonlinear stability analysis technique that will always determine stability for a given nonlinear system under consideration. Instead, special techniques are developed, with no one technique applicable to all nonlinear systems. For many systems containing several nonlinearities, stability may be determined only through simulation.

7.1.1 Definition for Nonlinear System

A nonlinear system is a system to which the principle of superposition does not apply. Hence, nonlinear differential equations are used to describe the input-output relationships of nonlinear systems. Some examples of nonlinear system equation are:

$$\dot{c}(t) + c^2(t) = r(t)$$
$$\dot{c}(t) + \sin[c(t)] = r(t)$$
$$\ddot{c}(t) + 3\dot{c}(t) + 2\dot{c}(t)c(t) = r(t)$$

The first equation is nonlinear because of the squared variable; the second, because of the sine of a variable; and the third, because of the product of a variable with its derivative. In general, the Laplace transform cannot be used to solve nonlinear differential equations of any type.

Although the analysis and design of linear control systems have been well developed, their counterparts for nonlinear systems are usually quite complex. In many cases it may be difficult to write the differential equations describing the input-output relationship of nonlinear systems. Hence, block diagrams are often used in the study of nonlinear systems, and the nonlinearities can be obtained through

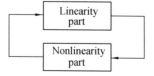

Figure 7.1 Nonlinear system with basic configuration

experimentation. Moreover, using block simplification, the linearity part and nonlinearity part can be reduced respectively. A system is called a nonlinear system with basic configuration if its linearity part and nonlinearity part can be separated as shown in Fig. 7.1. Some important methods for nonlinear systems are applicable only to those with basic configuration. In a block diagram describing nonlinear systems, the linearity part is usually represented by the transfer function.

Nonlinearities can be classified as dynamic and static. The nonlinear energy-stored device, which input and output may be related through a differential equation, is called a dynamic nonlinearity. On the other hand, the nonlinear device that does not involve a differential equation is called a static nonlinearity. The static nonlinearity is usually, for convenience, represented by a characteristic curve. Fig. 7.2 shows a block diagram of a nonlinear system, in which a nonlinear amplifier is shown as a curve representing a saturation characteristic.

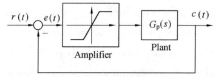

Figure 7.2 System with static nonlinearity

7.1.2 Some Properties of Nonlinear Systems

Since nonlinear systems do not obey the principle of superposition, they have many distinctive properties different from that of linear systems.

① The response of a nonlinear system will not be linear exponential or time weighted exponential as is the case for a linear system. The response shape to same type of inputs may be markedly different, in some cases it may converge to an equilibrium point and in other cases it may become unbounded; in some cases it may converge to an equilibrium point with decayed oscillation and in other cases without oscillation; moreover, it may converge to an equilibrium point with different frequencies. Fig. 7.3 shows step responses of a nonlinear system for different input amplitudes.

② Some nonlinear systems may be stable for certain sets of initial conditions, but may be unstable for different sets of initial conditions.

For example, consider a first-order nonlinear system with differential equation

$$\dot{x}(t) = -x(1-x)$$

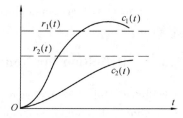

Figure 7.3 Step responses of a nonlinear system for different input amplitudes

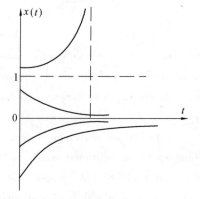

Figure 7.4 Response of a nonlinear system for different initial conditions

With the initial condition $x(0) = x_0$, the solution of equation is

$$x(t) = \frac{x_0 e^{-t}}{1 - x_0 + x_0 e^{-t}}$$

Fig. 7.4 plots the response curves for different initial conditions. It can be seen from the figure that the system stability is related to the initial conditions.

③ Multiple equilibrium points may exist for no input, so that when a steady state is attained it may not be same for all initial conditions. As shown in Fig. 7.4, the system has two equilibrium points, $x = 0$ is a stable equilibrium point and $x = 1$ is an unstable equilibrium point.

④ The response of a nonlinear system may converge to a periodic oscillation and form a closed trajectory that is called limit cycle in the nonlinear systems analysis. A periodic oscillation in a linear time-invariant system is sinusoidal and its amplitude is a function of both the amplitude of the system excitation and the initial conditions, as shown in Fig. 7.5(a). In a nonlinear system the same periodic oscillation or limit cycle can result from many different initial conditions, as shown in Fig. 7.5(b). Moreover, in general, limit cycles are nonsinusoidal.

⑤ A nonlinear system with a periodic input may exhibit a periodic output whose frequency is either a subharmonic or a harmonic of the input frequency.

⑥ In some cases, for a given amplitude and frequency of an input sinusoidal, a nonlinear system may have more than one stable periodic state; it is thus known as multimode system. A slight change in the input amplitude or frequency can cause the system to change states, a phenomenon referred to as jump resonance. A jump phenomenon is illustrated in Fig. 7.6.

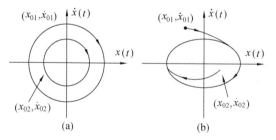
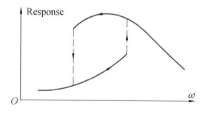

Figure 7.5　　　　　　　　　　Figure 7.6　Jump resonance in a frequency response

7.1.3　Common Nonlinearities

In most control systems we cannot avoid the presence of certain types of nonlinearities. System nonlinearities are often classified as inherent or intentional nonlinearities. Inherent nonlinearities are those that exist in the 'presumably linear' components selected to perform a function. Often the designer would be happier if they did not exist. On the other hand, intentional nonlinearities are those that are deliberately introduced into the system to perform a specific nonlinear function. They have been chosen during the design because of their economic, reliability, performance or other advantages.

1. Saturation

Saturation is one of the most common inherent nonlinearities. Examples are electronic amplifiers that saturate, mechanical stops in systems with translational or rotational motion, such as rudders in aircrafts and ships, and so forth.

The static characteristic of saturation is shown in Fig. 7.7(a). The output x is proportional to the input e only for a limited range of the input. As the magnitude of the input exceeds the range, the output approaches a constant. Although, for a practical saturation nonlinearity, the change from one range to the other is usually gradual, it is often sufficiently accurate in most cases to approximate the curve by a set of straight lines, as shown.

Considering the gain of a saturated element, it can be seen, as shown in Fig. 7.7(b), the equivalent gain is decreased as the input is increased. Consequently, the equivalent open-loop gain of the overall control system is decreased.

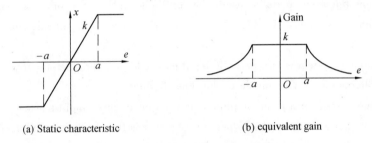

(a) Static characteristic   (b) equivalent gain

Figure 7.7   Saturation nonlinearity

2. Dead zone

In many physical devices the output is zero until the magnitude of the input exceeds a certain value. For example, while developing the mathematical model for a dc servomotor, we had assumed that any voltage applied to the armature windings would cause the armature to rotate, if the field current is maintained constant. In practice, rotation will result only if the torque produced by the motor is sufficient to overcome the static friction. As a result, if we plot the relationship between the steady-state angular velocity and the applied voltage, we get the

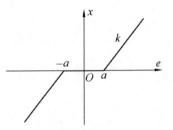

Figure 7.8   Static characteristic of dead-zone

characteristic shown in Fig. 7.8, which exhibits the dead-zone phenomenon. Many other devices exhibit similar characteristics.

Generally, the dead-zone nonlinearity will deduce the sensitivity of control system, especially the sensitivity of the error-sensing devices. Hence, the dead-zone is also called non-sensitive zone.

3. Relay

A relay is often used in control systems as it provides a large-power amplification rather inexpensively. The characteristic of an idea relay is shown in Fig. 7.9(a), where a change in the sign of the input causes an abrupt change in the output.

In the practical realization of a relay, the current in an iron-cored coil exerts magnetic force to move an arm so as to make the contact in one direction or another. Since, in practice, the current must exceed a certain value before the arm can be moved, an actual relay will exhibit a dead-zone as shown in Fig. 7.9(b). Furthermore, due to the phenomenon of magnetic hysteresis, a large current is needed to close the relay than the current at which the contacts open. Hence, in practice, most relays exhibit a dead-zone characteristic as well as hysteresis, as shown in Fig. 7.9(c). Another case is a relay with pure hysteresis as shown in Fig. 7.9(d).

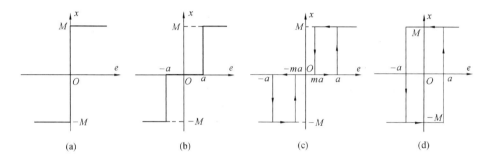

Figure 7.9 Static characteristics of relays

4. Backlash

An extremely complex form of nonlinearity to model is the backlash that exists in gears, because gears cannot mesh perfectly. When the drive gear changes direction, the play in the gears must be taken up before the load gear can begin moving again.

This nonlinearity results in time delay and the resultant hysteresis. Fig. 7.10 shows the static characteristic of backlash, in which the play width is 2ε. From the viewpoint of dynamics, the effect of a backlash is equivalent to introduce a time-delay element. On the other hand, from the viewpoint of steady-state performance, the effect of a backlash is equivalent to introduce a dead-zone.

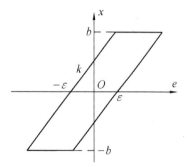

Figure 7.10 Static characteristic

7.1.4 Methods of Studying Nonlinear Systems

The performance of nonlinear systems can be studied by hardware, simulation, and/or analysis. Strictly speaking, the latter two cases consider the performance of the system model. Measurement and testing of the actual system, which certainly will provide the most reliable information on how it will behave, is normally the final stage of design. Building a system and test it without analysis or simulation can be both dangerous and expensive.

An important tool for studying nonlinear control systems is simulation. With computers one can undertake exhaustive simulation of complex systems. However, simulation without preliminary analysis can often be wasteful as well as ineffective, since each computer run merely provides a solution for the given parameters and inputs. Theory is needed to provide a frame of reference for understanding what is happening, and to show how changes can be made to improve performance. The study can then be completed by simulation on a computer in an intelligent manner, since it will be based on the preliminary analysis.

Although a number of methods have been employed for the study of nonlinear systems, we shall consider here two well-known methods. The first one is the describing function method, and another popular approach is the phase-plane method, which provides a graphical technique for obtaining the solution of nonlinear differential equations of the second-order system. Although the latter is not applicable to higher-order systems, quite often such systems can be represented with approximate second-order models.

7.2 Describing Function Method

Describing function method is an attempt to extend the familiar frequency response approach to nonlinear systems. Although this method is based on a simplifying approximation and is hence inexact, its main advantage is its basic simplicity. It often gives the designer an appreciation of the system behavior and indicates the modification that should be made to the system for satisfactory performance.

7.2.1 Definition of Describing Function

The describing function, as considered here, is applicable to the systems with basic configuration as shown in Fig. 7.11. Generally, the linear part, $G(s)$, can also include the transfer functions of the compensator and sensor.

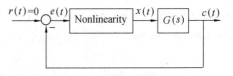

Figure 7.11 Nonlinear system with basic configuration

In order to develop the describing function method, let us consider the case that, in Fig. 7.11, the system input is zero and the input of nonlinearity is sinusoidal, i.e.,

$$e(t) = A\sin \omega t \qquad (7.1)$$

[A system with a zero input is called a free system. If, in addition, the system is time-invariant, the system is called an autonomous system.] Then, in the steady state, its output $x(t)$ is periodic but not, in general, sinusoidal. Fourier analysis of the nonsinusoidal waveform leads to a fundamental component and some harmonics as follows

$$x(t) = A_0 + \sum_{n=1}^{\infty}(A_n \cos n\omega t + B_n \sin n\omega t) = A_0 + \sum_{n=1}^{\infty} X_n \sin(n\omega t + \varphi_n) \qquad (7.2)$$

where

$$A_0 = \frac{1}{2\pi}\int_0^{2\pi} x(t)\,\mathrm{d}(\omega t) \tag{7.3}$$

$$A_n = \frac{1}{\pi}\int_0^{2\pi} x(t)\cos n\omega t\,\mathrm{d}(\omega t) \tag{7.4}$$

$$B_n = \frac{1}{\pi}\int_0^{2\pi} x(t)\sin n\omega t\,\mathrm{d}(\omega t) \tag{7.5}$$

$$X_n = \sqrt{A_n^2 + B_n^2}, \quad \varphi_n = \arctan\frac{A_n}{B_n} \tag{7.6}$$

In this chapter we restrict ourselves to consider only those nonlinearities that have odd symmetrical static characteristics, i. e. we will consider only the case of $A_0 = 0$.

If we further assume that $G(s)$ in Fig. 7.11 is of the low-pass nature with respect to the harmonics, i. e., $|G(j\omega)|$ is small for all other components of $x(t)$ compared to its value for the fundamental component. Thus the output $c(t)$ can be expressed as

$$c(t) = A_c \sin(\omega t + \theta) \tag{7.7}$$

This assumption is the foundation of describing function analysis. The harmonics in $x(t)$ are then not important, since they have very little effect on $c(t)$. The harmonics fed back to the nonlinearity input can be neglected, and $x(t)$ can be approximated as

$$x(t) \approx x_1(t) = A_1 \cos \omega t + B_1 \sin \omega t = X_1 \sin(\omega t + \varphi_1) \tag{7.8}$$

To facilitate the analysis of a nonlinear system as described above, the describing function of a nonlinear element is defined as the complex ratio of the fundamental component of its output to the sinusoidal input. From Eq. (7.8) we see that $x(t)$ can be approximated as a sinusoid of the same frequency as $e(t)$ but not the same amplitude and phase; thus the nonlinearity can be replaced with a complex gain of

$$N(A,\omega) = \frac{B_1 + jA_1}{A} \tag{7.9}$$

This equivalent gain is called the describing function. The describing function in general is a function of both the amplitude and the frequency of the input sinusoid. In this chapter, we only consider the static nonlinearities. In this case, the describing functions are only amplitude depended and may be denoted as $N(A)$.

Furthermore, for the single-value and odd symmetrical nonlinearities, we have

$$A_1 = 0, \quad \varphi_1 = 0$$

and

$$N(A) = \frac{B_1}{A} \tag{7.10}$$

Note the assumption of the definition of the describing function. This definition has meaning only if

① the input to the nonlinearity is a sinusoid;

② the nonlinearities have odd symmetrical static characteristics, and

③ the linear part following the nonlinearity is sufficiently low-pass so as to attenuate all harmonics to such values as to be neglected.

7.2.2 Describing Function of Common Nonlinearites

1. Saturation

The ideal saturation characteristic and its output waveform are shown in Fig. 7.12, where the static nonlinearity of saturation is given by

$$x(t) = \begin{cases} kA\sin \omega t & 0 \leqslant \omega t \leqslant \alpha_1 \\ ka = b & \alpha_1 < \omega t < (\pi - \alpha_1) \\ kA\sin \omega t & (\pi - \alpha_1) \leqslant \omega t \leqslant \pi \end{cases} \quad (7.11)$$

where $\alpha_1 = \arcsin \dfrac{a}{A}$.

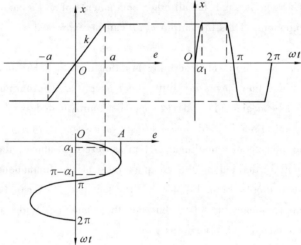

Figure 7.12 Saturation characteristic and its output waveform

Since the ideal saturation is a single valued and odd symmetrical nonlinearity, we have

$$A_1 = 0$$

$$B_1 = \frac{1}{\pi}\int_0^{2\pi} x(t)\sin \omega t\, d(\omega t) = \frac{4}{\pi}\int_0^{\pi/2} x(t)\sin \omega t\, d(\omega t) =$$

$$\frac{4}{\pi}\left[\int_0^{\alpha_1} kA\sin \omega t \sin \omega t\, d(\omega t) + \int_{\alpha_1}^{\pi/2} ka \sin \omega t\, d(\omega t)\right] =$$

$$\frac{4kA}{\pi}\left\{\left[\frac{1}{2}\omega t - \frac{1}{4}\sin 2\omega t\right]_0^{\alpha_1} + \frac{a}{A}[-\cos \omega t]_{\alpha_1}^{\pi/2}\right\} =$$

$$\frac{4kA}{\pi}\left(\frac{1}{2}\alpha_1 - \frac{1}{4}\sin 2\alpha_1 + \frac{a}{A}\cos \alpha_1\right) =$$

$$\frac{2kA}{\pi}\left(\arcsin \frac{a}{A} + \frac{a}{A}\sqrt{1 - \left(\frac{a}{A}\right)^2}\right)$$

Thus, the describing function of saturation characteristic is

$$N(A) = \frac{2k}{\pi}\left(\arcsin \frac{a}{A} + \frac{a}{A}\sqrt{1 - \left(\frac{a}{A}\right)^2}\right), \quad A \geqslant a \quad (7.12)$$

2. Dead-zone

As shown in Fig. 7.13, the static nonlinearity of dead-zone is given by

$$x(t) = \begin{cases} 0 & 0 \leqslant \omega t \leqslant \alpha_1 \\ kA\sin(\omega t - a) & \alpha_1 < \omega t < \pi - \alpha_1 \\ 0 & \pi - \alpha_1 \leqslant \omega t \leqslant \pi \end{cases} \quad (7.13)$$

where $\alpha_1 = \arcsin \dfrac{a}{A}$.

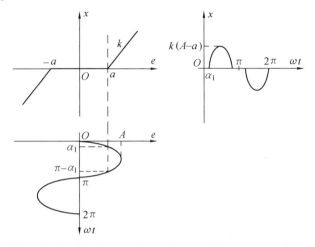

Figure 7.13 Dead-zone characteristic and its output waveform

The ideal dead-zone is a single valued and odd symmetrical nonlinearity, hence we have

$$A_1 = 0$$

$$B_1 = \frac{1}{\pi}\int_0^{2\pi} x(t)\sin \omega t \, d(\omega t) = \frac{4}{\pi}\int_0^{\pi/2} x(t)\sin \omega t \, d(\omega t) =$$

$$\frac{4}{\pi}\left[\int_{\alpha_1}^{\pi/2} kA(\sin \omega t - a)\sin \omega t \, d(\omega t)\right] =$$

$$\frac{2kA}{\pi}\left(\frac{\pi}{2} - \arcsin\frac{a}{A} - \frac{a}{A}\sqrt{1 - \left(\frac{a}{A}\right)^2}\right)$$

Thus, the describing function of dead-zone characteristic is

$$N(A) = \frac{2k}{\pi}\left(\frac{\pi}{2} - \arcsin\frac{a}{A} - \frac{a}{A}\sqrt{1 - \left(\frac{a}{A}\right)^2}\right), \quad A \geqslant a \quad (7.14)$$

3. Relay

Consider the relay nonlinearity shown in Fig. 7.14. The static nonlinearity is defined by

$$x(t) = \begin{cases} 0 & 0 \leqslant \omega t \leqslant \alpha_1 \\ M & \alpha_1 < \omega t < \pi - \alpha_2 \\ 0 & \pi - \alpha_2 \leqslant \omega t \leqslant \pi \end{cases} \quad (7.15)$$

where $\alpha_1 = \arcsin \dfrac{e_0}{A}$ and $\alpha_2 = \pi - \arcsin \dfrac{me_0}{A}$.

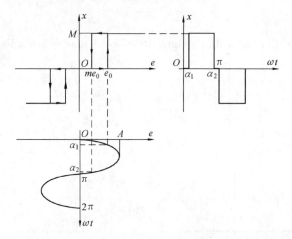

Figure 7.14 Relay characteristic and its output waveform

For the describing function we have

$$A_1 = \frac{1}{\pi}\int_0^{2\pi} x(t)\cos\omega t\, d(\omega t) = \frac{2}{\pi}\int_{\alpha_1}^{\alpha_2} M\cos\omega t\, d(\omega t) = \frac{2Me_0}{\pi A}(m-1)$$

$$B_1 = \frac{1}{\pi}\int_0^{2\pi} x(t)\sin\omega t\, d(\omega t) = \frac{2}{\pi}\int_{\alpha_1}^{\alpha_2} M\sin\omega t\, d(\omega t) =$$

$$\frac{2M}{\pi}\left(\sqrt{1-\left(\frac{me_0}{A}\right)^2} + \sqrt{1-\left(\frac{e_0}{A}\right)^2}\right)$$

Hence, we get

$$N(A) = \frac{2M}{\pi A}\left(\sqrt{1-\left(\frac{me_0}{A}\right)^2} + \sqrt{1-\left(\frac{e_0}{A}\right)^2}\right) + j\frac{2Me_0}{\pi A^2}(m-1), \quad A \geqslant e_0 \qquad (7.16)$$

Note that in this case the describing function is complex, and this do introduce phase shift.

Particularly, in the case of $e_0 = 0$, the described function for the ideal relay is given by

$$N(A) = \frac{4M}{\pi A}, \quad A \geqslant e_0 \qquad (7.17)$$

In the case of $m = 1$, the describing function for the relay with dead-zone is given by

$$N(A) = \frac{4M}{\pi A}\sqrt{1-\left(\frac{e_0}{A}\right)^2}, \quad A \geqslant e_0 \qquad (7.18)$$

As for the case of $m = -1$, the describing function for the relay with hysteresis is given by

$$N(A) = \frac{4M}{\pi A}\sqrt{1-\left(\frac{me_0}{A}\right)^2} + j\frac{2Me_0}{\pi A^2}(m-1), \quad A \geqslant e_0 \qquad (7.19)$$

4. Combined nonlinearities

In many systems nonlinearities occur at more than one position in the system. If the combination of nonlinear and linear elements is such that the input to all the nonlinearities will be approximately sinusoidal, for the situation under investigation each of these nonlinearities can be replaced by its describing function.

(1) Nonlinearities in series

The usual procedure for analyzing a series combination of nonlinearities is to obtain the overall equivalent nonlinearity, as illustrated in Fig. 7.15, and then evaluate the describing function of the new equivalent nonlinearity. Note that the describing function of a series combination is not equal to the product of the describing functions for the separate nonlinearities, because the effect of harmonics, produced by the first nonlinearity, on the second nonlinearity can not be ignored, and both nonlinearities can not be replaced by the describing functions.

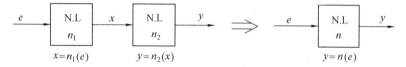

Figure 7.15 Series nonlinear combination

For example, consider combination of the dead-zone and saturation in series as shown in Fig. 7.16. The equivalent nonlinearity is a saturation characteristic with dead-zone.

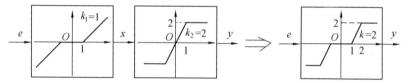

Figure 7.16 Example of series nonlinear combination

(2) Nonlinearities in parallel

A parallel combination of two nonlinearities is shown in Fig. 7.17. Obviously, since both nonlinearities have the same input $e(t) = \sin \omega t$, the overall output is the sum of the separated nonlinearities. Hence, if the linear part following the nonlinearity is sufficiently low-pass, the overall describing function is

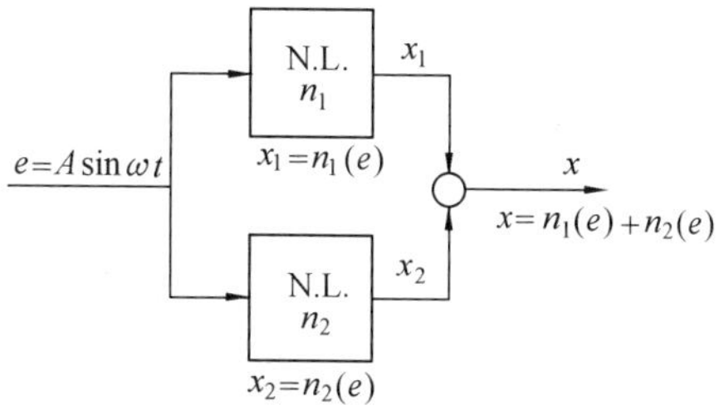

$$N(A) = N_1(A) + N_2(A) \qquad (7.20)$$

Figure 7.17 Parallel nonlinear combination

Although the above configuration of nonlinearities occurs infrequently in practical systems, the concepts are extremely useful in the synthesis of nonlinearities for simulation purpose and in evaluating describing functions. For example the describing function of dead-zone can be obtained to illustrate the procedure for obtaining describing functions.

We see from Fig. 7.18 that in this case the dead-zone characteristic can be obtained with

$n_1(e)$: linear gain of unit magnitude, and

$-n_2(e)$: saturation of unit gain and limit level a.

Therefore, the describing function of the dead-zone is

$$N(A) = 1 - \frac{2}{\pi}\left(\arcsin\frac{a}{A} + \frac{a}{A}\sqrt{1-\left(\frac{a}{A}\right)^2}\right) = \frac{2}{\pi}\left(\frac{\pi}{2} - \arcsin\frac{a}{A} - \frac{a}{A}\sqrt{1-\left(\frac{a}{A}\right)^2}\right)$$

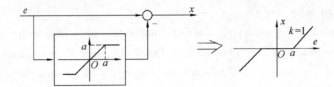

Figure 7.18 Dead-zone synthesis

The describing functions just derived, plus some other commonly used ones, are given in Table 7.1.

Table 7.1 **Nonlinearities and their describing functions.**

Nonlinearity	Describing function $N(A)$
(saturation)	$\dfrac{2k}{\pi}\left(\arcsin\dfrac{a}{A} + \dfrac{a}{A}\sqrt{1-\left(\dfrac{a}{A}\right)^2}\right), A \geq a$
(dead-zone)	$\dfrac{2k}{\pi}\left(\dfrac{\pi}{2} - \arcsin\dfrac{a}{A} - \dfrac{a}{A}\sqrt{1-\left(\dfrac{a}{A}\right)^2}\right), A \geq a$
(ideal relay)	$\dfrac{4M}{\pi A}$
(relay with dead-zone)	$\dfrac{4M}{\pi A}\sqrt{1-\left(\dfrac{e_0}{A}\right)^2}, A \geq a$
(relay with hysteresis)	$\dfrac{4M}{\pi A}\sqrt{1-\left(\dfrac{e_0}{A}\right)^2} - j\dfrac{4Me_0}{\pi A^2}, A \geq a$
(relay with dead-zone and hysteresis)	$\dfrac{2M}{\pi A}\left(\sqrt{1-\left(\dfrac{me_0}{A}\right)^2} + \sqrt{1-\left(\dfrac{e_0}{A}\right)^2}\right) + j\dfrac{2Me_0}{\pi A^2}(m-1), A \geq a$

Table 7.1

Nonlinearity	Describing function $N(A)$
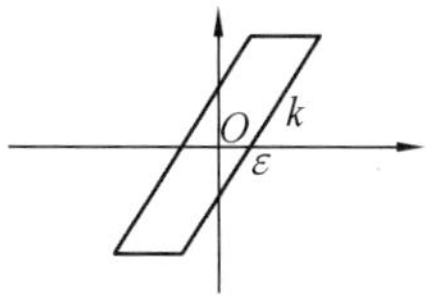 (hysteresis with slope k, width ε)	$\dfrac{k}{\pi}\left(\dfrac{\pi}{2} + \arcsin\left(1 - \dfrac{2\varepsilon}{A}\right) + 2\left(1 - \dfrac{2\varepsilon}{A}\right)\sqrt{\dfrac{\varepsilon}{A}\left(1 - \dfrac{2\varepsilon}{A}\right)}\right) +$ $j\dfrac{4k\varepsilon}{\pi A}\left(\dfrac{\varepsilon}{A} - 1\right), A \geqslant \varepsilon$
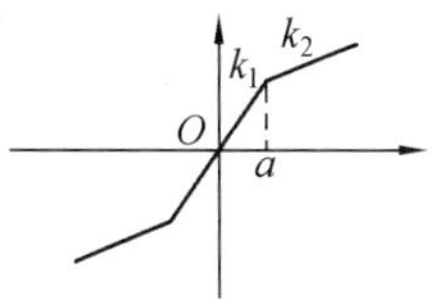 (piecewise linear with slopes k_1, k_2, break at a)	$k_2 + \dfrac{2}{\pi}(k_1 - k_2)\left[\arcsin\dfrac{a}{A} + \dfrac{a}{A}\sqrt{1 - \left(\dfrac{a}{A}\right)^2}\right], A \geqslant a$

7.2.3 Stability Analysis of Nonlinear Systems

Consider the system shown in Fig. 7.11. Assuming that the describing function analysis is applicable to this system, we get the equivalent system of Fig. 7.19, where $N(A)$ is the describing function of the nonlinearity, and $G(s)$ is the transfer function of the linearity part.

The closed-loop frequency response of the system is given by

$$\Phi(j\omega) = \dfrac{N(A)G(j\omega)}{1+N(A)G(j\omega)} \quad (7.21)$$

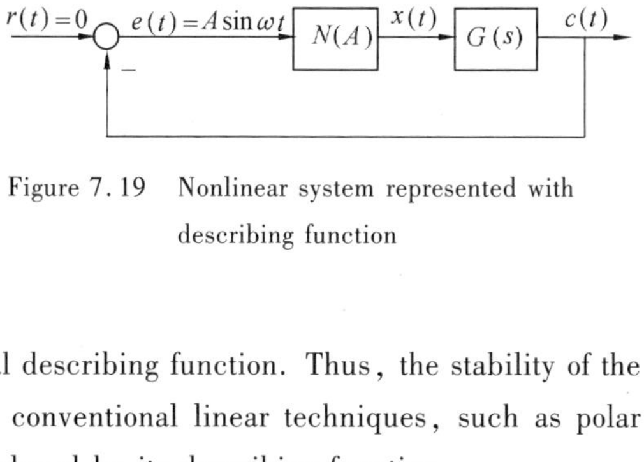

and the corresponding characteristic equation is

$$1 + N(A)G(j\omega) = 0$$

i.e.,

$$G(j\omega) = -\dfrac{1}{N(A)} \quad (7.22)$$

Figure 7.19 Nonlinear system represented with describing function

where $-1/N(A)$ is called the negative reciprocal describing function. Thus, the stability of the nonlinear system can be investigated using any conventional linear techniques, such as polar plot and Nichols chart, with the nonlinearity replaced by its describing function.

When the frequency response $G(j\omega)$ and the negative reciprocal describing function $-1/N(A)$ are plotted together on same complex plane, so called the Nyquist plane, any intersection, as shown in Fig. 7.20(a), will be a periodic solution of Eq. (7.22), which means that there exists an oscillation corresponding to the intersection in the control system. Since in general $G(j\omega)$ is a complex function of ω and $N(A)$ is a complex function of A, the solution of Eq. (7.22) gives both the frequency and amplitude of the oscillation. The oscillation may be stable or unstable, and the stable oscillation is called the self-oscillation due to zero-input. The stability of the oscillation is discussed further later in this section.

On the other hand, if no intersection of the locus $G(j\omega)$ and $N(A)$ exists, the stability of the system is assessed using the normal Nyquist criterion with respect to any point on the $-1/N(A)$ locus rather than the point $(-1, j0)$. Assuming that the linearity part of the system is

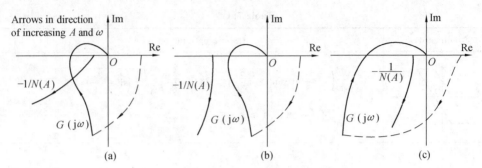

Figure 7.20 Illustration of stability

minimum phase-shift leads to that the nonlinear system is stable if $-1/N(A)$ locus is not encircled by $G(j\omega)$ plot as shown in Fig. 7.20(b), and the nonlinear system is unstable if $-1/N(A)$ locus is encircled by $G(j\omega)$ plot as shown in Fig. 7.20(c).

In many control systems, the describing function characterizes the effect of the nonlinearities quite accurately. The reason is that the plant, especially the power-output device, of a control system usually attenuates the higher harmonics of a periodic waveform strongly. Hence the effect of higher harmonics is small. On the other hand, it should be remembered, however, that the describing function method is an approximated technique. It is sometimes difficult to determine whether or not one can legitimately neglect the higher harmonics, for example in the case of that $-1/N(A)$ locus and $G(j\omega)$ plot are almost tangent at their intersection.

7.2.4 Stability of Limit Cycle

When a periodic solution to Eq. (7.22) is obtained does it correspond to a stable or unstable oscillation condition? If following a small change in amplitude the oscillation returns to the original solution state, the oscillation is said stable, whereas if the amplitude continue to move away from its equilibrium state, the oscillation is unstable. Unstable periodic movement is difficult to observe, since any disturbance in the system will prevent the unstable periodic movement from maintaining a steady-state oscillation, and the unstable periodic movement will converge, or diverge, or transfer to another periodic movement.

The stability of a periodic movement can be investigated using the describing function approach. Consider, for example, the system shown in Fig. 7.21. It is seen that two intersections occur between the locus $G(j\omega)$ and $-1/N(A)$. A periodic movement of $A_a \sin \omega_a t$ is predicted for the first intersection labeled a. However, note that an increase in amplitude, for example the operating point is moved to point c, causes the system to become unstable and thus the amplitude continues to increase; a decrease in amplitude, for example the

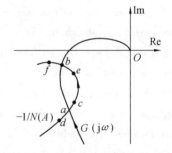

Figure 7.21 Illustration of periodic movement stability

operating point is moved to point d, causes the system to become stable and thus the amplitude continues to decrease. Hence, the periodic movement is unstable. As for the periodic movement of $A_b \sin \omega_b t$ predicted for the second intersection labeled b, an increase in amplitude, for example the operating point is moved to point f, causes the system to become stable and thus the amplitude will decrease; a decrease in amplitude, for example the operating point is moved to point e, causes the system to become unstable and thus the amplitude will increase. Hence, the periodic movement is stable.

A similar graphical criterion is more convenient for determining the stability of periodic movements. The Nyquist plane can be divided into two parts: stable region and unstable region, as shown in Fig. 7.22. For a stable limit cycle, along with an increase of amplitude A the $-1/N(A)$ locus should be moved from the unstable region into the stable region, as the intersection b in Fig. 7.22. As for an unstable periodic movement, along with an increase in amplitude A the $-1/N(A)$ locus should be moved from the stable region into the unstable region, as

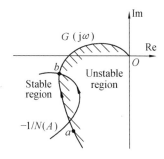

Figure 7.22 Stable region and unstable region in Nyquist plane

the intersection a in Fig. 7.22. Moreover, in the case of no intersection between $G(j\omega)$ and $-1/N(A)$, assuming that $G(s)$ is minimum phase-shift leads to that the nonlinear system is stable if $-1/N(A)$ locus is located in the stable region; the nonlinear system is unstable if $-1/N(A)$ locus is located in the unstable region.

Example 7.1 Analyze the stability of the system shown in Fig. 7.23(a).

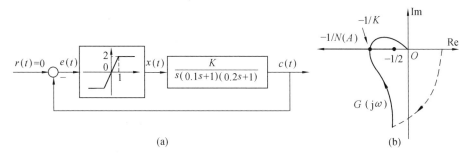

Figure 7.23 System and Nyquist plane for the Example 7.1

Solution For this system, from Table 7.1, the describing function of the saturation characteristic is

$$N(A) = \frac{2k}{\pi}\left(\arcsin\frac{a}{A} + \frac{a}{A}\sqrt{1-\left(\frac{a}{A}\right)^2}\right), \quad A \geqslant a$$

Letting $u = a/A$ and taking derivative of $N(u) = N(a/A)$ with respect to u yields

$$\frac{dN(u)}{du} = \frac{2k}{\pi}\left(\frac{1}{\sqrt{1-u^2}} + \sqrt{1-u^2} - \frac{u^2}{\sqrt{1-u^2}}\right) = \frac{4k}{\pi}\sqrt{1-u^2}$$

Note that $u = a/A < 1$ when $A > a$, hence $N(u)$ is an increasing function with respect to u, i. e., $N(A)$ is a decreasing function with respect to A. For $A = 1$, which is the limiting value for linear operation,

$$-1/N(A) = -1/2$$

and for $A \to \infty$,

$$-1/N(A) = -\infty$$

The locus of $-1/N(A)$ is then shown in Fig. 7.23(b). Also shown in this figure is a plot of $G(j\omega)$. By inspection, there is an intersection between $G(j\omega)$ and $-1/N(A)$ locus, and this intersection corresponds to a stable periodic movement, i. e., there exists a self-oscillation for this system.

Letting the phase-crossing frequency of the linear part be ω_g yields

$$-90° - \arcsin 0.1\omega_g - \arcsin 0.2\omega_g = -180°$$

i. e.,

$$\omega_g = \sqrt{50} \text{ rad/s}$$

Noting

$$|G(j\omega_g)| = \frac{K}{\sqrt{50} \cdot \sqrt{0.5+1} \cdot \sqrt{2+1}} = \frac{K}{15}$$

we see that a self-oscillation is predicted if $K > 7.5$.

Suppose, for example, that K is given a value of 15. Then, the frequency of the predicted self-oscillation is $\omega = \sqrt{50}$ rad/s and the amplitude is $A = 2.47$.

Example 7.2 Analyze the stability of the system shown in Fig. 7.24(a), where the nonlinearity is a dead zone.

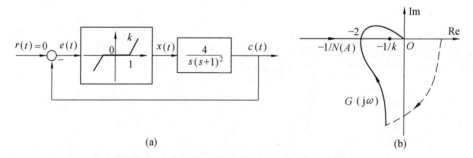

Figure 7.24 System and Nyquist plane for the Example 7.2

Solution From Table 7.1, the describing function of the nonlinearity is

$$N(A) = \frac{2k}{\pi}\left(\frac{\pi}{2} - \arcsin\frac{1}{A} - \frac{1}{A}\sqrt{1-\left(\frac{1}{A}\right)^2}\right), \quad A \geq 1$$

Comparing the describing functions of the saturate and dead-zone characteristics, it is easily seen that $N(A)$ is now an increasing function with respect to A. Since

$$\lim_{A \to 1}[-1/N(A)] = -\infty, \quad \lim_{A \to \infty}[-1/N(A)] = -1/k$$

we have $-1/N(A) \leq -1/k$. For k greater than 0.5, the describing function analysis is shown in Fig. 7.24, where the phase crossing frequency of $G(j\omega)$ is $\omega_g = 1$ rad/s and $|G(j\omega_g)| = 2$. The intersection between $-1/N(A)$ and $G(j\omega)$ corresponds to a periodic movement. It can be seen that an increase in A causes the system to become unstable and thus A continues to increase; a decrease in A causes the system to become stable and thus A continues to decrease. Hence the periodic movement is unstable.

This unstable periodic movement can be explained from physical consideration. For small signals with $|e(t)| \leq 1$, no signal transmitted through the dead zone, and the system is certainly stable. For somewhat larger signals, the effective gain of the nonlinearity is small, and the system is still stable. For very large signals, the dead zone effect is negligible, and the nonlinearity appears as a linear gain. For this linear gain, the system is unstable, and the system response will grow without limit.

Example 7.3 Consider the nonlinear system shown in Fig. 7.25(a). Analyze the system stability, using the describing function method, for the following cases:

(a) $G_c(s) = 1$;

(b) $G_c(s) = \dfrac{0.25s+1}{8.3(0.03s+1)}$.

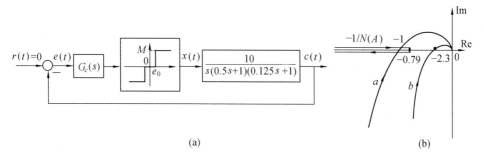

(a) (b)

Figure 7.25 System for the Example 7.3

Solution From Table 7.1, the describing function of the relay with a dead-zone is

$$N(A) = \frac{4M}{\pi A}\sqrt{1-\left(\frac{e_0}{A}\right)^2}, \quad A \geq e_0$$

Letting $u = e_0/A$ and taking derivative of $N(u) = N(e_0/A)$ with respect to u yields

$$\frac{dN(u)}{du} = \frac{4M}{\pi e_0}\left(\sqrt{1-u^2} - \frac{u^2}{\sqrt{1-u^2}}\right) = \frac{4M}{\pi e_0} \cdot \frac{1-2u^2}{\sqrt{1-u^2}}$$

Solving $d[N(u)]/du = 0$ results in

$$u_m = \frac{e_0}{A_m} = \frac{1}{\sqrt{2}}$$

Since $d[N(u)]/du > 0$ for $e_0 \leq A \leq A_m$ and $d[N(u)]/du < 0$ for $A > A_m$, both $N(A)$ and $-1/N(A)$ have the maximum values as $A = A_m$. Thus, we have

$$-\frac{1}{N(A)} = -\frac{\pi e_0}{2M} = -0.785$$

Noting that

$$\lim_{A \to e_0} -\frac{1}{N(A)} = -\infty, \quad \lim_{A \to \infty} -\frac{1}{N(A)} = -\infty$$

the locus of $-1/N(A)$ is then shown in Fig. 7.25(b).

Case a: $G_c(s) = 1$. The polar plot of linearity part is, as curve a, shown in Fig. 7.25(b), where the gain crossover frequency is $\omega_g = 4$ rad/s and the corresponding magnitude is $|G(j\omega_g)| = 1$. By inspection, there are two intersections between $G(j\omega)$ and $-1/N(A)$ locus. From Eq. (7.22), we get the periodic solutions

$$\begin{cases} \omega_1 = 4 \text{ rad/s} \\ A_1 = 1.1 \end{cases}, \quad \begin{cases} \omega_2 = 4 \text{ rad/s} \\ A_2 = 2.3 \end{cases}$$

By the stability criterion, the periodic movement with amplitude A_2 is stable, but not the periodic movement with amplitude A_1. Hence we can get the conclusion that there is no self-oscillation in the system if the initial condition or disturbance makes $A < A_1$; that there exists a self-oscillation, $e(t) = 2.3\sin 4t$, if the initial condition or disturbance makes $A > A_1$.

Case b: $G_c(s) = \dfrac{0.25s+1}{8.3(0.03s+1)}$. The polar plot of linearity part is now, as curve b, shown in Fig. 7.25(b), where the gain crossover frequency is $\omega_g = 11.97$ rad/s and the corresponding magnitude is $|G(j\omega_g)| = 0.226$. By inspection, the system is stable.

7.3 Phase-Plane Method

7.3.1 Introduction

The phase-plane method is basically a graphical method and applicable primarily to second-order nonlinear systems. For a nonlinear system $\ddot{x} = f(x, \dot{x})$, the phase-plane analysis is simply the plotting of many different system trajectories in the $[x, \dot{x}]$-plane, i.e., the so-called phase plane. Given an initial condition, this method can be utilized to plot a trajectory in the phase plane. A family of trajectories in the phase plane for a number of initial conditions is called a phase portrait, which provides information about stability and the existence of limit cycles.

For example, Fig. 7.26 is the phase portrait of a second system described by $\ddot{x} + \omega_n^2 x = 0$. The arrows show the direction of the variables along the trajectories for increasing time. It is useful to note that if $\dot{x} > 0$ then x must be increasing in value, i.e., if the trajectory is in the upper half-plane then the net movement in the trajectory must be from left to right. Similarly, if $\dot{x} < 0$ then x must be decreasing in value,

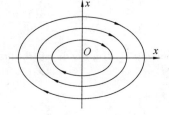

Figure 7.26　Phase portrait

i.e., if the trajectory is in the lower half-plane then the net movement in the trajectory must be from right to left.

The main advantage of this approach is that the trajectories can be sketched without analytical solution of the differential equation. It is also particularly suited to the analysis of systems with linear segmented nonlinear characteristics as the phase plane can be divided into some regions corresponding to operation on a particular linear segment of the nonlinearity.

7.3.2 Sketching Phase Portrait

1. Analytical method

The phase trajectory equations for simpler systems are not too difficult mathematically, and tr actable analytical solutions often can be obtained without problems. Then the phase portraits can be sketched through analytical method, taking direct integration or eliminating intermediate variable, as illustrated in the following example.

Example 7.4 Consider the satellite shown in Fig. 7.27(a). The purpose of an attitude-control system for this satellite is to maintain the attitude angle x at some specified value by firing the thrusters, and thus the system can be described by

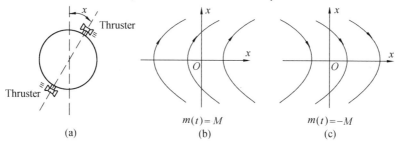

Figure 7.27 Satellite and its phase portrait

$$\ddot{x} = \frac{\tau}{J} = m, \quad x(0) = x_0, \quad \dot{x}(0) = \dot{x}_0$$

where τ is the torque generated by the thrusters, J is the satellite's inertia moment, and $m(t)$ is the thrust. It is assumed that when the thrusters fire the thrust is constant, and thus $m(t) = \pm M$.

Consider first the case of $m(t) = M$. From the system's dynamic equation, we have

$$\ddot{x} = \frac{d\dot{x}}{dt} = \frac{d\dot{x}}{dx} \cdot \frac{dx}{dt} = \dot{x}\frac{d\dot{x}}{dx} = M$$

i.e.,

$$\dot{x}d\dot{x} = Mdx$$

Since the variables $\dot{x}$ and x are separated, taking integration of this equation yields the phase trajectory equation

$$\frac{\dot{x}^2}{2} = Mx + C_1$$

where C_1 is an integral constant and determined by the initial conditions, i.e.,

$$C_1 = -Mx_0 + \frac{\dot{x}_0^2}{2}$$

The phase trajectory equation describes a family of parabolas in the $[x, \dot{x}]$-plane, and the phase portrait is plotted in Fig. 7.27(b).

For $m(t) = -M$, in a same way, we can get the phase trajectory equation
$$\frac{\dot{x}^2}{2} = - Mx + C_2$$
where C_2 is an integral constant and determined by the initial conditions. The phase portrait for this case is shown in Fig. 7.27(c).

Another analytical method of plotting the phase trajectory is to eliminate t from the expressions of $x(t)$ and $\dot{x}(t)$ to obtain an analytical expression for the trajectory.

2. Method of isoclines

The practical graphical method for obtaining trajectories in the state plane is the method of isoclines. Consider the second-order system described by
$$\ddot{x} = f(x, \dot{x}) \tag{7.23}$$
where $f(x, \dot{x})$ may be a linear or nonlinear analytical function. In the $(x, \dot{x})$-plane, at any point, the slope of a trajectory is
$$\frac{d\dot{x}}{dx} = \frac{d\dot{x}}{dt} \bigg/ \frac{dx}{dt} = \frac{\ddot{x}}{\dot{x}} = \frac{f(x, \dot{x})}{\dot{x}} \tag{7.24}$$
Eq. (7.24) is called slope equation of trajectory. Letting
$$\frac{d\dot{x}}{dx} = \alpha$$
for a given point in the phase plane, α is the slope of the trajectory that passes through that point. Eq. (7.24) can be rewritten as
$$\dot{x} = \frac{f(x, \dot{x})}{\alpha} \tag{7.25}$$
For a given value of α, Eq. (7.25) describes a curve in the phase plane. This curve is called an isocline, since a phase trajectory will always have the same slope α as it crosses this curve. By constructing several isoclines in the phase plane, we can sketch a system phase portrait. Of cause, the construction is must simpler if Eq. (7.25) is a family of straight lines, for example in the case of linear second-order systems.

Using the method of isoclines, the phase portrait can be sketched, as shown in Fig. 7.28, by using the following procedure:

① Draw a number of isoclines reasonably close together. (Generally, an included angle about 5° ~ 10° between two adjacent isoclines is recommended.)

② From the starting point A, draw a line having slope $(\alpha_1 + \alpha_2)/2$ and extending from the isocline for α_1 to point B at the isocline for α_2.

③ Continue from the point B using the procedure of step ②.

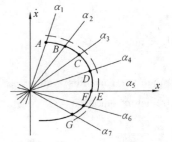

Figure 7.28　Construction of a trajectory using isoclines

④ Connect the points with a smooth curve.

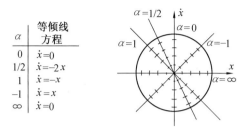

α	等倾线方程
0	$\dot{x}=0$
1/2	$\dot{x}=-2x$
1	$\dot{x}=-x$
-1	$\dot{x}=x$
∞	$\dot{x}=0$

Figure 7.29 Isocline method for a linear oscillator

Example 7.5 Consider a linear oscillator described by
$$\ddot{x} + x = 0$$
The slope equation is
$$\alpha = \frac{d\dot{x}}{dx} = -\frac{x}{\dot{x}}$$
and isocline equation is
$$\dot{x} = -\frac{x}{\alpha}$$

Thus, for this case, the isoclines are a family of straight lines that pass the origin. A table of some isocline equations, the phase plane showing isoclines, and a typical phase trajectory are shown in Fig. 7.29. Note that in this case the phase trajectories are circles.

3. The delta method

As the method of isoclines, the delta method is applicable for both linear and nonlinear systems. However, it is more suitable to nonlinear systems, for which the isoclines of the phase trajectories are usually no longer straight lines. A trajectory is approximated in this method by small arcs of circles.

At first rewrite Eq. (7.23) as
$$\ddot{x} + \omega^2 x = \omega^2 \delta(x,\dot{x}) \tag{7.26}$$
where $\delta(x,\dot{x})$ is defined as
$$\delta(x,\dot{x}) = \frac{f(x,\dot{x}) + \omega^2 x}{\omega^2} \tag{7.27}$$
and the term $\omega^2 x$ is selected so that $\delta(x,\dot{x})$ is not too large or too small in the ranges considered for x and $\dot{x}$. Then remain $\delta(x,\dot{x})$ constant and equal to $\bar{\delta}$ for small changes in x and $\dot{x}$. Thus, Eq. (7.26) becomes
$$\ddot{x} + \omega^2(x - \bar{\delta}) = 0 \tag{7.28}$$
The slope equation of phase trajectory is given by
$$\frac{d\dot{x}}{dx} = -\frac{\omega^2(x - \bar{\delta})}{\dot{x}} \tag{7.29}$$
Then, integrating the slope equation yields
$$\dot{x}^2 + \omega^2 x^2 - 2\omega^2 \bar{\delta} x = C \tag{7.30}$$

where C is an integral constant. Rearranging this equation, we get the phase trajectory equation

$$\left(\frac{\dot{x}}{\omega}\right)^2 + (x - \bar{\delta})^2 = R^2 \tag{7.31}$$

which is a circle.

A graphical interpretation is given in Fig. 7.30. For a given point $P_i(x_i, \dot{x}_i)$, the phase trajectory is perpendicular to P_iQ_i and can be approximated by a small circular arc drawn with center at $Q_i(\bar{\delta}_i, 0)$ and radius $R = P_iQ_i$.

By drawing the circular arc we obtain the next point P_{i+1} on the trajectory, where the angle $\angle P_iQ_iP_{i+1}$, denoted as $\Delta\theta$, is small. At the point P_{i+1} we can again calculate the value of $\bar{\delta}_{i+1}$ and, after locating the new center Q_{i+1}, draw the next circular arc.

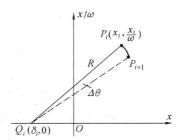

Figure 7.30 Graphical interpretation for the delta method

7.3.3 Basic Phase Portrait of Linear Systems

The phase plane method is particular suited to the analysis of systems with linear segmented nonlinear characteristics. Moreover, quite often many higher-order systems can be represented with approximate first-or second-order models. It is important to be familiar with the phase portrait of linear first-and second-order systems.

1. Phase trajectory of first-order system

Consider a linear first-order system described by

$$\dot{x} + Tx = 0 \tag{7.32}$$

The trajectory equation is

$$\dot{x} = -\frac{1}{T}x \tag{7.33}$$

Assuming $x(0) = x_0$ results in $\dot{x}(0) = \dot{x}_0 = -x_0/T$. As shown in Fig. 7.31, the phase trajectory is simply a straight line, passing through the origin, with a slope of $-1/T$. Along this line the phase trajectory is converge to the origin if $T > 0$, or diverge to infinite if $T < 0$.

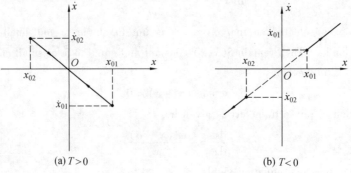

Figure 7.31 Phase trajectory of first-order system

2. Phase portrait of second-order system

Consider a linear second-order system described by

$$\ddot{x} + 2\zeta\omega_n \dot{x} + \omega_n^2 x = 0 \tag{7.34}$$

The trajectory slope equation is

$$\frac{d\dot{x}}{dx} = \frac{-2\zeta\omega_n \dot{x} - \omega_n^2 x}{\dot{x}} \tag{7.35}$$

The phase portrait can be sketched by using analytical or graphical method.

In the analysis of the phase portrait we are particular interested in those points at which $d\dot{x}/dx = 0/0$, i.e., the slope of a trajectory is undefined. These points are known as singular points or singularities. Obviously, at singular points the system can be in equilibrium. For the linear system given by Eq. (7.34) the only singular point is at the origin.

The behavior of a second-order system is obviously related to its characteristic roots. In what follows the phase portraits of Eq. (7.34) and other special second systems in the vicinity of singular points will be investigated. The singular points are classified according to patterns of trajectories in their vicinity.

(1) Two negative real roots

In the case of $\zeta \geqslant 1$ and $\omega_n > 0$, for Eq. (7.34), both characteristic roots

$$s_{1,2} = -\zeta\omega_n \pm \omega_n \sqrt{\zeta^2 - 1}$$

are negative real.

The phase portrait for the overdamped systems, $\zeta > 1$, is shown in Fig. 7.32(a). It can be seen that all trajectories approach and terminate at the origin, i.e. singular point, in the form of nonoscillatory attenuation. Two particular trajectories in the phase portrait, two straight lines with slope s_1 and s_2 respectively, are the asymptotes for all other trajectories. Near the origin the trajectories are tangent to one of the asymptotes. When the system is critically damped, $\zeta = 1$, as shown in Fig. 7.32(b), the phase portrait is similar to that in (a), except that there is only one asymptote. The singular point for an overdamped (or critical damped) response is called a stable node.

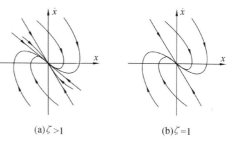

(a) $\zeta > 1$ (b) $\zeta = 1$

Figure 7.32 Phase portrait for stable node

(2) Complex roots with negative real parts

In the case of $0 < \zeta < 1$ and $\omega_n > 0$, for Eq. (7.34), the characteristic roots

$$s_{1,2} = -\zeta\omega_n \pm j\omega_n \sqrt{1 - \zeta^2} \tag{7.36}$$

are complex with negative real parts. From the time-domain analysis, the zero-input response presents a form of damped oscillation.

The phase portrait for the underdamped second-order systems is shown in Fig. 7.33. The characteristic is that all trajectories are converged logarithmic spirals. Motion along the trajectories is clockwise toward the origin for increasing time. The singular point for an underdamped response is called a stable focus.

(3) Imaginary roots

In the case of $\zeta = 0$ and $\omega_n \neq 0$, for Eq. (7.34), the characteristic roots are

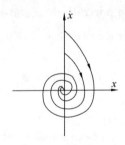

Figure 7.33   Phase portrait for stable focus

$$s_{1,2} = \pm j\omega_n$$

and the zero-input response is of undamped oscillation.

The phase portrait is shown in Fig. 7.34. All the trajectories are elliptical, and the singular point is called a center.

(4) Two positive real roots

In the case of $\zeta \leqslant -1$ and $\omega_n > 0$ (negative damping) for Eq. (7.34), both characteristic roots

$$s_{1,2} = -\zeta\omega_n \pm \omega_n\sqrt{\zeta^2 - 1}$$

are negative real, and the system is unstable.

The phase portrait is shown in Fig. 7.35. In this case all the trajectories depart from the origin and go to infinity, and the singular point is called an unstable node.

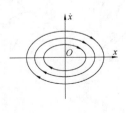

Figure 7.34   Phase portrait for center

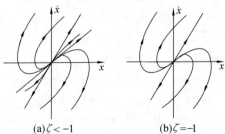

(a) $\zeta < -1$ (b) $\zeta = -1$

Figure 7.35   Phase portrait for unstable node

(5) Complex roots with positive real parts

For negative damping of $-1 < \zeta < 0$ in Eq. (7.34), the characteristic roots

$$s_{1,2} = \zeta\omega_n \pm j\omega_n\sqrt{1 - \zeta^2}$$

are complex conjugates with positive real parts. The system is unstable.

The phase portrait is shown in Fig. 7.36. In this case, all the trajectories are logarithmic spirals diverged from the origin. The singular point is called an unstable focus.

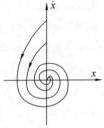

Figure 7.36   Phase portrait for unstable node

(6) Real roots of opposite sign

Consider the second-order system described by
$$\ddot{x} + 2\zeta\omega_n \dot{x} - \omega_n^2 x = 0 \tag{7.37}$$

The characteristic roots
$$s_{1,2} = -\zeta\omega_n \pm \omega_n\sqrt{\zeta^2 + 1}$$

are real but of opposite sign, as shown in Fig. 7.37 (a). Of course, the system is unstable.

The phase portrait is shown in Fig. 7.37(b). Almost all the trajectories are hyperbolas diverged from the origin. Two particular trajectories in the phase portrait, two straight lines with slope s_1 and s_2 respectively, are the asymptotes for all other trajectories. The singular point for this case is called a saddle point.

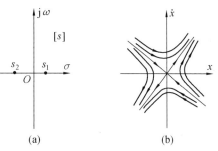

Figure 7.37 Phase portrait for saddle point

(7) One zero and one nonzero root

Consider the system described by
$$\ddot{x} + T\dot{x} = 0 \tag{7.37}$$

The characteristic roots are
$$s_1 = 0, \quad s_2 = -\frac{1}{T}$$

From the trajectory slope equation
$$\frac{d\dot{x}}{dx} = \frac{\ddot{x}}{\dot{x}} = \frac{-T\dot{x}}{\dot{x}}$$

the trajectory equation is given by
$$\dot{x}\left(\frac{d\dot{x}}{dx} + T\right) = 0 \tag{7.38}$$

This equation means that the trajectories, corresponding to the equation $d\dot{x}/dx = -T$, are a family of straight lines with a slope of $-T$, and x axis is apart of the phase portrait. The phase portrait is shown in Fig. 7.38, where every trajectory starts from its initial point $(x_0, \dot{x}_0)$ and ends at the x axis, i.e. the line $\dot{x} = 0$. In this case every point at x axis can be considered as a singular point, or the line $\dot{x} = 0$ is called a singular line.

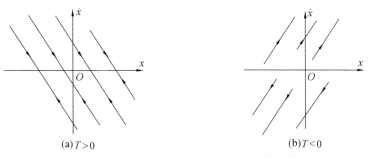

(a) $T>0$ (b) $T<0$

Figure 7.38 Phase portrait for equation $\ddot{x} + T\dot{x} = 0$

7.3.4 Phase Plane Analysis

A control system can often be approximated by a second-order differential equation and then studied using the phase plane method. In addition, when nonlinear phenomena occurring in such a system can be represented by linear segmented characteristics the phase plane analysis may be considerably simplified. The procedure is to divide the phase plane into various regions each of which corresponds to motion on a particular linear segment of the nonlinearity. The motion is then described by different linear differential equations in the various regions of the phase plane and solutions for particular initial values are obtained by matching the conditions at the region boundaries. Singular points for trajectories in a particular region may thus lie outside that region, in which case the singularity is known as a virtual singular point. The approach is illustrated with some examples as follows.

Example 7.6 The satellite, discussed in Example 7.3, is now placed in a feedback configuration in order to maintain the attitude angle x at $0°$. This attitude control system is shown in Fig. 7.39.

When x is other than $0°$, the appropriate thrusters will fire to force x toward $0°$. Note that switching of $m(t)$ occurs at $x = 0$. Thus the line $x = 0$ is called the switching line. A typical trajectory for the system is illustrated with Fig. 7.40(a). The system response is then a limit circle. The phase portrait is shown in Fig. 7.40(b). Hence, this control system is not acceptable.

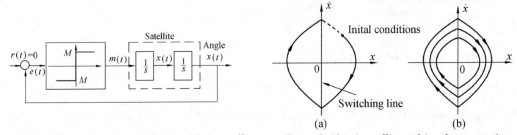

Figure 7.39 Attitude control system for a satellite   Figure 7.40 A satellite and its phase portrait

Example 7.7 The satellite control system is studied again in this example. Suppose that we add rate feedback, using a rate gyro with a gain of β as the sensor, to the control system, as shown Fig. 7.41.

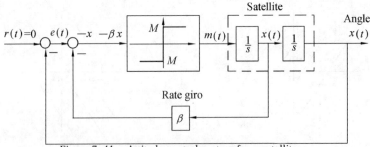

Figure 7.41 Attitude control system for a satellite

Note that the input to the satellite is still $\pm M$, and thus the response remains the family of parabolas shown in Fig. 7.27(b) and (c). However, the switching of $m(t)$ is now different.

Switching occurs when $e = 0$, i.e.,
$$x + \beta \dot{x} = 0$$
This equation indicates that the switching line is a straight line with slope of $d\dot{x}/dx = -1/\beta$. A typical response for this system is shown in Fig. 7.42. It can be seen that the system response is greatly improved with the addition of the rare feedback. Furthermore, the origin is now asymptotically stable.

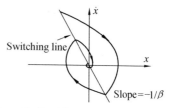

Figure 7.42 Phase portrait of a satellite

Problems

P7.1 A nonlinear system is shown in Fig. P7.1, where $G(s)$ is the linearity part. Determine the equivalent nonlinearity N of the nonlinearities N_1, N_2 and N_3.

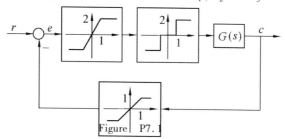

Figure P7.1

P7.2 Determine whether each system shown in Fig. P7.2 is stable and analyze the oscillation characteristics corresponding to each intersection between the $G(j\omega)H(j\omega)$ plot and $-1/N(A)$ locus. It is assumed that $G(s)H(s)$ is minimum phase-shift.

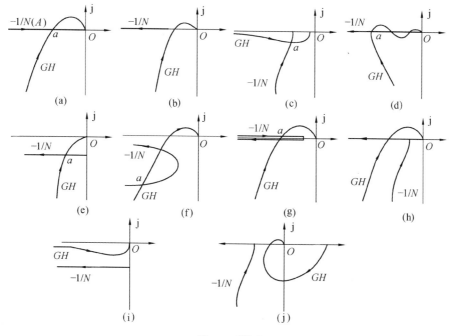

Figure P7.2

P7.3 Consider a nonlinear system shown in Fig. P7.3. (a) Analyze the system stability for $k=10$, using the describing function method. (b) Find the critical value of k so that the system is stable.

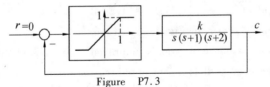

Figure P7.3

P7.4 Determine the self-oscillation parameters, amplitude and frequency, of the system shown in Fig. P7.4.

P7.5 Consider the system shown in Fig. P7.5. (a) Analyze the system stability, using the describing function method. (b) Determine how to adjust the parameters e_0 and M so that the system is stable.

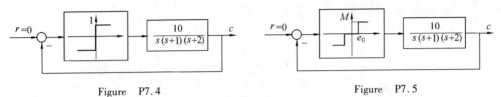

Figure P7.4   Figure P7.5

P7.6 Analyze the stability of the system shown in Fig. P7.6, and determine the amplitude and frequency of self-oscillation.

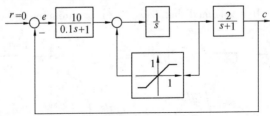

Figure P7.6

P7.7 Find the singularity points of the following equations and determine their kinds.
(a) $2\ddot{x}+\dot{x}^2+x=0$
(b) $\ddot{x}-(1-x^2)\dot{x}+x=0$

References

[1] 多尔夫.现代控制系统(英文影印版)[M].北京:科学出版社,2002.

[2] 多尔西.连续与离散控制系统(英文影印版)[M].北京:电子工业出版社,2002.

[3] 德赖斯.线性控制系统工程(英文影印版)[M].北京:清华大学出版社,2002.

[4] JOHN J D´AZZO, CONSTANTINE H HOUPIS. Linear Control System Analysis and Design [M]. New York:McGraw-Hill,1988.

[5] BENJAMIN C KUO. Automatic Control Systems [M]. New Jersey:Prentice-Hall,1982.

[6] 绪方胜彦.现代控制工程[M].北京:电子工业出版社,2000.

[7] 郑大钟.线性系统理论[M].北京:清华大学出版社,2002.

[8] 李友善.自动控制原理[M].北京:国防工业出版社,1989.

[9] 李文秀.自动控制原理[M].哈尔滨:哈尔滨工程大学出版社,2001.

[10] 胡寿松.自动控制原理[M].北京:科学出版社,2002.

[11] 孙虎章.自动控制原理[M].北京:中央广播电视大学出版社,1994.

[12] 吴麒.自动控制原理[M].北京:清华大学出版社,1991.

[13] 戴忠达.自动控制理论基础[M].北京:清华大学出版社,1991.

[14] 陈启宗.线性系统理论与设计[M].北京:科学出版社,1988.

[15] 王诗宓.自动控制理论例题习题集[M].北京:清华大学出版社,2002.

[16] LAUGHTON M A, SAY M G. Electrical Engineer´s Reference Book[M]. Scotland:Butterworths, 1985.

第1章 控制系统概述

1.1 概 述

1.1.1 控制工程和自动化

作为科学和数学的原理在诸如机器、装置、过程等系统设计、制造和运行方面的应用，控制工程关注的是了解和控制自然界的物质和自然力为人类造福。控制系统工程师关注的则是了解和控制与他们有关的系统为人类社会提供经济实用的产品。控制工程以反馈理论和线性系统分析作为基础，并且把网络理论和通信理论的概念结合在一起。因此，控制工程不是局限于任何一门工程学科，而是同样地适用于航天、化学、机械、环境以及电气等工程。例如，一个控制系统常常会包括电气、机械和化学的部分。而且，随着人们对于企业、社会以及行政系统动态行为了解的增加，控制这些系统的能力也将与日俱增。

一个装置或过程采用自动而不是人工的方式进行控制称为自动化；或者，我们也可以说，自动化是机器、装置和过程等的自动技术。自动化用于改善生产力和获取高质量的产品。

1.1.2 自动控制的历史

使一个装置或过程能够实现自动控制最简单的方法是通过传统的反馈控制。利用反馈控制一个系统已经有了很长的历史。历史上，在控制的发展过程中向前迈出的关键一步发生在工业革命时期。当时，机器的发展极大地增强了把原材料转化为产品而造福于人类社会的能力。相关的机器，尤其是蒸汽机，涉及大量的动力。而且不久人们就认识到，如果系统要安全有效地运行，这些动力就需要以一种协调和统一的方式加以控制。这一时期的一个重大进展是瓦特发明的飞球调速装置，这种装置通过控制蒸汽的流量来调节蒸汽机的速度。

世界大战也使控制工程得到了很大的发展。有些发展与武器制导系统和防空系统有关，而另一些发展则与因战争的消耗而增强了对生产的需求有关。

在20世纪60和70年代，人类进入太空也是依赖于控制的发展。之后，这些研究成果又经久不息地返回到人们的生活消费品，以及商业、环境和医学方面的应用。

到了20世纪末期，控制已经成为现代社会无处不在（但在很大程度上看不见）的要素。事实上，我们所接触的每一个系统都是以复杂的控制系统作为基础的。实例涵盖的范围可以从简单的家用产品（空调器的温度调节器、热水器中的温度调节装置等）到大规模系统（诸如化工装置、飞行器和加工制造过程）。

除了这些工业上的例子外，反馈调节机制主要集中在生物系统、通信网络、国民经济的运行,甚至于人类的相互影响。确实,只要仔细想一下,在我们生活的各个方面都可以发现这种或者那种形式的控制。

由此可见,控制工程是一个包括多种学科而令人动心的领域,具有广阔的实际应用范围。而且,在可以预见的未来,人们对于控制的兴趣不太可能减小。正好相反,由于市场全球化程度的增大和对于环境问题的日益关注,控制极有可能变得前所未有的重要。

市场正在越来越全球化,这种状况意味着制造业正在不可避免地把越来越多的努力放在质量和效益的问题上。这使人们的注意力集中到开发改进的控制系统,从而使得装置和生产过程能以最佳的可能方式运行。尤其是,改进后的控制是使技术支撑可能成为关键。

此外,要达到关心有限的自然资源和保护我们脆弱的环境这些目标时,控制工程是能使技术付诸实现的核心问题。

1.1.3 控制理论的一些历史时期

上面我们已经看到,控制工程在历史上至关重要的事件中已经向前迈出了几大步,其中每一步都伴随着基本控制理论方面相应的突破性进展。

在早期应用反馈概念时,工程师们有时会遇到预料不到的结果。之后,这就成为进行严谨分析的促进因素。

整个第二次世界大战时期控制的发展也伴随着控制理论方面重要的发展。其结果是产生使用简单的图解方法,可以用于分析单变量的反馈控制问题。现在专业术语通常把这些方法称为经典控制理论。

20世纪60年代是控制的另一种方法,即状态空间法研究成功的时期。这种方法随着维纳、卡尔曼(以及其他人)在最优估计和最优控制方面成果的发表而问世。这一成果使得多变量问题能够以一种统一的方式处理。而在经典控制理论的框架内,这类问题的解决如果不是不可能的,也是非常困难的。这一系列进展的结果术语称为现代控制理论。

关于自动控制的理论是一门非常有用的工程学科,它的内容广泛且动人心弦。人们很容易就能理解为什么要学习自动控制理论。

1.1.4 控制系统未来的发展

控制系统经久不衰的目标是提供广泛的灵活性和高度的自主性。当今的工业机器人被理解为高度自治的,一旦给机器人输进了程序,一般情况下不需要更多的干涉。由于感觉的限制,这些机器人系统在适应工作环境变化方面灵活性还有限,这也是计算机视觉研究的动机所在。其实,控制系统的适应能力是很强的,但要依靠人的监督管理。高级的机器人系统正在通过增强的感觉反馈为实现任务的适应性而努力。集中在人工智能、传感器集成、计算机视觉,以及离线计算机辅助设计/计算机辅助制造编程领域的研究将使系统通用性更好,并且开支更为节省。控制系统正在走向作为加强人类控制的自主运行。在监控、减少操作人员负担的人-机接口方法,以及计算机数据库管理方面的研究,目的就是改善操作人员的工作效率。许多研究活动对于机器人和控制系统都是共同的,目标都是减少实现的成本和扩大应用的领域。这包括改进的通信方法和高级编程语言。

1.1.5 控制系统

控制系统是一些相互作用的部件的有机集合,它们组合在一起通过物质、能量和信息的操纵和控制达到一些指定的目的。在研究自动控制理论时,我们关注的是系统内的信号流而不是物质流和能量流。

控制系统本质上是动态的,而且在一个控制系统的组成部分之间存在着因果关系。因此,一个系统及其组成部分可以用图示的方法,用所谓的方框图如图1.1所示。方框内是相应的系统或其组成部分的名称、功能或数学模型,带有箭头的直线指示信号流的方向。在一个控制系统中,需要被控制的装置或过程称为被控对象,选出用以表征系统行为的物理量称为被控量或输出。对应于被控量建立有期望值或指令输入。输入-输出关系表示系统的一种因果关系。扰动是一种(有别于输入指令的)外部作用,这种作用往往驱使被控量偏离其期望值。扰动-输出关系是系统的另一种因果关系。

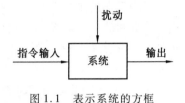

图 1.1 表示系统的方框

1.2 开环控制和闭环控制

通常一个自动控制系统由被控对象及其控制装置组成。基于控制作用如何产生,即控制作用的产生是否与实际输出有关,控制系统可以分为开环控制系统和闭环控制系统。

1.2.1 开环控制

开环控制系统是没有反馈的系统,而且在开环系统内从输入到输出的信号流是单向传输的。在开环控制的情况下,控制作用的生成与实际输出无关,即控制作用的生成仅仅取决于指令输入和/或扰动。

1. 按指令输入进行操纵

图1.2为开环控制的一种方式,即按指令输入进行控制的功能方框图。控制器接受指令输入并操纵被控对象获得期望的输出。输出的期望值有可能要改变,这时将需要改变输入来调整被控对象的运行。

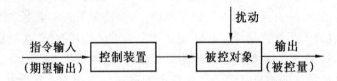

图 1.2 按指令输入进行控制的功能方框图

图1.3为用于使光盘以恒定转速旋转的转盘,是一个开环系统的例子。该系统使用电池提供一个与期望转速成正比的参考电压。这一电压经放大后施加于直流电机,而直流电机则可提供与所施加的电压成正比的转速。在这一系统中,如果由于某种原因造成转盘转速偏离期望值的变化,则没有办法改变施加于直流电机的电压以维持期望的转速。

在这种情况下,也可以说输出对于输入没有影响。

按输入进行操纵的另一个例子是图 1.4 所示的发电机-负载系统。由发电机提供的电压与励磁机的励磁电压成正比,励磁电压则可以用一个电位计调节。

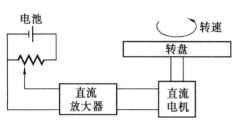

图 1.3　转盘速度的开环控制　　　　图 1.4　发电机-负载系统

显然,负载两端的电压将受到扰动的影响,例如负载的波动或者某些元部件参数的变化,这时实际输出则将偏离期望值,而且不能自行返回原先的状态。

2. 按扰动进行补偿

开环控制的另一种方式是图 1.5 所示的按扰动进行补偿的方框图。在这种情况下,如果扰动信号是可以测量的,控制器接受扰动信号并产生一个额外的控制作用来补偿扰动对于系统的影响。

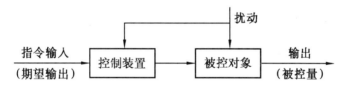

图 1.5　按扰动进行补偿的方框图

这一开环控制方式的一个例子还用于发电机-负载系统,如图 1.6 所示,但是励磁机绕组的一部分接在发电机-负载回路中。现在,如果负载电压由于负载的增加而降低的话,流过负载和励磁绕组的电流将增大,这将导致发电机端电压的增高。

虽然构成开环系统比较简单、花费也比较少,但是为了确定对应于所需输出的输入值,需要详细了解开环系统的各组成部分。而且,系统参数的变化和/或外部的扰动可能会对控制的精度有着不利的影响。

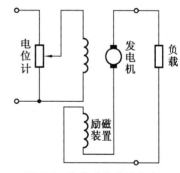

图 1.6　按扰动补偿的发电机-负载系统

1.2.2　闭环控制

图 1.7 为闭环控制系统的功能方框图。与开环控制系统相反,闭环控制系统是利用测量得到的实际输出结果与代表输出期望值的输入相比较。在这里实际输出的测量结果称为反馈信号。反馈控制系统是通过输出实际值与期望值的比较并利用比较的差异作为控制的手段,使系统的一个变量维持与另一个变量的预定关系。

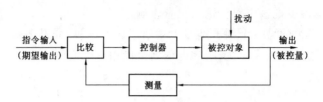

图 1.7 闭环控制系统的功能方框图

图 1.8 为一个转盘转速的闭环控制系统。图中测速发电机用于测量实际转速并输出一个与转速成正比的电压。该系统利用输出和输入的关系控制被控对象。输入与反馈信号之间的差异用于控制系统,使得差异连续不断地减小。反馈的概念已经是控制系统分析和设计的基础。

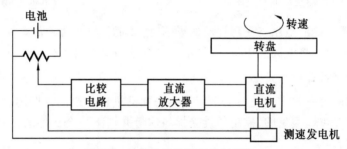

图 1.8 转盘转速的闭环控制系统

闭环系统显著的特点之一是由于反馈的作用而能如实地重现指令输入。由于系统是基于输出偏离输入的情况进行调节,控制作用持续产生足够的附加输出量使输出和输入趋于一致。不幸的是,反馈也要为系统中的振荡趋势负责。闭环控制系统的另一个与开环系统直接形成对比的重要特点是,即使在存在非线性特性的情况下,闭环控制系统通常也能正确地进行控制。

1.2.3 反馈系统的一般功能结构

每一个反馈控制系统都是由执行特定功能的元部件组成的。表示控制系统这一功能特征时,方框图是一种既方便又有用的方法。本质上,这是一种表示系统中所执行操作的手段和信号在整个系统中流动的方式。方框图关注的不是任何特定系统的物理特征,而仅仅是系统中各个部分之间的功能关系。反馈控制系统一般的功能配置和结构如图 1.9 所示。

由控制系统的工作原理图可以画出它的功能方框图。首先,根据系统将要完成的任务,可以确定被控对象、被控量以及可能的扰动;然后可以根据信号传递顺序确定各个功能方框。

由于反馈控制系统是基于偏差运行的,即控制信号就是实际输出和期望输出之间的差值,在每一个反馈控制系统中都一定有一个比较环节。

测量环节直接或间接地测量实际的被控量,并为比较环节提供一个反馈信号。输入环节提供功能转换使指令输入转换为适当的模式与反馈信号进行比较,结果得到的是系统的参考输入。

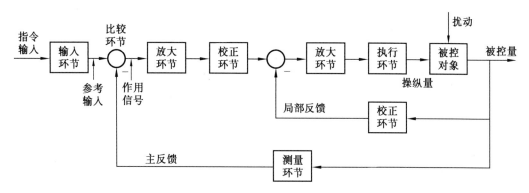

图 1.9　反馈回路的功能配置和结构

放大环节用于使偏差信号放大为功率信号,如果有必要还用于使偏差信号转换为合适的物理形式。

最后,执行环节执行控制作用操纵被控对象。

除了以上功能方框外,几乎所有的控制系统都还需要校正环节改善系统的性能。

1.2.4　控制系统的性能要求

一个合适的控制系统应当具有以下性质:

① 系统应当是稳定的,并且对于输入指令呈现出可以接受的响应,即被控量应当能以适当的速度跟随参考输入的变化,并且没有过大的振荡或超调。

② 系统应当以尽可能小的误差运行。

③ 系统应当减小不希望有的扰动的影响。

第 2 章　控制系统的数学模型

2.1　引　言

在控制系统的分析中,最为重要的步骤之一是该系统的数学描述和模型的建立。数学模型这一术语是指描述一个系统内各变量之间关系的数学表达方式。在经典控制理论中,数学模型通常是指联系一个系统或其组成部分输出与输入的数学关系。控制系统的数学模型是一个重要的问题,因为它可以按照系统各组成部分之间的因果关系清楚地了解系统。

由于所考虑的系统本质上是动态的,描述它们的数学方程通常是微分方程。而且,如果这些方程是线性的或者可以线性化的,就可以应用拉普拉斯变换简化方程的求解,同时由此得到的传递函数将成为极有价值的工具,用于系统的分析和设计。在实际应用中,由于控制系统的复杂性以及对于所有相关因素的认识不足,这就需要引入一些与系统运行有关的假设。因此,经常发现,作一些必要的假设并使系统线性化将有助于考虑物理系统的问题。然后,利用描述等效线性系统的物理规律,可以获得一组线性微分方程。最后,利用诸如拉普拉斯变换之类的数学工具,可以得到描述系统运行状况的解。

一般而言,也可以用描绘系统各组成部分之间相互关系和内部连接的原理图表示物理系统。在控制系统理论中,方框图经常用于描绘各种类型的系统。

在这一章,将讨论一些常见物理系统的数学模型。但是应当指出,本章只不过是作为系统建模的一个入门介绍,并不打算写成关于该主题全面的论述。

应当认识到,没有一个物理系统的数学模型是绝对精确的。可以通过增加方程的复杂程度提高模型的精度,但绝不可能得到完全精确的模型。一般情况下,努力要做到的是建立一个适合于正在解决的问题的模型,同时又要避免模型过分复杂。

2.2　微分方程和传递函数

2.2.1　引　言

从数学的观点出发,微分方程可以用于描述系统的动态行为。控制系统的各组成部分在本质上是各不相同的,它们所包括的器件有可能是电的、机械的、热力的、液压的,等等。描述这些器件的微分方程可以利用一些基本的物理规律获取,这些规律包括力、能量和质量的平衡。

考虑一个线性时不变连续系统由以下 n 阶微分方程描述

$$a_0 \frac{d^n c(t)}{dt^n} + a_1 \frac{d^{n-1} c(t)}{dt^{n-1}} + \cdots + a_{n-1} \frac{dc(t)}{dt} + a_n c(t) =$$
$$b_0 \frac{d^m r(t)}{dt^m} + b_1 \frac{d^{m-1} r(t)}{dt^{m-1}} + \cdots + b_{m-1} \frac{dr(t)}{dt} + b_m r(t) \tag{2.1}$$

式中，$c(t)$ 为输出量；$r(t)$ 为输入量；系数 $a_0,a_1,\cdots,a_n$ 和 $b_0,b_1,\cdots,b_m$ 均为常数，而且 $n \geq m$。这一微分方程代表了系统输入 $r(t)$ 和输出 $c(t)$ 之间的完全描述。一旦指定了系统的输入和初始条件，就可以通过求解方程获得输出响应。不过有一点是很清楚的，描述系统的微分方程法虽然是基本的方法，但也是一种相当麻烦的方法，尤其是在控制系统的设计中，高阶微分方程实际上很少使用。更重要的事实是，尽管可以使用有效的软件在计算机上求解高阶微分方程，但是线性控制理论中重要的研究工作却是依赖于无需实际求解系统微分方程的分析和设计方法。

使用传递函数使得一种比较方便的描述线性时不变系统的方法成为可能。为了获取方程(2.1)所表示的线性系统的传递函数，设初始条件为零，并对该方程的两侧取拉普拉斯变换，则可以得到

$$(a_0 s^n + a_1 s^{n-1} + \cdots + a_{n-1} s + a_n) C(s) =$$
$$(b_0 s^m + b_1 s^{m-1} + \cdots + b_{m-1} s + b_m) R(s) \tag{2.2}$$

该系统的传递函数定义为 $C(s)$ 与 $R(s)$ 之比，即

$$\frac{C(s)}{R(s)} = \frac{b_0 s^m + b_1 s^{m-1} + \cdots + b_{m-1} s + b_m}{a_0 s^n + a_1 s^{n-1} + \cdots + a_{n-1} s + a_n} \tag{2.3}$$

概括传递函数的一些性质可以有：

① 只有线性时不变系统才定义有传递函数。

② 一个系统输入和输出之间的传递函数定义为输出的拉普拉斯变换与输入的拉普拉斯变换之比，同时该系统的所有初始条件均假设为零。

③ 传递函数与输入激励作用无关。

传递函数在表征线性时不变系统的特性时起着重要的作用。传递函数与方框图结合在一起，构成了经典控制理论中表示线性时不变系统输入 – 输出关系的基本方法。

下面将应用有关的物理定律建立一些简单物理系统的数学模型。

2.2.2 简单网络

1. 电路

列写电网络微分方程的经典方法是由基尔霍夫的两个定律确定的回路法和节点法。

例 2.1 作为第一个例子，考虑图 2.1 中的简单 RC 电路，图中输入电压 v_i 作用于一个 RC 电路。

根据基尔霍夫电压定律，输出电压 v_o 通过以下微分方程与输入作用 v_i 相联系

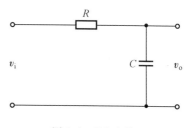

图 2.1 RC 电路

$$v_i(t) = RC \frac{dv_o(t)}{dt} + v_o(t)$$

该方程可以排列成一种规范形式

$$RC\frac{\mathrm{d}v_o(t)}{\mathrm{d}t} + v_o(t) = v_i(t) \tag{2.4}$$

设初始条件为零,取拉普拉斯变换后有

$$(RCs + 1)V_o(s) = V_i(s)$$

由上面的方程可以得到传递函数

$$\frac{V_o(s)}{V_i(s)} = \frac{1}{RCs + 1} \tag{2.5}$$

例 2.2 作为另一个例子,考虑图 2.2 中的 RLC 电路,图中 v_i 为输入电压,v_o 为输出电压。

设电容器上初始电压为零,由基尔霍夫电压定律我们可以得到

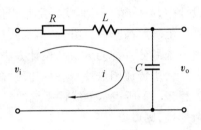

图 2.2 RLC 电路

$$v_i(t) = Ri(t) + L\frac{\mathrm{d}i(t)}{\mathrm{d}t} + v_o(t)$$

$$v_o(t) = \frac{1}{C}\int i(t)\mathrm{d}t$$

其中,回路电流 i 是一个中间变量。消去中间变量 i 结果为

$$LC\frac{\mathrm{d}^2 v_o(t)}{\mathrm{d}t^2} + RC\frac{\mathrm{d}v_o(t)}{\mathrm{d}t} + v_o(t) = v_i(t) \tag{2.6}$$

对该方程取拉普拉斯变换得到传递函数

$$\frac{V_o(s)}{V_i(s)} = \frac{1}{LCs^2 + RCs + 1} \tag{2.7}$$

2. 机械系统

(1) 平移系统

平移运动定义为沿着一条直线进行的运动。用于描述平移运动的变量有力、位移、速度和加速度。线性机械平移系统的基本元件可以是质量元件、阻尼器和弹簧元件。图 2.3 为平移运动的基本元件符号。

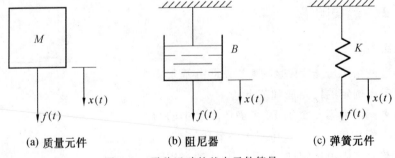

(a) 质量元件　　　　(b) 阻尼器　　　　(c) 弹簧元件

图 2.3 平移运动的基本元件符号

在图 2.3(a) 中,$f(t)$ 表示作用力,$x(t)$ 表示位移,而 M 则表示质量。这样,由牛顿第二定律有

第 2 章 控制系统的数学模型

$$f(t) = M \frac{d^2 x(t)}{dt^2} \tag{2.8}$$

这里假设质量元件是刚性的。

对于剩下的两种元件,顶部的连接点可以相对于底部的连接点运动。首先考虑图 2.3(b) 所示的阻尼器。与油、空气等有关的粘滞摩擦是一种可以物理观察到的阻尼现象,而汽车上的减震器则是一种物理器件。阻尼器的数学模型为

$$f(t) = B \frac{dx(t)}{dt} \tag{2.9}$$

式中,B 为黏性摩擦系数。

由胡克定律,图 2.3(c) 所示弹簧的定义方程为

$$f(t) = K x(t) \tag{2.10}$$

式中,K 为弹簧系数。

例 2.3 图 2.4 为一个简单的弹簧 – 质量 – 阻尼器系统,系统受到外力 $f(t)$ 的作用,而运动则用位移 $x(t)$ 表示。

在这一例子中,用一个阻尼器作为壁间黏性摩擦的模型,即摩擦力的大小线性正比于质量元件的速度。应用牛顿第二定律得到

$$f(t) - Kx(t) - B \frac{dx}{dt} = M \frac{d^2 x(t)}{dt^2}$$

即

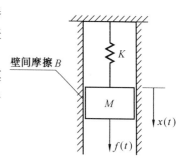

图 2.4 弹簧 – 质量 – 阻尼器系统

$$M \frac{d^2 x(t)}{dt^2} + B \frac{dx}{dt} + Kx(t) = f(t) \tag{2.11}$$

设初始条件为零并取拉普拉斯变换,得到

$$(Ms^2 + Bs + K) X(s) = F(s)$$

结果得到传递函数为

$$\frac{X(s)}{F(s)} = \frac{1}{Ms^2 + Bs + K} \tag{2.12}$$

(2) 旋转系统

一个物体的旋转运动可以定义为围绕一条固定轴线的运动,通常用于描述旋转运动的变量为力矩、角位移、角速度和角加速度。线性旋转系统与线性平移系统相似(微分方程具有相同的形式),而且用于列写线性平移系统方程的方法也可以用于旋转系统。

线性旋转运动的三种基本元件画,如图 2.5 所示。

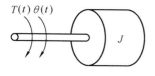

(a) 转动惯量

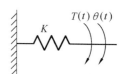

(c) 扭簧

图 2.5 线性旋转运动的三种基本元件

图 2.5(a)中的第一种元件是转动惯量,其关系式定义为

$$T(t) = J\frac{d^2\theta(t)}{dt^2} \tag{2.13}$$

其中,$T(t)$为外加的力矩;J为转动惯量;$\theta(t)$为旋转的角度。这一方程与平移系统中质量元件的方程相似。

图 2.5(b)中的第二种元件是黏性摩擦,它的定义方程为

$$T(t) = B\frac{d\theta(t)}{dt} \tag{2.14}$$

式中,B为阻尼系数。至于图 2.5(c)中所示的扭簧,定义方程则为

$$T(t) = K\theta(t) \tag{2.15}$$

其中,K为扭簧系数。

例 2.4 考虑图 2.6 中的扭摆。摆锤的转动惯量用 J 表示,摆锤与空气之间的摩擦用 B 表示,而吊杆的弹性作用则用 K 表示。假设力矩施加在摆锤上。

将作用在摆锤上的力矩相加,可以有

$$J\frac{d^2\theta(t)}{dt^2} + B\frac{d\theta(t)}{dt} + K\theta(t) = T(t) \tag{2.16}$$

然后很容易推导出传递函数为

$$\frac{\theta(s)}{T(s)} = \frac{1}{Js^2 + Bs + K} \tag{2.17}$$

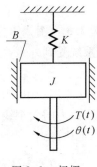

图 2.6 扭摆

(3) 相似系统

容易就可以注意到方程(2.6)、(2.11)和(2.16)的对等关系,其中电压 $v_i(t)$、力 $f(t)$ 和力矩 $T(t)$ 为对等的变量,通常称为相似量,而这些系统则称为相似系统。

对于电气、机械、热力以及流体系统,存在着具有类似解的相似系统。相似系统的存在为分析人员提供了把一个系统的解运用到具有相同微分方程描述的所有相似系统的可能性。因此,关于电气系统分析和设计所获悉的结果可以立即运用于了解机械、热力以及流体等系统。

2.2.3 齿轮传动机构

齿轮传动机构是一种机械装置,它们以力、力矩、速度和位移发生变化的方式使能量从系统的某一部分传输到另一部分。这种装置也可以看作是用于获取最大能量传递的匹配装置。

齿轮传动机构的一种等效表示方法,如图 2.7 所示。图中 $T(t)$ 为外加的力矩,$\theta_1(t)$ 和 $\theta_2(t)$ 为角位移,$T_1(t)$ 和 $T_2(t)$ 为齿轮传递的力矩,J_1 和 J_2 为齿轮的转动惯量,N_1 和

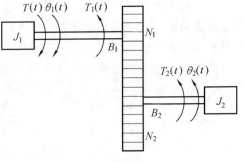

图 2.7 齿轮传动机构

N_2 为齿数,B_1 和 B_2 为黏性摩擦系数。

从动齿轮的力矩方程为

$$T_2(t) = J_2 \frac{d^2\theta_2(t)}{dt^2} + B_2 \frac{d\theta_2(t)}{dt} \tag{2.18}$$

驱动齿轮一侧的力矩方程为

$$T(t) = J_1 \frac{d^2\theta_1(t)}{dt^2} + B_1 \frac{d\theta_1(t)}{dt} + T_1(t) \tag{2.19}$$

力矩 T_1 和 T_2 之间的关系为

$$\frac{T_1}{T_2} = \frac{N_1}{N_2} \tag{2.20}$$

而转速 ω_1 和 ω_2 之间的关系为

$$\frac{\omega_1}{\omega_2} = \frac{d\theta_1/dt}{d\theta_2/dt} = \frac{N_2}{N_1} \tag{2.21}$$

因此我们有

$$T_1(t) = \frac{N_1}{N_2} T_2(t) = \left(\frac{N_1}{N_2}\right)^2 J_2 \frac{d^2\theta_1(t)}{dt^2} + \left(\frac{N_1}{N_2}\right)^2 B_2 \frac{d\theta_1(t)}{dt} \tag{2.22}$$

该式表明,可以把齿轮传动机构一侧的转动惯量、力矩、速度和位移折算到另一侧。将式(2.22)代入式(2.19)得到

$$T(t) = J_{1e} \frac{d^2\theta_1(t)}{dt^2} + B_{1e} \frac{d\theta_1(t)}{dt} \tag{2.23}$$

式中 $J_{1e} = J_1 + \left(\frac{N_1}{N_2}\right)^2 J_2$,$B_{1e} = B_1 + \left(\frac{N_1}{N_2}\right)^2 B_2$。

2.2.4 电枢控制直流电动机

直流电动机是一种将直流电能转换为旋转能向负载递送能量的功率执行装置。图2.8为一种电枢控制直流电动机的工作原理图。在这种情况下,电枢电压 $v_a(t)$ 是该系统的输入,而电动机轴的转角 $\theta_m(t)$ 则看作输出。电枢线圈的电阻和电感分别为 R_a 和 L_a。

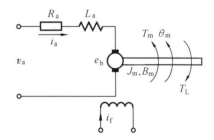

图 2.8 电枢控制直流电动机的工作原理图

电枢电流与施加于电枢的输入电压的关系为

$$v_a(t) = L_a \frac{di_a(t)}{dt} + R_a i_a(t) + e_b(t) \tag{2.24}$$

式中,$e_b(t)$ 为正比于电动机转速 $\omega_m(t)$ 的反电势。因此我们有

$$e_b(t) = C_e \omega_m(t) = C_e \frac{d\theta_m(t)}{dt} \tag{2.25}$$

式中,C_e 为电动机的反电势常数。

电动机产生的电磁转矩为

$$T_m(t) = C_m i_a(t) \tag{2.26}$$

式中,C_m 为电动机的转矩常数。由电动机电枢上的合力矩可以推导出

$$T_m(t) = J_m \frac{d^2\theta_m(t)}{dt^2} + B_m \frac{d\theta_m(t)}{dt} + T_L(t) \tag{2.27}$$

式中,J_m 为电动机的转动惯量;B_m 为摩擦引起的阻尼系数;T_L 为负载力矩。

由以上方程我们有

$$J_m L_a \frac{d^3\theta_m}{dt^3} + (B_m L_a + J_m R_a) \frac{d^2\theta_m}{dt^2} + (B_m R_a + C_e C_m) \frac{d\theta_m}{dt} =$$

$$C_m v_a - L_a \frac{dT_L}{dt} - R_a T_L \tag{2.28}$$

如果电枢电感可以忽略不计,则式(2.28)变为

$$J_m R_a \frac{d^2\theta_m}{dt^2} + (B_m R_a + C_e C_m) \frac{d\theta_m}{dt} = C_m v_a - R_a T_L \tag{2.29}$$

取拉普拉斯变换,我们得到传递函数

$$\frac{\theta_m(s)}{V_a(s)} = \frac{C_m}{s(J_m R_a s + B_m R_a + C_e C_m)} = \frac{K_m}{s(T_m s + 1)} \tag{2.30}$$

$$\frac{\theta_m(s)}{T_L(s)} = -\frac{R_a}{s(J_m R_a s + B_m R_a + C_e C_m)} = -\frac{K_L}{s(T_m s + 1)} \tag{2.31}$$

式中 $T_m = \frac{JR_a}{fR_a + C_e C_m}$, $K_m = \frac{C_m}{fR_a + C_e C_m}$, $K_L = \frac{R_a}{fR_a + C_e C_m}$。

2.2.5 位置控制系统

例 2.5 图 2.9 为直流位置控制系统即伺服机构,图中输出轴的位置要求跟随输入轴的位置。正比于输入轴位置 θ_i 和输出轴位置 θ_o 两者之差的误差电压 v_s 作用于放大器。放大器的输出连接到直流伺服电机的电枢如图所示。电机轴通过一个减速比为 $i = Z_2/Z_1$ 的减速箱连接到输出轴。

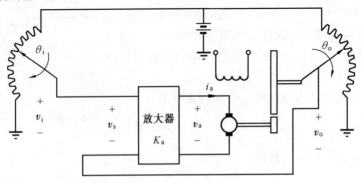

图 2.9 直流位置控制系统

很容易推导出误差检测装置和放大器的关系式,即

$$v_s(t) = v_i(t) - v_o(t)$$
$$v_i(t) = K_1\theta_i(t)$$
$$v_o(t) = K_1\theta_o(t)$$
$$v_a(t) = K_a v_s(t) \tag{2.32}$$

其中,K_1 是转换装置(这里是电位器)的转换系数;K_a 是放大器的电压放大系数。

假设负载力矩为零,则电机的电枢电压 v_a 和角位移 θ_m 之间的关系为

$$JR_a\frac{d^2\theta_m}{dt^2} + (BR_a + C_eC_m)\frac{d\theta_m}{dt} = C_m v_a \tag{2.33}$$

其中,J 为等效转动惯量;B 为等效阻尼系数。至于减速箱,我们有

$$\theta_o(t) = \frac{\theta_m(t)}{i} \tag{2.34}$$

最终有

$$J\frac{d^2\theta_o(t)}{dt^2} + F\frac{d\theta_o(t)}{dt} + K\theta_o(t) = K\theta_i(t) \tag{2.35}$$

式中 $F = \frac{BR_a + C_mC_e}{iR_a}$,$K = \frac{C_mK_aK_1}{iR_a}$。取拉普拉斯变换得到

$$\frac{\theta_o(s)}{\theta_i(s)} = \frac{K}{Js^2 + Fs + K} \tag{2.36}$$

2.3 非线性系统的线性近似

对于我们来说,只要有可能就用线性时不变微分方程作为物理系统的模型是很有用的。在这一章已经介绍了几类物理系统的线性模型。

绝大部分物理系统在变量的某一范围内是线性的。例如例2.3中的弹簧-质量-阻尼器系统,只要质量元件偏离量 $x(t)$ 不大,该系统就是线性的并可以用线性微分方程描述。但是,如果 $x(t)$ 连续不断地增大,弹簧最终将拉伸过度并断裂。所以,每一个系统都必须考虑线性的问题和线性的适用范围。

一个由关系式 $y = x^2$ 表示其运动特性的系统不是线性的,因为叠加性不能得到满足。由关系式 $y = ax + b$ 描述的系统也不是线性的,因为它不满足齐次性。但是第二个系统在工作点 (x_0, y_0) 附近对于小的变化 Δx 和 Δy 可以认为是线性的。当 $x = x_0 + \Delta x$ 和 $y = y_0 + \Delta y$ 时,我们有

$$y_0 + \Delta y = a(x_0 + \Delta x) + b \tag{2.37}$$

因此

$$\Delta y = a\Delta x \tag{2.38}$$

该关系式满足线性系统的叠加性和齐次性。

严格地讲,所有物理系统在本质上都是非线性的。因此,当使用一个物理系统的线性模型时,是在应用某种形式的线性化。很多机械系统和电系统在变量的一个合理范围内可以认为是线性的。对于热力和流体环节情况却并不总是这样,它们在更多的情况下符

合非线性。值得庆幸的是,在假设是小偏差信号的情况下,常常可以使非线性环节线性化。这是用于获取电路和晶体管的线性等效电路的常规方法。在很多情况下,线性化模型可以得到精确的结果,即线性模型可以精确地描述系统的特性。

在另一些情况下,线性化模型表示物理系统的近似效果很不好,没有丝毫的把握可以使用。在这些情况下应当采用其他的分析方法。对于有些至今还没有找到有效分析方法的系统,就必须通过仿真确定系统的特性。

非线性增益

为了引入线性化的概念,我们考虑图 2.10 的非线性增益特性,该特性的输入为 x,输出为 $y = f(x)$。假设在我们感兴趣的范围内 $y = f(x)$ 是一个光滑的函数,而且如图所示增益曲线上的工作点在输入值 x_0 处。

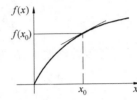

图 2.10 非线性增益特性

假设输入 x 出现了一个小的变化 Δx,那么非线性特性的输入为 $x = x_0 + \Delta x$。$f(x)$ 在工作点 (x_0, y_0) 附近的泰勒级数展开式为

$$y = f(x_0) + \frac{df(x)}{dx}\bigg|_{x=x_0} (x - x_0) + \frac{1}{2!} \frac{d^2 f(x)}{dx^2}\bigg|_{x=x_0} (x - x_0)^2 + \cdots \quad (2.39)$$

记 $\Delta x = x - x_0$,该展开式可以改写为

$$y - y_0 = \Delta y = \frac{df(x)}{dx}\bigg|_{x=x_0} \cdot \Delta x + \frac{1}{2!} \frac{d^2 f(x)}{dx^2}\bigg|_{x=x_0} \cdot (\Delta x)^2 + \cdots \quad (2.40)$$

增益特性在工作点上切线的斜率为

$$\frac{df(x)}{dx}\bigg|_{x=x_0}$$

它能很好地近似表示偏离工作点小范围 $(x - x_0)$ 内的曲线。这样,作为一种合理的近似,忽略 $f(x)$ 展开式中的高次导数项,式(2.40)变为

$$\Delta y \approx \frac{df(x)}{dx}\bigg|_{x=x_0} \cdot \Delta x \quad (2.41)$$

为了方便起见,式(2.41)经常改写为

$$y = Kx \quad (2.42)$$

其中

$$K = \frac{df(x)}{dx}\bigg|_{x=x_0}$$

但是应当注意,这一方程中的两个变量 x 和 y 指的都是增量。

Δy 和 Δx 之间的关系是线性的,因为函数在工作点上计算得到的导数是常数。图 2.10 的非线性增益在工作点 (x_0, y_0) 附近线性化了。式(2.41)的精度与 Δx 的大小以及 $y = f(x)$ 在工作点 (x_0, y_0) 附近的平滑程度有关。

对于有两个输入的平滑非线性函数 $y = f(x_1, x_2)$,线性化有着类似的方法。$y = f(x_1, x_2)$ 关于工作点 $y_0 = f(x_{10}, x_{20})$ 的泰勒级数展开式为

$$y = f(x_{10}, x_{20}) + \left[\frac{\partial f}{\partial x_1}\bigg|_{\substack{x_1=x_{10}\\x_2=x_{20}}} \cdot (x_1 - x_{10}) + \frac{\partial f}{\partial x_2}\bigg|_{\substack{x_1=x_{10}\\x_2=x_{20}}} \cdot (x_2 - x_{20}) \right] +$$

$$\frac{1}{2!}\left[\left.\frac{\partial^2 f}{\partial x_1^2}\right|_{\substack{x_1=x_{10}\\x_2=x_{20}}} \cdot (x-x_{10})^2 + 2\left.\frac{\partial^2 f}{\partial x_1 \partial x_2}\right|_{\substack{x_1=x_{10}\\x_2=x_{20}}} \cdot (x-x_{10})(x-x_{20}) + \right.$$

$$\left.\left.\frac{\partial^2 f}{\partial x_2^2}\right|_{\substack{x_1=x_{10}\\x_2=x_{20}}} \cdot (x-x_{20})^2\right] + \cdots \tag{2.43}$$

忽略表达式 $y=f(x_1,x_2)$ 中的高阶导数项得到

$$y = f(x_{10},x_{20}) + \left[\left.\frac{\partial f}{\partial x_1}\right|_{\substack{x_1=x_{10}\\x_2=x_{20}}}(x_1-x_{10}) + \left.\frac{\partial f}{\partial x_2}\right|_{\substack{x_1=x_{10}\\x_2=x_{20}}}(x_2-x_{20})\right] \tag{2.44}$$

即

$$y - y_0 = K_1(x_1-x_{10}) + K_2(x_2-x_{20}) \tag{2.45}$$

或者简记为

$$y = K_1 x_1 + K_2 x_2 \tag{2.46}$$

式中

$$K_1 = \left.\frac{\partial f}{\partial x_1}\right|_{\substack{x_1=x_{10}\\x_2=x_{20}}}, \quad K_2 = \left.\frac{\partial f}{\partial x_2}\right|_{\substack{x_1=x_{10}\\x_2=x_{20}}}$$

例 2.6 图 2.11 为可控硅三相桥式整流电路的特性曲线，图中输入为控制角 α，输出为整流电压 E_d，而正常的工作点则为 A 点，即 $(E_d)_0 = E_{d0}\cos\alpha_0$，其中 E_{d0} 为 $\alpha=0$ 时的整流电压。确定该可控硅整流电路的线性化数学模型。

解 E_d 和 α 之间的关系为

$$E_d = 2.34 E_2 \cos\alpha = E_{d0}\cos\alpha \tag{2.47}$$

式中，E_2 为交流相电压的有效值。

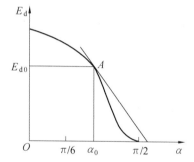

图 2.11 可控硅三相桥式整流电路的特性曲线

如果控制角 α 在一个小范围内变化，这一个非线性特性就能够线性化。由式（2.47）可有

$$E_d - E_{d0}\cos\alpha_0 = K_s(\alpha - \alpha_0) \tag{2.48}$$

或者

$$\Delta E_d = K_s \Delta\alpha \tag{2.49}$$

其中

$$K_s = \left.\frac{dE_d}{d\alpha}\right|_{\alpha=\alpha_0} = -E_{d0}\sin\alpha_0$$

2.4 方 框 图

由于简易和通用的原因，方框图经常用于描绘所有各种类型的控制系统。方框图可以只是用于简单地表示一个系统的组成和内部连接；或者，它也可以与传递函数一起用于表示整个系统的动态因果关系。

如果系统所有环节的数学关系和功能关系都已经知道，方框图就可以用作该系统解析求解或者计算机求解的一个参考。而且，如果系统的所有环节都假定是线性的，则可以借助于方框图代数获取整个系统的传递函数。

2.4.1 方框图的概念

方框图由一些单向传递信号、具有运算功能的方框组成,方框表示我们所感兴趣的变量之间的传递函数。按照定义,一个线性时不变系统或者元部件的传递函数是其输出量的拉普拉斯变换与输入量的拉普拉斯变换之比。令 $R(s)$ 为输入量,$C(s)$ 为输出量,$G(s)$ 为传递函数,则图 2.12 所示的方框图是另一种以图形方式表示的代数方程

$$C(s) = G(s)R(s) \qquad (2.50)$$

的方法。这一方框的输出等于方框中给出的传递函数乘以输入。如图所示,由信号线的箭头方向明确地规定了输入和输出。

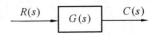

图 2.12 输入 - 输出关系的方框图表示

有时候用方框图表示一个代数方程时还需要增加的元素是相加点,这可以用表示方程

$$C(s) = G_1(s)R_1(s) \pm R_2(s) \qquad (2.51)$$

的图 2.13 说明。在方框图中,相加点用一个小圆圈表示。按照规定,离开相加点的信号等于进入该点各信号的代数和,各个分量的符号根据放在靠近该分量箭头旁的正负号确定。注意,尽管一个相加点可以有任意多个输入,但是只有一个输出。

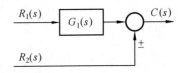

图 2.13 有相加点的方框图

为了表示一个比较复杂的系统,要用到方框的相互连接。例如,图 2.14 中所示系统有两个输入量和两个输出量。利用传递函数关系,我们可以写出输出量的联立方程为

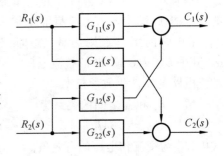

图 2.14 双输入 - 双输出系统的方框图表示

$$\begin{aligned}C_1(s) &= G_{11}(s)R_1(s) + G_{12}(s)R_2(s) \\ C_2(s) &= G_{21}(s)R_1(s) + G_{22}(s)R_2(s)\end{aligned} \qquad (2.52)$$

式中,$C_{ij}(s)$ 是联系第 i 个输出量和第 j 个输入量的传递函数。表示这一方程组的方框图见图 2.14,图中另一个有用的方框图元素是记号为"·"的分支点。

通常,一个方框图由一些方框、带有箭头的信号线、相加点以及分支点组成。

2.4.2 建立方框图的例子

现在给出几个简单的例子,用图说明如何根据一个物理系统的动态方程组建立它的方框图。

例 2.7 图 2.15 中所示的为 RC 无源网络,图中 $v_i(t)$ 为输入,$v_o(t)$ 为输出。利用同一个图中指定的回路电流和节点电压,一组独立的方程为

$$v_i = R_1 i_1 + v_1$$

$$v_1 = \frac{1}{C_1}\int (i_1 - i_2)\,dt$$

$$v_1 = R_2 i_2 + v_o$$

$$v_o = \frac{1}{C_2}\int i_2\,dt$$

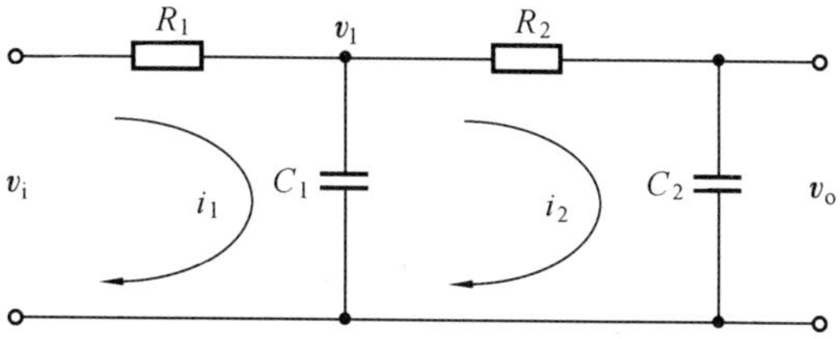

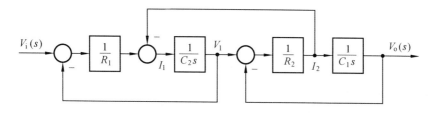

图 2.15　RC 无源网络

取拉普拉斯变换,并将变换结果写成因果关系形式得到

$$I_1(s) = \frac{1}{R_1}[V_i(s) - V_1(s)]$$

$$V_1(s) = \frac{1}{C_1 s}[I_1(s) - I_2(s)]$$

$$I_2(s) = \frac{1}{R_2}[V_1(s) - V_o(s)]$$

$$V_o(s) = \frac{1}{C_2 s}I_2(s)$$

把变量 $V_i(s), I_1(s), V_1(s), I_2(s), V_o(s)$ 自左至右按照顺序排列,则可构成 RC 无源网络的方框图,如图 2.16 所示。

图 2.16　RC 无源网络的方框图

例 2.8　由方程(2.24)~(2.27),一台电枢控制直流电机的动态行为可以描述为

$$v_a(t) = L_a \frac{di_a(t)}{dt} + R_a i_a(t) + e_b(t)$$

$$T_m(t) = C_m i_a(t)$$

$$e_b(t) = C_e \omega_m(t)$$

$$T_m(t) = J_m \frac{d\omega_m(t)}{dt} + B_m \omega_m(t) + T_L(t)$$

$$\omega_m(t) = \frac{d\theta_m(t)}{dt}$$

其中，v_a 为输入电枢电压；θ_m 为输出角位移；i_a 为电枢电流；e_b 为反电势；ω_m 为电机转速；T_m 为电磁转矩；T_L 为负载力矩。取拉普拉斯变换得到

$$V_a(s) = (L_a s + R_a)I_a(s) + E_b(s)$$

$$T_m(s) = C_m I_a(s)$$

$$E_b(s) = C_e \omega_m(s)$$

$$T_m(s) = (J_m s + B_m)\omega_m(s) + T_L(s)$$

$$\omega_m(s) = s\theta_m(s)$$

将以上方程改写为因果关系形式，得到

$$I_a(s) = \frac{1}{L_a s + R_a}[V_a(s) - E_b(s)]$$

$$T_m(s) = C_m I_a(s)$$

$$E_b(s) = C_e \omega_m(s)$$

$$\omega_m(s) = \frac{1}{J_m s + B_m}[T_m(s) - T_L(s)]$$

$$\theta_m(s) = \frac{1}{s}\omega_m(s)$$

考虑到 $E_b(s)$ 是反电势，并将变量 $V_a(s)$，$I_a(s)$，$T_m(s)$，$\omega_m(s)$，$\theta_m(s)$ 自左至右按照顺序排列，则可构成该电机的方框图如图 2.17 所示。

虽然直流电机是一个开环系统，图 2.17 中的方框图表明该电枢控制电机有一个由反电势造成的"内置"反馈回路。

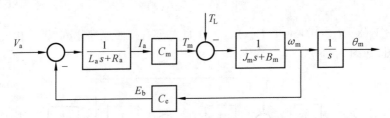

图 2.17　电枢电压控制直流伺服电机的方框图

2.4.3　方框图的简化

为了分析一个复杂的反馈控制系统或者要获得该系统总的传递函数，通过适当的等效变换重新安排方框图的结构，使得方框图能够容易分析或者使系统的方框图简化成一个方框都是常用的方法。

1. 方框图代数

应用方框图代数时，在下面所讨论的三种情况下，方框图可以直接简化成一个方框。

ⅰ. 串联方框。这种结构由两个（或更多个）方框组成如图 2.18 所示，其中每一个方框的传递函数都是已知的。在这种情况下，第一个方框的输出是第二个方框的输入。两个以这样的方式连接的方框称为串联方框。这样，总的传递函数就是

$$\frac{C(s)}{R(s)} = \frac{C(s)}{X(s)} \cdot \frac{X(s)}{R(s)} = G_1(s)G_2(s) \tag{2.53}$$

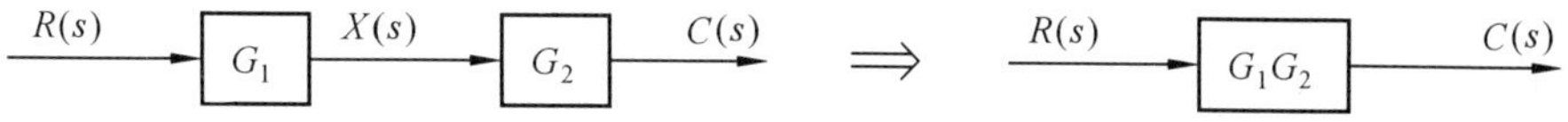

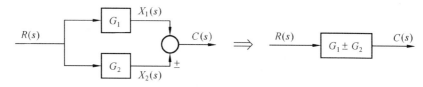

图 2.18　串联方框的简化

根据这条规则,串联连接的方框可以合并成一个方框,而总的传递函数只是各个方框传递函数的乘积。这一规则可以扩大到两个以上串联的方框。

ⅱ. 并联方框。另一种要考虑的结构如图 2.19 所示。各个方框的输入相同,而系统的输出则为各个方框输出的总和。这些方框称为是并联的,其总的传递函数就是

$$\frac{C(s)}{R(s)} = \frac{X_1(s) \pm X_2(s)}{R(s)} = G_1(s) \pm G_2(s) \tag{2.54}$$

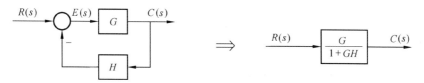

图 2.19　并联方框图的简化

根据这一规则,并联连接的方框可以合并成一个方框,而总的传递函数则是各个方框传递函数的总和。我们也可以看到,这一规则可以直接扩大到两个以上并联的方框。

ⅲ. 典型反馈连接。如图 2.20 所示,这一系统包含一条传递函数为 $G(s)$ 的前向通路和一条传递函数为 $H(s)$ 的反馈通路。由方框图可知

$$E(s) = R(s) - H(s)C(s) \tag{2.55}$$

$$C(s) = G(s)E(s) \tag{2.56}$$

将式(2.55)代入式(2.56)得到

$$C(s) = G(s)[R(s) - H(s)C(s)] = G(s)R(s) - G(s)H(s)C(s) \tag{2.57}$$

该式整理后结果为

$$[1 + (G(s)H(s))]C(s) = G(s)R(s) \tag{2.58}$$

即

$$\frac{C(s)}{R(s)} = \frac{G(s)}{1 + G(s)H(s)} \tag{2.59}$$

因此,该方框图可以简化成单一的方框,如图 2.20 所示。这一个闭环传递函数特别重要,因为它代表了许多实际存在的控制系统。

图 2.20　典型反馈连接的简化

注意,如果反馈回路是正反馈而不是负反馈,则

$$\frac{C(s)}{R(s)} = \frac{G(s)}{1 - G(s)H(s)} \tag{2.60}$$

2. 方框移动操作规则

为了简化一个方框图,有时候把个别的方框移动到相加点或分支点的另一侧是很明智的。这样做的时候,另外几条规则将会很有用。这些规则归纳在图 2.21 中,它们是:

ⅳ. 相加点移动到方框的前面。
ⅴ. 相加点移动到方框的后面。
ⅵ. 分支点移动到方框的前面。
ⅶ. 分支点移动到方框的后面。

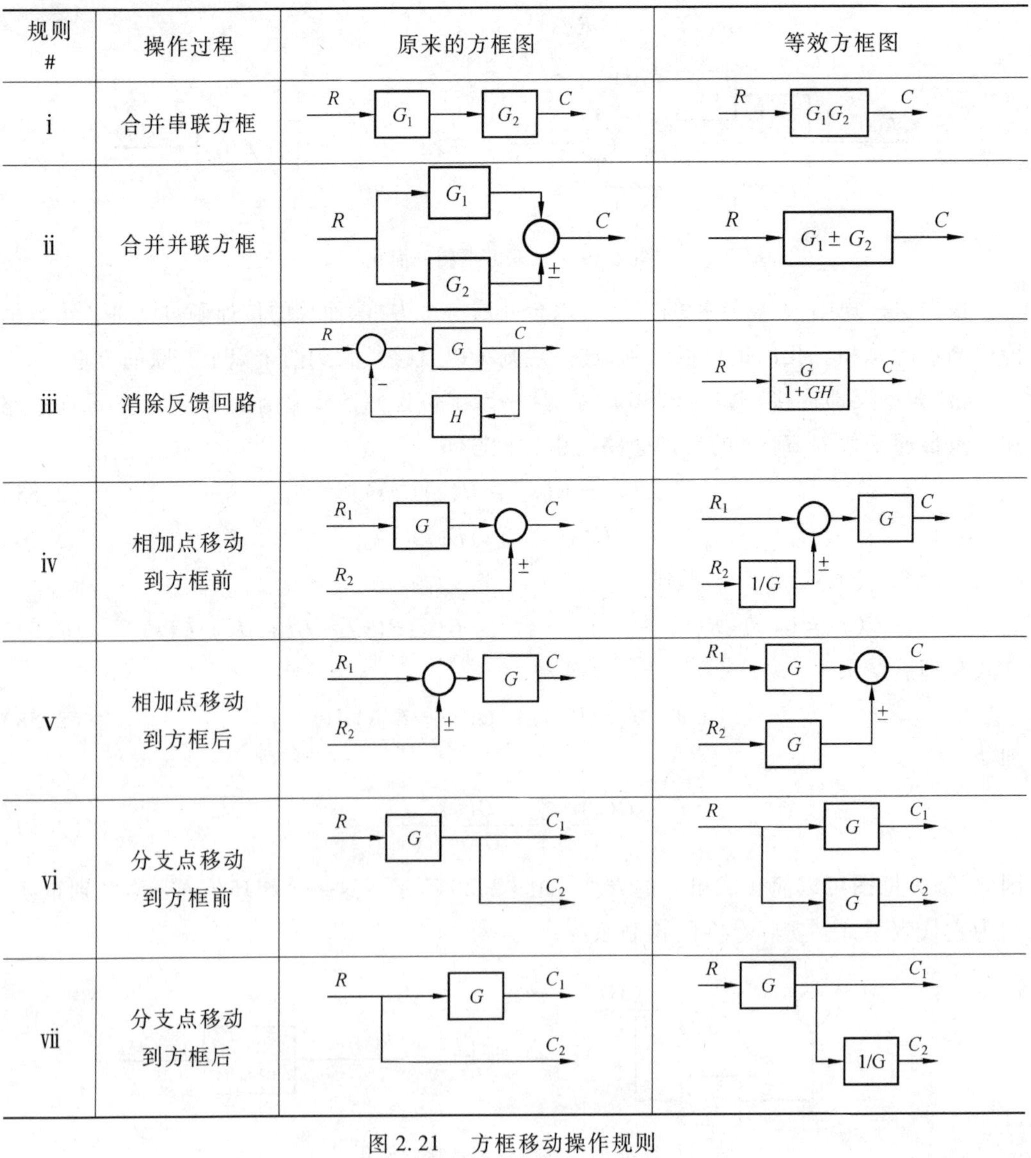

规则 #	操作过程	原来的方框图	等效方框图
ⅰ	合并串联方框	$R \to G_1 \to G_2 \to C$	$R \to G_1 G_2 \to C$
ⅱ	合并并联方框	R 分别经 G_1、G_2 汇合 $\pm$ 得 C	$R \to G_1 \pm G_2 \to C$
ⅲ	消除反馈回路	$R \to \ominus \to G \to C$,反馈经 H	$R \to \dfrac{G}{1+GH} \to C$
ⅳ	相加点移动到方框前	$R_1 \to G \to \pm \to C$,R_2 相加在后	$R_1 \to \pm \to G \to C$,$R_2 \to 1/G \to \pm$
ⅴ	相加点移动到方框后	$R_1 \to \pm \to G \to C$,R_2 相加在前	$R_1 \to G \to \pm \to C$,$R_2 \to G \to \pm$
ⅵ	分支点移动到方框前	$R \to G \to C_1$,分支 C_2 在后	$R \to G \to C_1$,$R \to G \to C_2$
ⅶ	分支点移动到方框后	$R \to G \to C_1$,分支 C_2 在前	$R \to G \to C_1$,$R \to 1/G \to C_2$

图 2.21 方框移动操作规则

在每一种情况下,等效的方框图都具有与原来方框图相同的输出方程。

一般情况下,一个复杂的方框图可以利用图 2.21 中的规则通过系统的变换完成方框图的简化。简化时按照以下顺序进行。

第 1 步:合并所有的串联方框(规则 ⅰ);

第 2 步:合并所有的并联方框(规则 ⅱ);

第 3 步:消除所有的内回路(规则 ⅲ);

第 4 步:相加点／分支点移动到方框的前面／后面(规则 ⅳ ～ ⅶ)。

例 2.9 简化图 2.22 中的方框图,并确定传递函数 $C(s)/R(s)$。

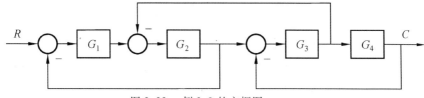

图 2.22 例 2.9 的方框图

解 观察图 2.22 可以看到,图中没有可以简化的串联方框、并联方框,也没有可以简化的原型反馈回路。简化从第 4 步开始,如图 2.23(b) 所示,将相加点 s_2 移动到它左侧方框的前面,并与相加点 s_1 交换位置。然后利用第 7 步,将分支点 p_2 移动到它右侧方框的后面,并交换 p_2 和 p_3 的位置。

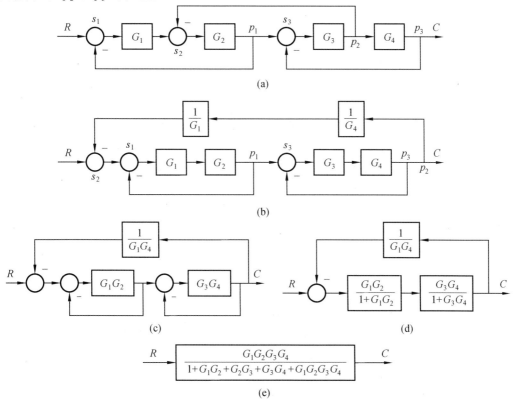

图 2.23 例 2.9 的简化过程

可以看出,由 $C_1(s)$ 和 $C_2(s)$ 组成的部分以及由 $C_3(s)$ 和 $C_4(s)$ 组成的部分分别是串联连接,并可以简化成如图 2.23(c) 的等效方框。

然后就可以看出有两个简单的反馈回路,并可以简化成如图 2.23(d) 的等效方框。

显然,由此我们就可以把给定的方框图简化成如图 2.23(e) 所示单一的方框。

最后,获得传递函数为

$$\frac{C(s)}{R(s)} = \frac{G_1 G_2 G_3 G_4}{1 + G_1 G_2 + G_2 G_3 + G_3 G_4 + G_1 G_2 G_3 G_4}$$

2.5 信号流图

方框图对于分析和设计控制系统是很有用的。但是,对于一个内在联系相当复杂的系统的传递函数,方框图的简化过程很麻烦,并且常常难以完成。控制系统的另一种表示方法是信号流图,它看上去有点像一个简化的方框图。信号流图法的主要优点是流图增益公式的实用性。该公式不需要流图的任何简化过程或者移动操作就可以给出系统变量之间的关系。

2.5.1 用于信号流图的一些定义

信号流图是由一些通过有向支路连接起来的节点组成的图,它是线性代数方程组的一种图形表示方法。例如,图 2.24 所示的图形就是以下方程组的信号流图,其中

$$x_2 = ax_1 + cx_3$$
$$x_3 = bx_2 + dx_4$$
$$x_5 = x_3$$

其中,x_1, x_2, x_3, x_4 和 x_5 为变量。

节点代表系统的变量,并用符号"●"表示。节点执行两个功能:① 对所有流入支路的信号求和。② 把节点总的信号(所有流入信号的和)传输到所有的流出支路。支路是连接在两个节点之间的单向线段,并起着信

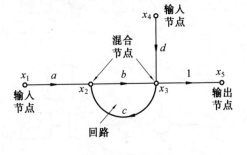

图 2.24　信号流图

号乘法器的作用,信号流动的方向由放置在支路上的箭头指示,而乘数(增益或传递函数)则放在靠近箭头的地方。此外,下面的术语对于使用流图增益公式是很有用的。

源点(输入节点):源点是只有流入支路的节点。例如,在图 2.24 中节点 x_1 和 x_4 是源点。

陷点(输出节点):陷点是只有流出支路的节点。例如,在图 2.24 中节点 x_5 是一个陷点。通过引入具有单位增益的支路,任何一个非输入节点都可以处理成陷点。

混合节点(一般节点):混合节点是既有流入支路又有流出支路的节点。

通路:通路是相同方向支路的任何一个顺序连接。

前向通路:前向通路是从输入节点出发到输出节点结束的通路,而且沿该通路有不多于一次的节点。例如,在图 2.24 中有一条从节点 x_1 到 x_5 的前向通路,另外还有一条从节

点 x_4 到 x_5 的前向通路。

回路：回路是一条起始和终止于同一节点的通路,而且沿该通路没有其他节点相遇不止一次。例如,在图 2.24 中只有一个回路。

不接触回路：有些回路如果它们没有任何公共节点则称为不接触回路。

通路增益：通路增益是通过一条通路时所遇到的支路增益的乘积。

前向通路增益：前向通路增益定义为前向通路的通路增益。

回路增益：回路增益定义为回路的通路增益。

2.5.2 绘制信号流图的例子

信号流图只是在表示一组代数方程时能够指示出各变量之间相互关系的一种图示方法。

例 2.10 仍然考虑例 2.7 的 RC 无源网络。图 2.15 中所示 RC 无源网络可以用以下写成因果关系形式的方程组描述

$$I_1(s) = \frac{1}{R_1}[V_i(s) - V_1(s)] = \frac{1}{R_1}V_i(s) - \frac{1}{R_1}V_1(s)$$

$$V_1(s) = \frac{1}{C_1 s}[I_1(s) - I_2(s)] = \frac{1}{C_1 s}I_1(s) - \frac{1}{C_1 s}I_2(s)$$

$$I_2(s) = \frac{1}{R_2}[V_1(s) - V_o(s)] = \frac{1}{R_2}V_1(s) - \frac{1}{R_2}V_o(s)$$

$$V_o(s) = \frac{1}{C_2 s}I_2(s)$$

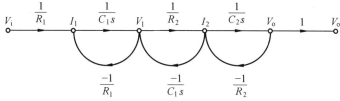

图 2.25 例 2.10 的信号流图

自左到右顺序安排变量 $V_i(s),I_1(s),V_1(s),I_2(s),V_o(s)$,可以画出该网络的信号流图如图 2.25 所示。

2.5.3 梅逊增益公式

一般情况下,一个源点(输入量)和一个陷点(输出量)之间总的增益(传递函数)可以使用梅逊增益公式计算

$$T = \sum_{k=1}^{n} \frac{T_k \Delta_k}{\Delta} \tag{2.61}$$

式中,T 为指定源点和陷点之间总的增益;T_k 为第 k 条前向通路的增益;n 为前向通路的总数;Δ 为流图的特征式;Δ_k 为流图特征式关于第 k 条前向通路的余子式。

流图的特征式为

$$\Delta = 1 - \sum L_{m1} + \sum L_{m2} - \sum L_{m3} + \cdots \tag{2.62}$$

其中,L_{mr} 是 r 个不接触回路的第 m 种可能组合的增益乘积,即 $\Delta = 1 - ($所有不同回路增益之和$) + ($所有两两互不接触回路增益乘积之和$) - ($所有三三互不接触回路增益乘积之和$) + \cdots$

余子式 Δ_k 是去掉与第 k 条前向通路相接触的回路后的特征式。

例 2.11 考虑例 2.10 的信号流图。利用增益公式确定传递函数 $V_o(s)/V_i(s)$。

解 通过观测,由图 2.25 的信号流图可以得到以下结论:

① 信号流图有三个回路

$$L_{11} = -\frac{1}{R_1 C_1 s}, \quad L_{21} = -\frac{1}{R_2 C_1 s}, \quad L_{31} = -\frac{1}{R_2 C_2 s}$$

② 回路 L_{11} 与 L_{31} 不相接触。因此,流图的特征式为

$$\Delta = 1 - \left(-\frac{1}{R_1 C_1 s} - \frac{1}{R_2 C_1 s} - \frac{1}{R_2 C_2 s}\right) + \left(-\frac{1}{R_1 C_1 s}\right)\left(-\frac{1}{R_2 C_2 s}\right) =$$

$$\frac{R_1 R_2 C_1 C_2 s^2 + (R_1 C_1 + R_2 C_1 + R_2 C_2)s + 1}{R_1 R_2 C_1 C_2 s^2}$$

③ 从 $V_i(s)$ 到 $V_o(s)$ 只有一条前向通路,而且该前向通路的增益为

$$T_1 = \frac{1}{R_1} \cdot \frac{1}{C_1 s} \cdot \frac{1}{R_2} \cdot \frac{1}{C_2 s} = \frac{1}{R_1 R_2 C_1 C_2 s^2}$$

④ 由于该前向通路与所有回路都接触,我们有 $\Delta_1 = 1$,所以,该网络的传递函数为

$$\frac{V_o(s)}{V_i(s)} = \frac{T_1 \Delta_1}{\Delta} = \frac{1}{R_1 R_2 C_1 C_2 s^2 + (R_1 C_1 + R_2 C_1 + R_2 C_2)s + 1}$$

由于方框图和信号流图之间的相似性,它们两者的输入 - 输出关系都可以用梅逊增益公式确定。给定一个线性系统的方框图后,前向通路增益、回路增益、特征式和余子式都可以直接由方框图获取。

例 2.12 确定图 2.26 中所示方框图的传递函数 $C(s)/R(s)$,$E(s)/R(s)$ 和 $X_1(s)/R(s)$。

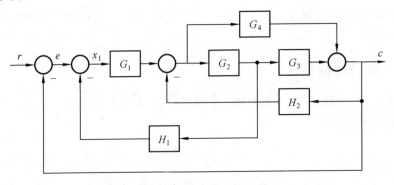

图 2.26 例 2.12 的方框图

解 (1) 确定传递函数 $C(s)/R(s)$。

① 图中有 5 个单独的回路，这些回路的传递函数分别为

$L_{11} = -G_1G_2H_1$，$L_{21} = -G_2G_3H_2$，$L_{31} = -G_1G_2G_3$，$L_{41} = -G_4H_2$，$L_{51} = -G_1G_4$

② 图中没有不相接触的回路，因此流图的特征式为

$$\Delta = 1 + G_1G_2H_1 + G_2G_3H_2 + G_1G_2G_3 + G_4H_2 + G_1G_4$$

③ 从 $R(s)$ 到 $C(s)$ 有两条前向通路，由此我们有

$$T_1 = G_1G_2G_3, \quad \Delta_1 = 1$$
$$T_2 = G_1G_4, \quad \Delta_2 = 1$$

所以，从 $R(s)$ 到 $C(s)$ 的传递函数为

$$\frac{C(s)}{R(s)} = \sum_{k=1}^{2} \frac{T_k\Delta_k}{\Delta} = \frac{G_1G_2G_3 + G_1G_4}{1 + G_1G_2H_1 + G_2G_3H_2 + G_1G_2G_3 + G_4H_2 + G_1G_4}$$

(2) 确定传递函数 $E(s)/R(s)$。注意，一个流图的特征式是唯一的。图中只有一条从 $R(s)$ 到 $E(s)$ 的前向通路，而且该通路的传递函数为

$$T_1 = 1$$

该前向通路与回路 L_{11}、L_{21} 以及 L_{41} 不相接触，由此该前向通路的余子式为

$$\Delta_1 = 1 + G_1G_2H_1 + G_2G_3H_2 + G_4H_2$$

所以，从 $R(s)$ 到 $E(s)$ 传递函数为

$$\frac{E(s)}{R(s)} = \frac{T_1\Delta_1}{\Delta} = \frac{1 + G_1G_2H_1 + G_2G_3H_2 + G_4H_2}{1 + G_1G_2H_1 + G_2G_3H_2 + G_1G_2G_3 + G_4H_2 + G_1G_4}$$

(3) 确定传递函数 $X_1(s)/R(s)$。从 $R(s)$ 到 $X_1(s)$ 只有一条前向通路，而且该通路的传递函数为

$$T_1 = 1$$

在这一情况下，前向通路与回路 L_{21} 和 L_{41} 不相接触，由此，该前向通路的余子式为

$$\Delta_1 = 1 + G_2G_3H_2 + G_4H_2$$

最终，从 $R(s)$ 到 $X_1(s)$ 的传递函数为

$$\frac{X_1(s)}{R(s)} = \frac{1 + G_2G_3H_2 + G_4H_2}{1 + G_1G_2H_1 + G_2G_3H_2 + G_1G_2G_3 + G_4H_2 + G_1G_4}$$

2.5.4 由方框图绘制信号流图

梅逊公式可以用于直接根据线性系统的方框图确定系统的输入 – 输出关系。但是，为了能够清楚地识别所有的回路和不相接触回路部分，如果在应用梅逊增益公式之前先画出方框图等效的信号流图，有时候可能还是有些用处的。

为了说明信号流图和方框图是如何相关的，图 2.27 给出了例 2.12 的方框图及其等效的信号流图。

由于信号流图中的节点可以解释为所有流进信号的相加点，在通常绘制信号流图时建议：

① 所有输入量和扰动都选为源点；

② 所有输出量都选为陷点；

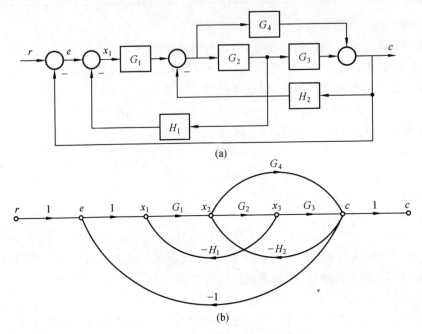

图 2.27 方框图及其等效的信号流图

③ 所有其他位于相加点后的变量和离开分支点的变量选为混合节点。

2.6 线性系统的传递函数

在熟悉了物理系统的数学建模之后,可以把建模的话题延伸到包括控制系统的特性,例如对输入测试信号的瞬态响应、稳态误差、扰动抑制等等。以下定义在反馈控制系统中的传递函数,在我们的学习中是很有用处的。

考虑画在图 2.28 中具有典型结构的控制系统,该系统只有一条从输入到输出的前向通路和一个回路。在图中 $R(s)$ 为输入,$C(s)$ 为输出,$D(s)$ 为作用于前向通路的扰动,$\varepsilon(s)$ 为偏差,$G_1(s)$ 和 $G_2(s)$ 分别为扰动作用点前后前向通路的传递函数,$H(s)$ 则为反馈传递函数。通常前向通路的传递函数记为 $G(s)$。显然,对于图 2.28 所示系统有

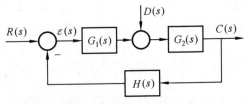

图 2.28 具有典型结构的控制系统

$$G(s) = G_1(s)G_2(s) \tag{2.63}$$

在许多情况下 $H(S)$ 等于 1 或者是其他的常数。

2.6.1 闭环传递函数

利用梅逊公式很容易获得图 2.29 所示系统的以下传递函数

$$\Phi(s) = \frac{C(s)}{R(s)} = \frac{G_1(s)G_2(s)}{1 + G_1(s)G_2(s)H(s)} = \frac{G(s)}{1 + G(s)H(s)} \qquad (2.64)$$

$$\Phi_d(s) = \frac{C(s)}{D(s)} = \frac{G_2(s)}{1 + G_1(s)G_2(s)H(s)} = \frac{G_2(s)}{1 + G(s)H(s)} \qquad (2.65)$$

$$\Phi_\varepsilon(s) = \frac{\varepsilon(s)}{R(s)} = \frac{1}{1 + G_1(s)G_2(s)H(s)} = \frac{1}{1 + G(s)H(s)} \qquad (2.66)$$

$$\Phi_{\varepsilon,d}(s) = \frac{\varepsilon(s)}{D(s)} = \frac{-G_2(s)H(s)}{1 + G_1(s)G_2(s)H(s)} = \frac{-G_2(s)H(s)}{1 + G(s)H(s)} \qquad (2.67)$$

其中 $\Phi(s)$ 称为输出对于输入的闭环传递函数并经常简称为闭环传递函数；$\Phi_d(s)$ 称为输出对于扰动的闭环传递函数；$\Phi_\varepsilon(s)$ 称为偏差对于输入的闭环传递函数；而 $\Phi_{\varepsilon,d}(s)$ 则称为偏差对于扰动的闭环传递函数。

如果控制信号 $R(s)$ 和扰动信号 $D(s)$ 同时作用于系统，那么，输出信号为

$$C(s) = \Phi(s)R(s) + \Phi_d(s)D(s) = \\ \frac{G_1(s)G_2(s)}{1 + G_1(s)G_2(s)H(s)}R(s) + \frac{G_2(s)}{1 + G_1(s)G_2(s)H(s)}D(s) \qquad (2.68)$$

类似地，在这种情况下偏差信号为

$$\varepsilon(s) = \Phi_\varepsilon(s)R(s) + \Phi_{\varepsilon,d}(s)D(s) = \\ \frac{1}{1 + G_1(s)G_2(s)H(s)}R(s) - \frac{G_2(s)H(s)}{1 + G_1(s)G_2(s)H(s)}D(s) \qquad (2.69)$$

观察这四个闭环传递函数可以发现，它们具有相同的分母 $1 + G(s)H(s)$，即它们具有相同的闭环极点。这意味着一个系统的闭环极点仅仅取决于方程

$$1 + G(s)H(s) = 0 \qquad (2.70)$$

该方程称为反馈控制系统的特征方程。

2.6.2 开环传递函数

在式(2.70)给出的特征方程中，$G(s)H(s)$ 定义为开环传递函数，而且这一传递函数在许多分析和设计方法中非常有用。对于图 2.28 中所示的单回路反馈系统，开环传递函数正好就是前向通路传递函数与反馈通路传递函数的乘积，即

$$G(s)H(s) = G_1(s)G_2(s)H(s) \qquad (2.71)$$

应当注意，一个闭环系统的开环传递函数既不是开环系统的传递函数，也不是闭环系统的回路传递函数。在图 2.28 中，开环系统的传递函数为 $G_1(s)G_2(s)$，而回路传递函数则为 $-G_1(s)G_2(s)H(s)$。

开环传递函数在许多分析和设计方法中非常有用，而且常常写成以下形式之一。

第一种是多项式形式

$$G(s)H(s) = \frac{M(s)}{N(s)} \qquad (2.72)$$

式中，$M(s)$ 和 $N(s)$ 两者都是用 s 表示的多项式。

第二种是零极点形式

$$G(s)H(s) = \frac{k\prod_{i=1}^{m}(s-z_i)}{\prod_{j=1}^{n}(s-p_j)} \qquad (2.73)$$

式中,$z_i(i=1,2,\cdots,m)$ 和 $p_j(j=1,2,\cdots,n)$ 分别为开环传递函数的零点和极点。

第三种是典型因子形式

$$G(s)H(s) = \frac{K\prod_{i=1}^{m}(\tau s+1)}{s^v \prod_{j=1}^{l}(Ts+1)\prod_{k=1}^{(n-v-l)/2}(\hat{T}^2 s^2 + 2\hat{T}\xi s+1)} \qquad (2.74)$$

式中,K 称为开环增益并定义为

$$K = \lim_{s\to 0} s^v G(s)H(s) \qquad (2.75)$$

2.7 线性系统的脉冲响应

2.7.1 脉冲响应的定义

一个线性系统的脉冲响应定义为输入是单位脉冲时系统的输出响应。因此,对于传递函数为 $G(s)$ 的系统,如果 $r(t) = \delta(t)$,那么系统输出的拉普拉斯变换就是该系统的传递函数,即

$$C(s) = G(s)L[\delta(t)] = G(s) \qquad (2.76)$$

因为单位脉冲函数的拉普拉斯变换为 1。

对式(2.76)的两侧取拉普拉斯反变换得到

$$c(t) = g(t) \qquad (2.77)$$

式中,$g(t)$ 为 $G(s)$ 的拉普拉斯反变换,是线性系统的脉冲响应(有时候也称为权函数)。因此我们可以说,脉冲响应的拉普拉斯变换是传递函数。

由于传递函数是表示线性系统特性的有效方法,在理论上这也就意味着,如果一个线性系统具有零初始状态,那么通过单位脉冲激励该系统并测量系统的输出就可以描述或者识别该系统。在实际应用中,虽然在物理上不能生成纯脉冲信号,但是通常脉宽很窄的脉冲可以提供适宜的近似信号。

2.7.2 脉冲响应的用途

式(2.3)中传递函数的推导是建立在对系统微分方程了解的基础上,而且根据式(2.3)得到解 $C(s)$ 时还假设了 $R(s)$ 和传递函数有解析的形式可供使用。但实际情况并不可能总是这样的,对于很多情况,输入信号 $r(t)$ 不能进行拉普拉斯变换或者只能以实验数据的形式利用。在这样的情况下,为了对系统进行分析,我们将不得不与脉冲响应 $g(t)$ 打交道。

有一类系统称为适应控制系统,在它们的分析和设计中,有时候线性系统脉冲响应的

测定是非常关键的。现实生活中,在一个较长的时间里大部分系统的动态特性都会在某种程度上发生变化。这种变化有可能是由于元部件的磨损、运行环境的缓慢变迁等原因造成的。有些系统只是参数以一种可预见或不可预见的方式随时间变化。例如,制导导弹在飞行中的瞬态特性将由于导弹质量的变化和大气环境的变化而随时间变化。另一方面,对于一个简单的包含质量和摩擦的机械系统,后者将或者由于"老化"或者由于表面状况而发生不可预见的变化。这样,在假设参数已知且固定不变的条件下设计的控制系统,有可能由于系统参数发生变化而得不到令人满意的响应。为了使系统也能具有根据变化的参数和环境自行修正和自行调整的能力,在系统运行期间就有必要连续地或者以适当的时间间隔辨识系统的瞬态特性。辨识的方法之一是测量系统的脉冲响应,使得设计参数可以相应地进行调整,从而在所有的时间都可以获得最佳的控制。

这样,在前面的讨论中已经介绍了线性系统传递函数和脉冲响应的定义。这两个定义通过拉普拉斯变换直接联系起来,实质上它们代表着关于一个系统的相同信息。但是必须强调,只有线性系统而且是初始条件假设为零时才定义有传递函数和脉冲响应。

习　题

题2.1　下面的微分方程代表线性时不变系统,其中$r(t)$表示输入,$c(t)$表示输出。找出各个系统的传递函数。

(a) $\dfrac{d^3c(t)}{dt^3} + 6\dfrac{d^2c(t)}{dt^2} + 11\dfrac{dc(t)}{dt} + 6c(t) = r(t)$

(b) $\dfrac{d^3c(t)}{dt^3} + 3\dfrac{d^2c(t)}{dt^2} + 4\dfrac{dc(t)}{dt} + c(t) = 2\dfrac{dr(t)}{dt} + r(t)$

题2.2　题2.2图所示是一个质量-弹簧-阻尼器系统。确定输入力f与输出位移x之间的微分方程。

题2.3　确定题2.3图所示系统的传递函数$X_2(s)/F(s)$。两个质量元件都在无摩擦的表面上滑动,且有$K = 1$ N/m。

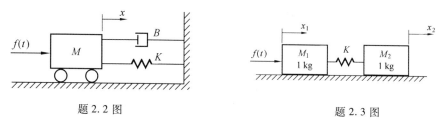

题2.2图　　　　　　　　　　　　题2.3图

题2.4　一个温度调节器对温度的响应用下式表示

$$R = R_0 e^{-0.1T}$$

式中R为电阻,T为温度(单位为℃),$R_0 = 10\,000\ \Omega$。对于工作在$T = 20℃$的调节器,找出小范围温度偏差时的线性化模型。

题2.5　求取题2.5图所示无源网络的传递函数$V_o(s)/V_i(s)$。

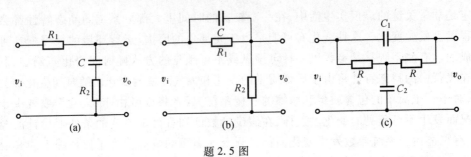

题 2.5 图

题 2.6 求取题 2.6 图所示有源网络的传递函数 $V_o(s)/V_i(s)$。

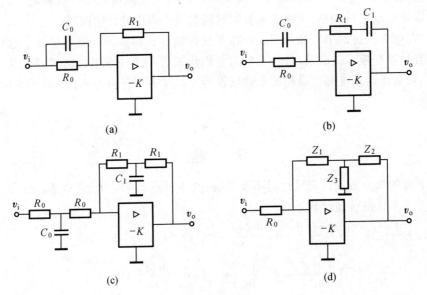

题 2.6 图

题 2.7 为系统绘制显示所有变量的方框图,描述系统的微分方程组为

$$x_1(t) = r(t) - c(t) - n_1(t)$$
$$x_2(t) = K_1 x_1(t)$$
$$x_3(t) = x_2(t) - x_5(t)$$
$$T\dot{x}_4 = x_3(t)$$
$$x_5(t) = x_4(t) - K_2 n_2(t)$$
$$\ddot{c} + \dot{c}(t) = K_0 x_5(t)$$

其中 $r(t)$ 为输入,$c(t)$ 为输出,$n_1(t)$ 和 $n_2(t)$ 是两个扰动,K_0、K_1、K_2 和 T 为常数。

题 2.8 某系统由以下方程组描述

$$X_1(s) = G_1(s)R(s) - G_1(s)[G_7(s) - G_8(s)]C(s)$$
$$X_2(s) = G_2(s)[X_1(s) - G_6(s)X_3(s)]$$
$$X_3(s) = G_3(s)[X_2(s) - G_5(s)C(s)]$$

$$Y(s) = G_4(s)X_3(s)$$

为该系统绘制显示所有变量的方框图,并求取传递函数 $C(s)/R(s)$。

题 2.9 题 2.9 图所示系统有两个输入和两个输出。确定传递函数 $C_1(s)/R_1(s)$,$C_1(s)/R_2(s)$,$C_2(s)/R_1(s)$ 和 $C_2(s)/R_2(s)$。写出 $C_1(s)$ 和 $C_2(s)$ 的表达式。

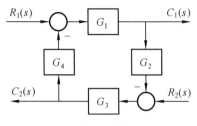

题 2.9 图

题 2.10 通过方框图简化确定题 2.10 图所示系统的传递函数。

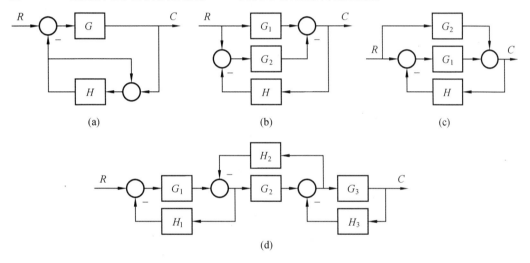

题 2.10 图

题 2.11 通过方框图简化推导题 2.11 图所示系统的传递函数 $C(s)/R(s)$ 和 $C(s)/N(s)$。

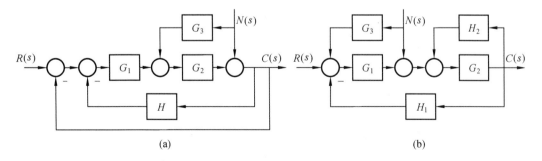

题 2.11 图

题 2.12　通过方框图简化求题 2.12 图所示系统的传递函数。

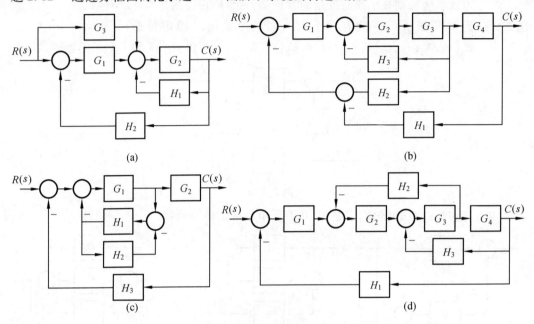

题 2.12 图

题 2.13　若要求图示系统的前向通路的传递函数满足

$$\frac{C(s)}{E(s)} = \frac{100(s+10)}{s(s+5)(s+20)}$$

确定该系统中的 K_1、K_2 和 $H(s)$。

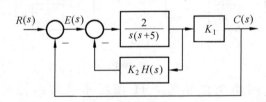

题 2.13 图

题 2.14　题 2.14(a) 和 (b) 图所示两个系统等价吗？说明理由。

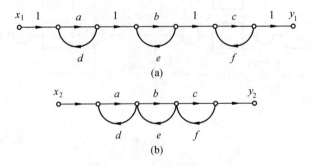

题 2.14 图

题 2.15　求题 2.15 图所示各系统总的传递函数。

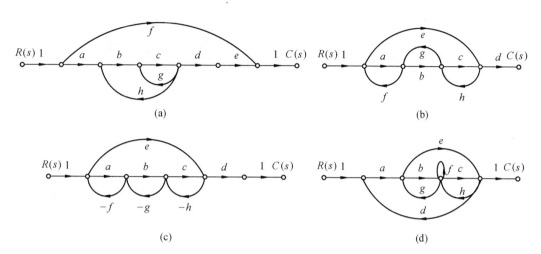

题 2.15 图

题 2.16　求题 2.16 图所示多回路交叉系统的传递函数。

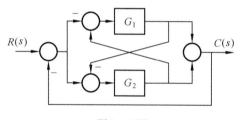

题 2.16 图

题 2.17　绘制题 2.12 图所示方框图等价的信号流图,并用梅逊公式求取各系统的传递函数。

第3章 控制系统的时域分析

3.1 引　言

在大部分控制系统中,评价系统的时间响应是一个令人感兴趣的主题。在分析问题时,输入信号施加于一个系统,而该系统的性能则通过研究时域内的响应进行评价。例如,如果控制系统的目标是使输出量尽可能地接近跟随输入信号,就有必要将输入和输出作为时间的函数进行比较。

3.1.1 瞬态响应和稳态响应

控制系统的时间响应通常分为两个部分:瞬态响应和稳态响应。如果一个时间响应记为 $c(t)$,则瞬态响应和稳态响应可以分别用 $c_t(t)$ 和 $c_{ss}(t)$ 表示。

在控制系统的应用中,当一个响应已经到达它的稳态时仍然可以随着时间变化。稳态响应只是时间趋向无穷大时固定下来的响应部分,即

$$c_{ss}(t) = \lim_{t \to \infty} c(t) \tag{3.1}$$

因此,一个正弦波可以认为是稳态响应,因为当时间趋向无穷大时,正弦波的表现都是固定的。同样地,如果一个响应是由 $c(t)=t$ 所描述的,那么它也可以判定为稳态响应。

瞬态响应定义为时间趋向无穷大时响应中变为零的部分。所以,瞬态响应 $c_t(t)$ 具有以下特性

$$\lim_{t \to \infty} c_t(t) = 0 \tag{3.2}$$

因此也可以说,稳态响应是过渡过程逐渐消逝后响应剩余下来的部分。

所有的控制系统在到达稳态之前都将不同程度地呈现出过渡过程现象。由于在物理系统中不可避免地存在质量、电容、电感等等,响应不能够瞬时地跟随输入的突然变化,因此通常就可以观察到过渡过程。

控制系统的瞬态响应很重要,因为它是系统的动态行为部分,在到达稳态之前,响应与输入或期望输出之间的差异必须密切加以观察。当稳态响应与一定的参考信号相比较时,可以给出系统最终精度的指示。如果输出的稳态响应与参考信号的稳态不能完全一致,我们就说系统存在稳态误差。

3.1.2 时间响应的典型测试信号

与众多电路和通讯系统不一样,许多实际控制系统的输入激励作用事先并不是已知的。在很多情况下,一个控制系统的实际输入有可能是以随机的方式随时间变化。例如,在一个雷达跟踪系统中被跟踪目标的位置和速度会以一种不可预测的方式变化,因而它

们不能用确定的数学表达式表示。这给设计者带来一个问题,因为要设计一个对任何输入信号都能令人满意的控制系统是一件非常困难的事情。对于分析和设计而言,就有必要假设一些基本类型的输入函数,使一个系统的性能能够关于这些测试信号进行评估。通过合适地选择这些基本的测试信号,不仅可以使问题的数学处理系统化,而且由这些输入得到的响应还可以预测系统在其他较复杂输入作用下的特性。在设计问题中,可以指定关于这些测试信号的性能指标,使得系统设计可以满足这些指标。

当线性时不变系统的响应在频域中进行分析时,采用频率可以变化的正弦输入。而为了方便,在时域分析中经常采用以下的测试信号。

1. 阶跃函数

阶跃函数表示输入中瞬时的变化。例如,如果输入是机械转轴的角位置,那么阶跃输入就表示转轴的突然旋转。阶跃信号有时候也称为位置信号。作为测试信号,阶跃函数是很有用的,因为它在起始瞬间幅值的跳变揭示了大量的关于系统快速做出响应的信息。而且从原理上讲,由于跳变的不连续性,阶跃函数的频谱具有很宽的频带,作为测试信号它等价于应用无数个频率范围很宽的正弦信号。

阶跃函数的数学表达式为

$$r(t) = \begin{cases} r_0 & t > 0 \\ 0 & t < 0 \end{cases} \tag{3.3}$$

或者

$$r(t) = r_0 \cdot 1(t) \tag{3.4}$$

阶跃函数在 $t=0$ 处没有定义。阶跃函数与时间的函数关系如图3.1(a)所示,可以获得阶跃函数的拉普拉斯变换为

$$R(s) = \frac{r_0}{s} \tag{3.5}$$

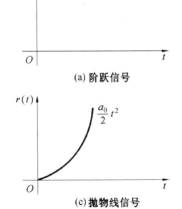

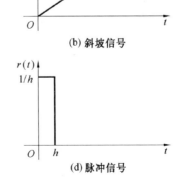

图 3.1 典型测试信号

2. 斜坡函数

在斜坡函数的情况下,信号是随时间线性增加的。斜坡函数可以看作阶跃函数的积分。如果输入量是属于转轴的角位移一类的信号,那么斜坡输入表示转轴恒速的旋转。斜坡信号有时候也称为速度信号。斜坡函数的变化要比阶跃函数快一个等级,它具有测试系统将如何对随时间线性变化的信号做出响应的能力。

斜坡函数在数学上表示为

$$r(t) = \begin{cases} v_0 t & t \geq 0 \\ 0 & t < 0 \end{cases} \tag{3.6}$$

或者简单地表示为

$$r(t) = v_0 t \cdot 1(t) \tag{3.7}$$

图 3.1(b) 所示为一个斜坡函数,可以获得斜坡函数的拉普拉斯变换为

$$R(s) = \frac{v_0}{s^2} \tag{3.8}$$

3. 抛物线函数

抛物线函数可以看作是斜坡函数的积分,可把图 3.1(c) 所示抛物线函数描述为

$$r(t) = \begin{cases} \dfrac{a_0}{2} t^2 & t > 0 \\ 0 & t < 0 \end{cases} \tag{3.9}$$

或者简记为

$$r(t) = \frac{a_0}{2} t^2 \cdot 1(t) \tag{3.10}$$

可以获得它的拉普拉斯变换为

$$R(s) = \frac{a_0}{s^3} \tag{3.11}$$

抛物线函数的变化要比斜坡函数快一个等级。抛物线信号有时候称为加速度信号。在实际应用中,很少发现有必要使用变化比抛物线函数更快的测试信号。这是因为如同我们稍后将证明的,为了跟踪或跟随变化较快的输入,系统的阶数就必须比较高,而这意味着系统的稳定性可能会遇到麻烦。

4. (理想) 脉冲函数

工程上所用实际的脉冲函数如图 3.1(d) 所示,它的数学表达式为

$$r(t) = \begin{cases} \dfrac{1}{h} & 0 < t \leq h \\ 0 & t < 0 \text{ 或 } t > h \end{cases} \tag{3.12}$$

式中,h 为脉冲的宽度。

在控制系统的分析中,我们经常使用理想脉冲函数 $\delta(t)$。当 $h \to 0$ 时,由表达式 (3.12) 的脉冲函数可以得到 $\delta(t)$,即

$$\delta(t) = \begin{cases} \infty & t = 0 \\ 0 & t \neq 0 \end{cases} \tag{3.13}$$

且
$$\int_{-\infty}^{\infty} \delta(t)\,dt = 1 \tag{3.14}$$

可以获得(理想)脉冲函数的拉普拉斯变换为
$$R(s) = 1 \tag{3.15}$$

所有这些测试信号都具有在数学上描述简单和实验室容易实现的共同特点。

3.1.3 瞬态响应的性能指标

控制系统时间响应的瞬态部分是时间趋向于无穷大时响应中变为零的部分。当然，瞬态响应只有对于稳定系统才有意义，因为对于不稳定系统而言，响应的瞬态部分不会减弱并将失控。

控制系统瞬态性能的特征通常利用阶跃响应表示。用以表示阶跃输入作用下瞬态响应特征的典型性能指标包括上升时间、峰值时间、超调量、调整时间等。图 3.2 说明了线性控制系统典型的阶跃响应。

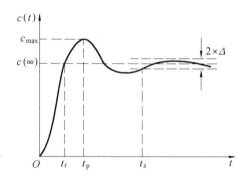

图 3.2 具有代表性的阶跃响应

(1) 上升时间 t_r

上升时间度量系统对于突然变化做出响应的速度，它可以按不同的方式定义。一般情况下，上升时间定义为阶跃响应第一次到达终值所需要的时间。

(2) 峰值时间 t_p

峰值时间也是系统响应速度的度量，是阶跃响应到达第一次超调最大值所需要的时间。

(3) 最大(百分比)超调量 σ_p

最大超调量是一个稳定系统的响应振荡特性的度量，它的大小也用于度量系统的相对稳定性，最大超调量经常表示成阶跃响应终值的百分比，即
$$\sigma_p = \frac{c_{\max} - c(\infty)}{c(\infty)} \times 100\% \tag{3.16}$$

(4) 调节时间 t_s

调节时间定义为阶跃响应调节到某一允许误差范围内所需的时间。如图 3.2 所示，允许误差范围 $\pm\Delta$ 通常取响应终值的 5% 或 2%。

一个控制系统的瞬态响应可以按两个要素进行描述：

① 用上升时间、峰值时间和调整时间表示的响应速度。
② 用超调量表示的响应与期望响应的接近程度。

由控制系统本身的特性所决定,这两个要素通常是矛盾的要求,因而必须做出折中的选择。

上面定义的这四个指标给出了阶跃响应瞬态特性直接的度量,当阶跃响应曲线已经绘制时,这些参数相对比较容易获取。但是,除了一些简单的情况外,用解析的方法确定这些参数是相当困难的。

许多将要遇到的控制系统主要是用最大(百分比)超调量和调节时间表征它们的瞬态响应。至于没有振荡的响应,上升时间可以定义为阶跃响应从其终值的10%上升到90%所需的时间,而峰值时间和超调量则没有定义。

3.2 一阶系统的时间响应

3.2.1 一阶系统

一个系统的阶数定义为该系统特征多项式的阶数。所研究的一种最简单的动态系统是用一阶微分方程描述的系统。这样的系统称为一阶系统。从前面的例子中可以看到,许多不尽相同的控制系统都可以用一阶形式的数学模型表示。对于一阶系统,输入 $R(s)$ 和输出 $C(s)$ 之间一般化的传递函数为

$$\frac{C(s)}{R(s)} = \frac{1}{Ts+1} \quad (3.17)$$

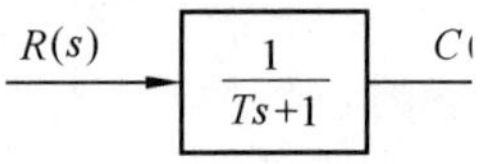

并可以用图 3.3 所画的方框图形式表示。将对这一类系统进行较为详尽的研究,确定它们对于一些典型输入的响应。

图 3.3 一阶系统

3.2.2 阶跃响应

如果输入为单位阶跃函数,即 $R(s) = 1/s$,则系统的输出为

$$C(s) = \frac{1}{Ts+1} \cdot \frac{1}{s} = \frac{\frac{1}{T}}{s\left(s+\frac{1}{T}\right)} = \frac{1}{s} - \frac{1}{s+\frac{1}{T}} \quad (3.18)$$

即

$$c(t) = 1 - e^{-\frac{1}{T}t} \quad (3.19)$$

输出的大小可以计算得到为

$$c(t)\big|_{t=0} = 1 - e^0 = 0, \qquad c(t)\big|_{t=T} = 1 - e^{-1} = 0.632$$
$$c(t)\big|_{t=2T} = 1 - e^{-2} = 0.865, \qquad c(t)\big|_{t=3T} = 1 - e^{-3} = 0.950$$
$$c(t)\big|_{t=4T} = 1 - e^{-4} = 0.982, \qquad c(t)\big|_{t\to\infty} = 1 - e^{-\infty} = 1$$

一阶系统也可以这样描述,一个时间常数后响应到达它们终值的63.2%,2T 后到达它们终值的86.5%,3T 后为95.0%,4T 后为98.2%。图 3.4 中画出的是一阶系统的阶跃

响应曲线。从实际应用的意义上讲,我们认为经过三个或四个时间常数后阶跃响应将到达终值,即一阶系统的调节时间为

$$t_s = \begin{cases} 3T & \Delta = 5\% \\ 4T & \Delta = 2\% \end{cases} \quad (3.20)$$

注意,阶跃响应曲线所具有的起始斜率为 $1/T$,即

$$\left.\frac{dc(t)}{dt}\right|_{t=0} = -\left.\frac{d}{dt}(e^{-t/T})\right|_{t=0} = \frac{1}{T} \quad (3.21)$$

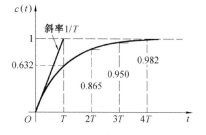

图 3.4 一阶系统的阶跃响应

这表明如果响应能以起始速率持续增大则将在 T s 后到达它的终值。因此,参数 T 称为一阶系统的时间常数,单位为 s。

3.2.3 脉冲响应

在单位脉冲的情况下,即 $R(s)=1$,输出变为

$$C(s) = \frac{1}{Ts+1} = \frac{\frac{1}{T}}{s+\frac{1}{T}} \quad (3.22)$$

即

$$c(t) = \frac{1}{T}e^{-\frac{1}{T}t} \quad (3.23)$$

系统的输出如图 3.5 所示。由于初始条件已经假设为零,输出必须从 $t=0_-$ 时的 0 瞬间变为 $t=0_+$ 时的 $1/T$。

可以注意到,由于脉冲函数是阶跃函数的导数,脉冲响应就是阶跃响应的导数。

3.2.4 斜坡响应

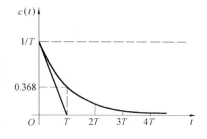

图 3.5 一阶系统的脉冲响应

如果输入信号为单位斜坡函数,即 $R(s)=1/s^2$,则有

$$C(s) = \frac{1}{Ts+1} \cdot \frac{1}{s^2} = \frac{1}{s^2} - \frac{T}{s} + \frac{T}{s+\frac{1}{T}} \quad (3.24)$$

时域响应则变为

$$c(t) = t - T + Te^{-\frac{1}{T}t} = t - T(1 - e^{-\frac{1}{T}t}) \quad (3.25)$$

图 3.6(a) 显示了这一情况下输入和输出之间的关系。系统的响应由瞬态部分和稳态部分组成。在稳态下输出滞后于输入,滞后时间等于时间常数 T。可以证明,如果输入不是单位斜坡而是 $r(t)=At$,那么如图 3.6(b) 所示,输入和稳态时输出之间相差的一个固定时间将为 AT。

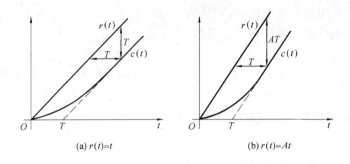

(a) $r(t)=t$　　　　(b) $r(t)=At$

图 3.6　一阶系统的斜坡响应

应当注意到,斜坡响应是阶跃响应的积分。

3.2.5　一阶反馈系统

既然已经对一阶系统的响应进行了详细的研究,下面就可以对反馈控制的一个主要好处作个说明。假设需要对一个已知是一阶的系统进行控制。最简单的方法将是实现如图 3.3 所说明的开环控制。如果时间常数假设为 $T=1$ s,那么系统对于单位阶跃的响应将如图 3.4 所示,响应形式为

$$c(t) = 1 - e^{-t} \tag{3.26}$$

在 1 s 后到达输出稳态值的 63.2%。

现在如图 3.7 所示,假设同一个一阶系统作为闭环系统的被控对象,闭环系统还有一个增益 K 可变的放大器作为控制器。这一次输入和输出之间的关系由闭环传递函数给出为

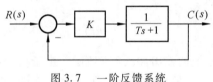

图 3.7　一阶反馈系统

$$\frac{C(s)}{R(s)} = \frac{KG(s)}{1 + KG(s)} = \frac{K}{Ts + K + 1} \tag{3.27}$$

如果输入为单位阶跃,则

$$C(s) = \frac{K}{Ts + K + 1} \cdot \frac{1}{s} = \frac{K/T}{s[s + (K+1)/T]} = \frac{K}{K+1} \cdot \frac{1}{s} - \frac{K}{K+1} \cdot \frac{1}{s + (K+1)/T} \tag{3.28}$$

因此时间响应为

$$c(t) = \frac{K}{K+1}(1 - e^{-\frac{K+1}{T}t}) \tag{3.29}$$

假设 $T=1$ 并任意设定 $K=1$,则可得到

$$c(t) = 0.5(1 - e^{-2t}) \tag{3.30}$$

这一输出(阶跃响应)如图 3.8 所示。图中也包括了开环系统的阶跃响应(用虚线表示)。

注意以下重要的观察结果:

① 反馈系统的稳态输出为 0.5,输入和稳态输出之间存在 0.5 的误差。

② 反馈系统的时间常数为 0.5，响应比开环系统的响应快。

因此已经使系统的瞬态响应改善得很好，不过输入和稳态输出之间的误差在这一情况下也相当可观，这对于系统的性能是个严重的问题。

但是现在注意，如果增益 K 增加到 10，那么输出变为

$$c(t) = 0.909(1 - e^{-11t}) \tag{3.31}$$

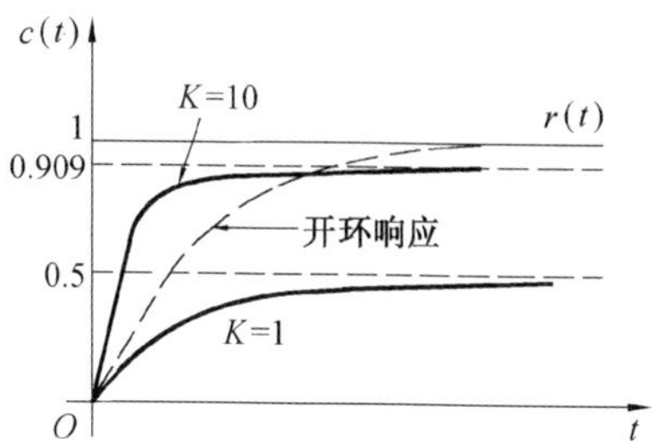

这一情况也画在了图 3.8 中，它表明输入和稳态输出之间的误差现在减少到小于 0.1，同时闭环系统的时间常数为 0.09 s，响应速度超过开环系统响应速度的 10 倍。

图 3.8 反馈系统的阶跃响应

这一例子清楚地表明，通过应用反馈有可能改善一个系统的动态性能；与此同时这种做法也可能会有一些不利的影响，例如在输入和稳态输出之间出现了误差。但是，通过使用较大的增益，不但确实使系统的响应变快，而且误差也变得很小。在实际情况下，增益不可能无限增大，因为发生的饱和现象将导致系统变为非线性。

3.2.6 一阶系统的极点和零点

考虑前面详细研究过的一阶系统的传递函数

$$G(s) = \frac{1}{Ts + 1} \tag{3.32}$$

可以看到，该开环系统没有零点，有一个极点在

$$s = -1/T \tag{3.33}$$

回想一下，该系统对于任何输入的瞬态响应都是以时间常数 T 表征的，或者更明确一些是以 $-1/T$，即该系统的极点表征的。这一特性并不限于一阶系统，而且可以在更一般的意义下陈述如下：

"一个系统的瞬态响应取决于传递函数的极点。使传递函数的分母为零得到称之为特征方程的表达式，通过求解特征方程就可以获得传递函数的极点。"

人们可以观察到，是极点决定了系统响应的特点，因为对于任何输入 $R(s)$ 而言，极点都将出现在输出 $C(s)$ 的分母内。零点的作用是在进行拉普拉斯反变换以获取时域响应之前确定部分分式的系数。在一般的反馈情况下

$$\frac{C(s)}{R(s)} = \frac{G(s)}{1 + G(s)H(s)}$$

这时特征方程的形式为

$$1 + G(s)H(s) = 0 \tag{3.34}$$

在一阶闭环系统的情况下，描述该系统的传递函数为

$$\frac{C(s)}{R(s)} = \frac{KG(s)}{1 + KG(s)} = \frac{K}{Ts + K + 1} \tag{3.35}$$

位于

$$s = -\frac{K+1}{T} = -\frac{1}{T'} \tag{3.36}$$

的极点称为闭环极点。以上式子表明,闭环极点与 K 密切相关,而且系统的瞬态响应也将因此而与这一参数密切相关,这证实了先前观察到的反馈对于一阶系统的影响。由式 (3.36) 可知, K 的取值较大时所得的 s 值就较大,较大的 s 则产生较小的等效时间常数 T' 和较快的响应速度。

3.3 二阶系统的时间响应

3.3.1 二阶系统

下面要进行研究的一种系统称为二阶系统,因为这些系统要用二阶微分方程描述。二阶系统很重要,因为它们的行为与一阶系统大不相同,会呈现出诸如有振荡的响应或超调之类的特点。而且一般地讲,二阶系统的研究有助于形成理解分析和设计方法的基础。

1. 典型二阶系统及其传递函数

考虑图 3.9 所示二阶系统的方框图,它可以表示一个位置控制系统,譬如第 2 章中的例 2.5,或者其他什么系统。由方框图可以获得传递函数为

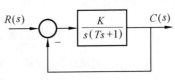

图 3.9 典型二阶系统的方框图

$$\frac{C(s)}{R(s)} = \frac{K}{Ts^2 + s + K} \tag{3.37}$$

但是,像一阶系统一样,通常传递函数的系数写成具有某种物理意义的方式。令

$$\zeta = \frac{1}{2\sqrt{KT}}, \quad \omega_n = \sqrt{\frac{K}{T}}$$

典型二阶系统的传递函数为

$$\frac{C(s)}{R(s)} = \frac{\omega_n^2}{s^2 + 2\zeta\omega_n s + \omega_n^2} \tag{3.38}$$

式中, ζ 无量纲,称为阻尼比; ω_n 称为自然无阻尼振荡频率。

2. 二阶系统的极点

由式 (3.38) 描述的二阶系统有两个极点

$$s_{1,2} = -\zeta\omega_n \pm \omega_n\sqrt{\zeta^2 - 1} \tag{3.39}$$

根据阻尼比 ζ 的取值,这两个极点可以是实数或者一对共轭复数,如图 3.10 的复平面上所示。在 s 平面上,闭环极点通常用实心的小圆点"·"表示。

在 $0 < \zeta < 1$ 的情况下,记住以下的关系是很有用的。

① 极点和原点之间连线的长度为

$$l = \sqrt{(\zeta\omega_n)^2 + (\omega_n \cdot \sqrt{1-\zeta^2})^2} = \omega_n \tag{3.40}$$

② 极点和原点之间连线与负实轴之间的夹角

$$\cos\theta = \frac{\zeta\omega_n}{\omega_n} = \zeta \tag{3.41}$$

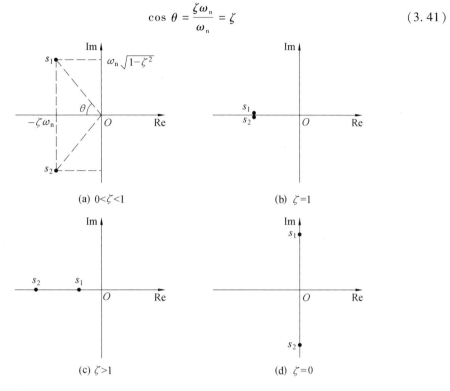

图 3.10 二阶系统的极点

3.3.2 阶跃响应

相当多关于二阶系统响应内在属性的信息可以通过研究单位阶跃输入作用下二阶系统的表现获得。在这种情况下,式(3.38)指出输出为

$$C(s) = \frac{\omega_n^2}{s^2 + 2\zeta\omega_n s + \omega_n^2} \cdot \frac{1}{s} = \frac{1}{s} - \frac{s + 2\zeta\omega_n}{s^2 + 2\zeta\omega_n s + \omega_n^2} \tag{3.42}$$

显然,由于 ζ 的不同取值,当系统的极点是实数或复数时,响应将呈现不同的特征。

1. $0 < \zeta < 1$ 的情况

在这种情况下,系统的两个极点是负实部的共轭复数

$$s_{1,2} = -\zeta\omega_n \pm j\omega_n\sqrt{1-\zeta^2} \tag{3.43}$$

输出为

$$C(s) = \frac{1}{s} - \frac{s + 2\zeta\omega_n}{(s+\zeta\omega_n)^2 + \omega_n^2(1-\zeta^2)}$$

令

$$\omega_d = \omega_n\sqrt{1-\zeta^2} \tag{3.44}$$

则

$$C(s) = \frac{1}{s} - \left[\frac{s + \zeta\omega_n}{(s+\zeta\omega_n)^2 + \omega_d^2} + \frac{\zeta\omega_n}{(s+\zeta\omega_n)^2 + \omega_d^2}\right] \tag{3.45}$$

取拉普拉斯反变换得到

$$c(t) = 1 - \left[e^{-\zeta\omega_n t}\cos\omega_d t + \frac{\zeta\omega_n}{\omega_d}e^{-\zeta\omega_n t}\sin\omega_d t \right] = 1 - e^{-\zeta\omega_n t}\left(\cos\omega_d t + \frac{\zeta}{\sqrt{1-\zeta^2}}\sin\omega_d t\right) =$$

$$1 - \frac{1}{\sqrt{1-\zeta^2}}e^{-\zeta\omega_n t}\sin(\omega_d t + \theta) \tag{3.46}$$

式中

$$\theta = \arccos\zeta \tag{3.47}$$

注意,式(3.46)的瞬态分量是一个衰减的振荡量。图 3.11 显示了阻尼比 ζ 不同的取值时阶跃响应与无因次时间 $\omega_n t$ 的函数关系。

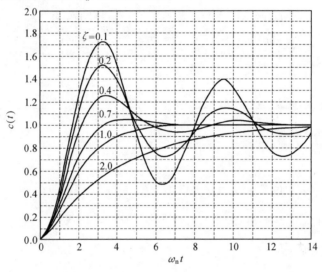

图 3.11 不同 ζ 时典型二阶系统的阶跃响应

在这里,认识到 $0 < \zeta < 1$ 时该图显示的一些特点是很重要的:

① 系统以频率 ω_d 进行振荡,ω_d 有时称为有阻尼振荡频率,而且它正好由极点的虚部给出;

② 振荡呈指数衰减,衰减速率则由常数 $\zeta\omega_n$ 决定,$\zeta\omega_n$ 称为衰减因子,而且正好由极点的实部给出。

在这种情况下,由于阶跃响应的欠阻尼特性,该系统称为是欠阻尼的。

2. $\zeta > 1$ 的情况

如果 $\zeta > 1$,那么两个极点为不相等的实数

$$s_{1,2} = -\zeta\omega_n \pm \omega_n\sqrt{\zeta^2 - 1} \tag{3.48}$$

而单位阶跃响应则为

$$C(s) = \frac{\omega_n^2}{(s-s_1)(s-s_2)} \cdot \frac{1}{s} =$$

$$\frac{1}{s} + \frac{\omega_n^2}{s_1(s_1-s_2)} \cdot \frac{1}{s-s_1} + \frac{\omega_n^2}{s_2(s_2-s_1)} \cdot \frac{1}{s-s_2} \tag{3.49}$$

取拉普拉斯反变换得到

$$c(t) = 1 + \frac{\omega_n^2}{s_1(s_1 - s_2)} e^{s_1 t} + \frac{\omega_n^2}{s_2(s_2 - s_1)} e^{s_2 t} \tag{3.50}$$

式(3.50)包含由两项衰减指数函数组成的瞬态分量。在 $\zeta > 1$ 的情况下，响应具有过阻尼特性，或者说系统是过阻尼的。

3. $\zeta = 1$ 的情况

当 $\zeta = 1$ 时，两个极点是相等的实数，即

$$s_{1,2} = -\omega_n \tag{3.51}$$

单位阶跃响应为

$$C(s) = \frac{1}{s} - \frac{s + 2\omega_n}{(s + \omega_n)^2} = \frac{1}{s} - \left[\frac{\omega_n}{(s + \omega_n)^2} + \frac{1}{s + \omega_n} \right] \tag{3.52}$$

取拉普拉斯反变换，得到结果为

$$c(t) = 1 - (\omega_n t + 1) e^{-\omega_n t} \tag{3.53}$$

该式也有两项衰减指数函数。在这种情况下，出于明显的原因，系统称为是临界阻尼的。

4. $\zeta = 0$ 的情况

在这种情况下，两个极点为虚数，即

$$s_{1,2} = \pm j\omega_n \tag{3.54}$$

输出为

$$C(s) = \frac{1}{s} - \frac{s}{s^2 + \omega_n^2} \tag{3.55}$$

取拉普拉斯反变换，其结果为

$$c(t) = 1 - \cos \omega_n t \tag{3.56}$$

在 $\zeta = 0$ 即零阻尼的情况下，响应是一个纯正弦量，系统则称为是无阻尼的。正如定义一样，自然无阻尼振荡频率 ω_n 对应于无阻尼正弦量的频率。

3.3.3 性能指标的估算

在实际应用中，典型二阶系统

$$\frac{C(s)}{R(s)} = \frac{\omega_n^2}{s^2 + 2\zeta\omega_n s + \omega_n^2} \tag{3.57}$$

性能指标的估算是很有意义的。

对于过阻尼($\zeta > 1$)或者临界阻尼($\zeta = 1$)的情况，阶跃响应中不存在超调。至于调节时间，要确定精确的数值很困难，但是在 $\zeta \geq 0.7$ 的情况下可以用以下的经验公式获取

$$t_s \approx \frac{1}{\omega_n}(6.45\zeta - 1.7), \quad \Delta = 5\% \tag{3.58}$$

下面将考虑 $0 < \zeta < 1$ 时式(3.57)所描述的二阶系统，它的单位阶跃响应表达式为

$$c(t) = 1 - \frac{1}{\sqrt{1 - \zeta^2}} e^{-\zeta\omega_n t} \sin(\omega_d t + \theta) \tag{3.59}$$

式中 $\omega_d = \omega_n\sqrt{1-\zeta^2}$,$\theta = \arccos\zeta$。

(1) 上升时间 t_r

由公式 (3.59),对于上升时间我们有 $c(t_r) = 1$,即

$$1 - \frac{1}{\sqrt{1-\zeta^2}} e^{-\zeta\omega_n t_r} \sin(\omega_d t_r + \theta) = 1$$

该式可以简化为

$$\sin(\omega_d t_r + \theta) = 0$$

或者

$$t_r = \frac{k\pi - \theta}{\omega_d}$$

式中 k 为正整数。由于上升时间是阶跃响应第一次到达其终值的时间,因此

$$t_r = \frac{\pi - \theta}{\omega_d} = \frac{\pi - \theta}{\omega_n\sqrt{1-\zeta^2}} \tag{3.60}$$

(2) 峰值时间 t_p

对式 (3.59) 求导并将求导结果置零可以获得峰值时间。由此得到

$$\left.\frac{dc(t)}{dt}\right|_{t=t_p} = -\frac{\zeta\omega_n e^{-\zeta\omega_n t_p}}{\sqrt{1-\zeta^2}}\sin(\omega_d t_p + \theta) + \frac{\omega_d e^{-\zeta\omega_n t_p}}{\sqrt{1-\zeta^2}}\cos(\omega_d t_p + \theta) =$$

$$\frac{\omega_n e^{-\zeta\omega_n t_p}}{\sqrt{1-\zeta^2}}[-\zeta\sin(\omega_d t_p + \theta) + \sqrt{1-\zeta^2}\cos(\omega_d t_p + \theta)] = 0$$

由于 $t_p \neq \infty$,该式可以简化为

$$-\zeta\sin(\omega_d t_p + \theta) + \sqrt{1-\zeta^2}\cos(\omega_d t_p + \theta) = 0$$

又由于 $\cos\theta = \zeta$ 和 $\sin\theta = \sqrt{1-\zeta^2}$ 有

$$\sin(\omega_d t_p) = 0$$

即

$$t_p = \frac{k\pi}{\omega_d}$$

式中,k 为正整数。显然,峰值时间发生在 $k = 1$ 时。因此,峰值时间为

$$t_p = \frac{\pi}{\omega_d} = \frac{\pi}{\omega_n\sqrt{1-\zeta^2}} \tag{3.61}$$

(3) 最大(百分比)超调量 σ_p

把式 (3.61) 代入式 (3.59) 可以获得最大超调量。由此可得

$$c(t_p) = 1 - \frac{1}{\sqrt{1-\zeta^2}} e^{-\zeta\omega_n \frac{\pi}{\omega_n\sqrt{1-\zeta^2}}} \sin\left(\omega_d \frac{\pi}{\omega_d} + \theta\right) =$$

$$1 - \frac{1}{\sqrt{1-\zeta^2}} e^{-\pi\zeta/\sqrt{1-\zeta^2}} \sin\theta = 1 - e^{-\pi\zeta/\sqrt{1-\zeta^2}}$$

由定义,最大(百分比)超调量为

$$\sigma_p = e^{-\pi\zeta/\sqrt{1-\zeta^2}} \times 100\% \tag{3.62}$$

注意，阶跃响应的最大超调量只是阻尼比的函数。典型二阶系统的最大超调量与阻尼比之间的关系，如图 3.12 所示。

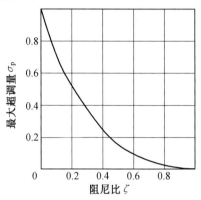

图 3.12 σ_p 与 ζ 的关系曲线

(4) 调节时间 t_s

由调节时间的定义我们有

$$|c(t) - c(\infty)| = \left|\frac{1}{\sqrt{1-\zeta^2}}e^{-\zeta\omega_n t}\sin(\omega_d t + \theta)\right| \leq \Delta, \quad t \geq t_s$$

要确定调节时间的精确数值是很困难的。但是当 $t \geq t_s$ 时我们可以看到，对于调节时间近似地有

$$\frac{1}{\sqrt{1-\zeta^2}}e^{-\zeta\omega_n t} \leq \Delta$$

或者

$$\zeta\omega_n t \geq \ln\frac{1}{\Delta\sqrt{1-\zeta^2}}$$

由上式可取得

$$t_s \approx \frac{1}{\zeta\omega_n}\ln\frac{1}{\Delta\sqrt{1-\zeta^2}} \tag{3.63}$$

对于较小的阻尼比，当 $0 < \zeta < 0.9$ 时调节时间可由近似公式给出为

$$t_s \approx \frac{3}{\zeta\omega_n}, \quad \Delta = 5\% \tag{3.64}$$

或者

$$t_s \approx \frac{4}{\zeta\omega_n}, \quad \Delta = 2\% \tag{3.65}$$

现在审视一下上升时间和调节时间之间的关系。可以看到，对于较小的 ζ 将可以得到较短的上升时间，但是较快的调节时间则需要较大的 ζ。因此，当一个设计问题中所有这些指标都要令人满意地得到满足时，ζ 的取值就需要做出折中的处理。同时考虑到最大超调量的要求时，一般情况下使综合性能令人满意的阻尼比可接受的范围为 0.5 ~ 0.8。

例 3.1 图 3.13 为一个被控对象固定不变的反馈控制系统,但控制器的增益及其极点的位置则由系统的设计者指定。确定极点的位置和 K 的数值,以满足以下阶跃响应设计要求:

(1) $\Delta = 5\%$ 时调节时间不超过 1 s;

(2) 最大(百分比)超调量不超过 10%。

在计算出合适的系统参数后,确定:

(a) 有阻尼振荡频率 ω_d;

(b) 峰值时间 t_p。

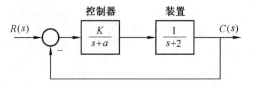

图 3.13 例 3.1 的反馈控制系统

解 由超调量的要求

$$e^{-\pi\zeta/\sqrt{1-\zeta^2}} \leqslant 0.1$$

求解此式得到 $\zeta = 0.6$,$\Delta = 5\%$ 时,由式(3.64)可得调整时间为 $t_s = \dfrac{3}{\zeta\omega_n} = 1$ s,因此 $\omega_n = 5$ rad/s。

系统的闭环传递函数采用以下形式

$$\frac{C(s)}{R(s)} = \frac{K}{s^2 + (2+a)s + (K+2a)}$$

与二阶系统的标准形式相比得到

$$\omega_n = \sqrt{K+2a}, \quad \zeta = \frac{2+a}{2\sqrt{K+2a}}$$

将 ζ 和 ω_n 代入方程可有

$$\sqrt{K+2a} = 5, \quad \frac{2+a}{2\sqrt{K+2a}} = 0.6$$

求解这些方程得到的结果为

$$a = 4, \quad K = 17$$

因此,令 $K = 17$ 和 $a = 4$ 则设计可以得到满足。

从前面所推导的闭环传递函数的分母,可以直接获得系统的特征方程为

$$s^2 + (2+a)s + (K+2a) = 0$$

即

$$s^2 + 6s + 25 = 0$$

特征方程的根为

$$s_{1,2} = -3 \pm j4$$

可见,有阻尼振荡频率为

$$\omega_d = 4 \text{ rad/s}$$

峰值时间则为

$$t_p = \frac{\pi}{\omega_d} = 0.785 \text{ s}$$

3.4 高阶系统的时间响应

3.4.1 高阶系统的阶跃响应

所有输入-输出关系用三阶或者更高阶数微分方程描述的系统都称为高阶系统。一个高阶系统的瞬态行为的特征仍然以它的阶跃响应来表示。

考虑一个 n 阶系统

$$\Phi(s) = \frac{C(s)}{R(s)} = \frac{b_0 s^m + \cdots b_{m-1} s + b_m}{s^n + a_1 s^{n-1} + \cdots + a_{n-1} s + a_n} = \frac{N(s)}{D(s)} \quad (3.66)$$

式中,$N(s)$ 为分子多项式;而 $D(s)$ 则为分母多项式。考虑到动态性能的要求,许多控制系统设计成呈现出衰减的振荡特性。由于 $\Phi(s)$ 的极点必定为实数或共轭复数,不失一般性,假设 $\Phi(s)$ 有 p 个实极点和 q 对共轭复极点,而且所有这些极点互异。这样,系统对于单位阶跃信号的响应为

$$C(s) = \frac{N(s)}{\sum\limits_{i=1}^{p}(s-s_i)\sum\limits_{j=1}^{q}(s^2 + 2\zeta_j \omega_{nj} s + \omega_{nj}^2)} \cdot \frac{1}{s} \quad (3.67)$$

将该式展开成部分分式的形式,响应可以表示为

$$C(s) = \frac{A_0}{s} + \sum_{i=1}^{p} \frac{B_i}{s-s_i} + \sum_{j=1}^{q} \frac{C_j s + D_j}{s^2 + 2\zeta_j \omega_{nj} s + \omega_{nj}^2} \quad (3.68)$$

式中,A_0,$B_i(i=1,2,\cdots,p)$,C_j 和 $D_j(j=1,2,\cdots,q)$ 是由部分分式展开式获得的系数。式 (3.68) 表明,如果所有这些极点均为负实数或负实部的共轭复数,则响应的瞬态部分包含 p 个衰减指数分量和 q 个衰减正弦分量。

取拉普拉斯变换不难得到时域内的响应表达式 $c(t)$。这样,由 $c(t)$ 的表达式或者相应的响应曲线就可以按照定义估算性能指标。

显然,高阶系统的动态分析比较复杂。而且,要推导出估算性能指标的解析表达式也很困难。不过,幸运的是对于许多高阶系统,利用主导极点法可以简化系统的分析和性能指标的估算。

3.4.2 主导极点法

1. 主导极点的概念

通过一个例子引入主导极点的概念,考虑以下三阶系统

$$\Phi(s) = \frac{C(s)}{R(s)} = \frac{50}{(s+1)(s+5)(s+10)}$$

它的单位阶跃响应为

$$C(s) = \frac{50}{(s+1)(s+5)(s+10)} \cdot \frac{1}{s} =$$

$$\frac{1}{s} - \frac{50}{36} \cdot \frac{1}{s+1} + \frac{50}{100} \cdot \frac{1}{s+5} - \frac{50}{450} \cdot \frac{1}{s+10}$$

即
$$c(t) = 1 - 1.39e^{-t} + 0.5e^{-5t} - 0.11e^{-10t}$$

而响应曲线及其各分量则画在图 3.14 内。

比较三个瞬态分量可以看到,各分量的衰减速率和初始值两者都与相应的极点到虚轴的距离密切相关。与 e^{-5t} 和 e^{-10t} 项相比,e^{-t} 项具有慢得多的衰减速率和大得多的初始幅值。因此,对于除了 $t \to 0$ 以外的所有时间,e^{-5t} 和 e^{-10t} 项在合成的时域响应中的贡献可以忽略不计。所以我们可以说,e^{-t} 项在响应中起着主导作用,相应地还可以说,$s = -1$ 是该系统的主导极点。

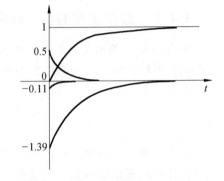

图 3.14　一个三阶系统的阶跃响应及其分量

这一结果可以推广到更为普遍的情况,对于没有有限零点的传递函数可以观察到以下现象:

① 极点距离虚轴越近,相应的响应分量衰减得越慢。

② 极点距离虚轴越近,相应分量的初始幅值越大。

因此,在很多情况下,如果除了一个或一对距离虚轴最近的极点外,所有其他的极点都位于距离虚轴足够远的地方,例如相距 5 倍以上,则系统的响应可以用距离虚轴最近的极点的响应近似表示。而且,这样的一个实极点或一对共轭复极点称为系统的主导极点。但是应当注意,如果系统的传递函数含有零点,而且该零点相当接近距离虚轴最近的极点,那么这些零点将对系统的瞬态响应有着相当大的影响,而距离虚轴最近的极点也有可能不再是系统的主导极点。

2. 高阶系统时间响应的近似表示

考虑式 (3.66) 中给出的系统,并假设系统具有一对主导极点

$$s_{1,2} = -\zeta\omega_n \pm j\omega_n\sqrt{1-\zeta^2} = -\zeta\omega_n \pm j\omega_d \tag{3.69}$$

那么,阶跃响应就变为

$$C(s) = \frac{N(s)}{D(s)} \cdot \frac{1}{s} \approx$$

$$\frac{A_0}{s} + \left[\frac{(s-s_1)N(s)}{D(s)} \cdot \frac{1}{s}\right]_{s=s_1} \cdot \frac{1}{s-s_1} + \left[\frac{(s-s_2)N(s)}{D(s)} \cdot \frac{1}{s}\right]_{s=s_2} \cdot \frac{1}{s-s_2} \tag{3.70}$$

即

$$c(t) \approx A_0 + 2\left.\frac{(s-s_1)N(s)}{sD(s)}\right|_{s=s_1} e^{-\zeta\omega_n t}\cos\left[\omega_d t + \angle\left.\frac{(s-s_1)N(s)}{sD(s)}\right|_{s=s_1}\right] \tag{3.71}$$

注意,在式 (3.71) 中非主导极点项被忽略不计,但是在主导极点项中反映了非主导极点对于响应的影响。

应用主导极点法,根据式 (3.71) 给出的阶跃响应的近似表达式可以推导出以下计算性能指标的公式。

第 3 章 控制系统的时域分析

(1) 峰值时间 t_p

根据定义推导,峰值时间可由下式估算

$$t_p = \frac{1}{\omega_d}\left(\pi - \sum_{i=1}^{m}\angle(s_1 - z_i) + \sum_{j=3}^{n}\angle(s_1 - s_j)\right) \tag{3.72}$$

式中,$z_i(i=1,2,\cdots,m)$ 为零点;s_1 为主导极点;$s_j(j=3,\cdots,n)$ 为非主导极点。

式(3.72)表明,闭环零点将使 t_p 缩短,而非主导极点则将使其增长。而且一般地讲,距离虚轴越近,它们的影响就越强。

(2) 最大(百分比)超调量 σ_p

根据定义推导,最大(百分比)超调量可由下式估算

$$\sigma_p\% = \frac{\prod_{i=1}^{m}|s_1 - z_i|}{\prod_{i=1}^{m}|z_i|} \cdot \frac{\prod_{j=3}^{n}|s_j|}{\prod_{j=3}^{n}|s_1 - s_j|} e^{-\zeta\omega_n t_p} \times 100\% \tag{3.73}$$

由式(3.73)可以看到,如果 $|s_1 - z_i| > |z_i|$,闭环零点对于超调量的影响是使之增大;而如果 $|s_1 - s_j| > |s_j|$,则非主导极点的影响是使之减小。

(3) 调节时间 t_s

调节时间可由下式估算

$$t_s = \frac{1}{\zeta\omega_n}\ln\left(\frac{2}{\Delta} \frac{\prod_{i=1}^{m}|s_1 - z_i|}{\prod_{i=1}^{m}|z_i|} \cdot \frac{\prod_{j=2}^{n}|s_j|}{\prod_{j=2}^{n}|s_1 - s_j|}\right) \tag{3.74}$$

由式(3.74)可以看到,如果 $|s_1 - z_i| > |z_i|$,闭环零点对于调节时间的影响趋向于使之增长;而如果 $|s_1 - s_j| > |s_j|$,则非主导极点的影响趋向于是使之缩短。

对于带零点的二阶系统,当零点距离虚轴足够远时,可以用由主导极点法推导出来的近似公式估算性能指标。

例 3.2 单位负反馈系统的开环传递函数为

$$G(s) = \frac{0.5s + 1}{s(s + 0.5)}$$

确定该系统的最大超调量 σ_p。

解 该系统的闭环传递函数为

$$\Phi(s) = \frac{0.5s + 1}{s^2 + s + 1}$$

闭环极点和零点分别为

$$s_{1,2} = -0.5 \pm j0.866, \quad z_1 = -2$$

由于 $|z_1| >> |\text{Re}(s_1)|$,可以用高阶系统的近似公式估算该系统的性能指标。由式 (3.72)~(3.74) 有

$$t_p = \frac{1}{\omega_d}[\pi - \angle(s_1 - z_i)] = \frac{1}{0.866}\left(\pi - \arctan\frac{0.866}{2 - 0.5}\right) = 3.02 \text{ s}$$

$$\sigma_p\% = \frac{|s_1 - z_1|}{|z_1|} e^{-\zeta\omega_n t_p} \times 100\% =$$

$$\frac{\sqrt{(2-0.5)^2 + 0.866^2}}{2} \times e^{-(0.5 \times 3.02)} \times 100\% = 19.1\%$$

3.4.3 降阶为低阶系统

还有一种使具有主导极点的高阶系统分析得以简化的方法是把它们降阶为低阶系统。这种方法和另一种近似方法一样,在获取性能指标时避免了过多的运算。

例如,考虑三阶系统

$$\frac{C(s)}{R(s)} = \frac{50}{(s+1)(s+5)(s+10)}$$

为了获得准确的对应于降阶系统的传递函数,首先需要把该传递函数表示成典型因式的形式(各因式的常数项均为1),即

$$\frac{C(s)}{R(s)} = \frac{1}{(s+1)(0.2s+1)(0.1s+1)}$$

略去非主导极点,现在降阶后的系统变为

$$\frac{C(s)}{R(s)} = \frac{1}{s+1}$$

在进行降阶之前,传递函数必须表示成典型因式的形式,否则非主导极点项删去后传递函数的增益会发生变化。

如果一个极点比另一个极点距离虚轴远得多,过阻尼二阶系统的调节时间常常也可以用这种主导极点法估算调节时间。

例3.3 估算以下二阶系统的调节时间 t_s。

$$\Phi(s) = \frac{2}{s^2 + 8s + 4}$$

解 由系统传递函数的分母多项式我们有

$$\omega_n^2 = 4, \quad 2\zeta\omega_n = 8$$

阻尼比和无阻尼自然振荡频率分别为

$$\zeta = 2, \quad \omega_n = 2$$

该系统是一个过阻尼二阶系统。传递函数的极点为

$$s_1 = -0.54, \quad s_2 = -7.46$$

由于 $|s_2| >> |s_1|$,极点 s_1 可以认为是主导极点,该二阶系统可以近似为一阶系统

$$\Phi(s) = \frac{2}{s^2 + 8s + 4} = \frac{2}{(s+0.54)(s+7.46)} = \frac{\frac{2}{0.54 \times 7.46}}{(\frac{1}{0.54}s+1)(\frac{1}{7.46}s+1)} \approx \frac{0.5}{\frac{1}{0.54}s+1}$$

调节时间为

$$t_s \approx 3 \times \frac{1}{0.54} = 5.56 \text{ s}, \quad \Delta = 5\%$$

3.4.4 三阶系统

设三阶系统的传递函数为

$$\frac{C(s)}{R(s)} = \frac{s_0 \omega_n^2}{(s+s_0)(s^2 + 2\zeta\omega_n s + \omega_n^2)}$$

其中 $s_{1,2} = -\zeta\omega_n \pm j\omega_n\sqrt{1-\zeta^2}$ 是一对负实部的共轭复极点,$s_3 = -s_0$ 则是一个负的实极点。该系统的单位阶跃信号的响应为

$$C(s) = \frac{s_0 \omega_n^2}{(s+s_0)(s^2 + 2\zeta\omega_n s + \omega_n^2)} \cdot \frac{1}{s}$$

将该式展开成部分分式的形式并记 $\omega_d = \omega_n\sqrt{1-\zeta^2}$,则可得到

$$C(s) = \frac{1}{s} + \frac{A}{s+s_0} + \frac{B}{s+\zeta\omega_n - j\omega_d} + \frac{C}{s+\zeta\omega_n + j\omega_d}$$

式中

$$A = \frac{-\omega_n^2}{s_0^2 - 2\zeta\omega_n s_0 + \omega_n^2}$$

$$B = \frac{s_0(2\zeta\omega_n - s_0) - js_0(2\zeta^2\omega_n - \zeta s_0 - \omega_n)/\sqrt{1-\zeta^2}}{2[(2\zeta^2\omega_n - \zeta s_0 - \omega_n)^2 + (2\zeta\omega_n - s_0)^2(1-\zeta^2)]}$$

$$C = \bar{B}$$

其中 $\bar{B}$ 为 B 的共轭复数。

取拉普拉斯变换且令 $\beta = s_0/\zeta\omega_n$,则得到

$$c(t) = 1 + Ae^{-s_0 t} + be^{-\zeta\omega_n t}\cos\omega_d t + ce^{-\zeta\omega_n t}\sin\omega_d t, \quad t \geq 0$$

式中

$$A = \frac{-\omega_n^2}{s_0^2 - 2\zeta\omega_n s_0 + \omega_n^2} = -\frac{1}{\beta\zeta^2(\beta-2) + 1}$$

$$b = 2\text{Re}B = -\frac{\beta\zeta^2(\beta-2)}{\beta\zeta^2(\beta-2) + 1}$$

$$c = -2\text{Im}B = -\frac{\beta\zeta[\beta\zeta^2(\beta-2) + 1]}{[\beta\zeta^2(\beta-2) + 1]\sqrt{1-\zeta^2}}$$

将这些系数代入 $c(t)$ 表达式并重新整理该式得到

$$c(t) = 1 - \frac{1}{\beta\zeta^2(\beta-2) + 1}e^{-s_0 t} - $$

$$\frac{1}{\beta\zeta^2(\beta-2) + 1}e^{-\zeta\omega_n t}\left[\beta\zeta^2(\beta-2)\cos\omega_d t + \frac{\beta\zeta[\beta\zeta^2(\beta-2) + 1]}{\sqrt{1-\zeta^2}}\sin\omega_d t\right], \quad t \geq 0$$

利用计算机仿真可以确定系统对于给定 ζ, ω_n 和不同 β 时的单位阶跃响应。$\zeta = 0.45$

和 $\omega_n = 1$ 时的结果见表 3.1,表中调节时间采用 2% 的允许误差带。

表 3.1　$\zeta = 0.45, \omega_n = 1$ 时第三个极点的影响

s_0	β	超调量/%	调节时间
0.444	0.99	0	9.63
0.666	1.47	3.9	6.3
1.111	2.47	12.3	8.81
2.5	5.56	18.6	8.67
20	44.4	20.5	8.37
∞	∞	20.5	8.24

$s_{1,2}$ 是否以及如何主导响应取决于 $\zeta\omega_n$ 和 s_0 的大小。在具有主导极点的情况下,可以用主导极点法估算系统的响应。

例如,取 $\zeta = 0.45, \omega_n = 1, \beta = 5.56$,结果有

$$s_0(=2.5) > 5\zeta\omega_n(=2.25)$$

可以认为 $s_{1,2}$ 是一对主导极点。

忽略 s_3 造成的瞬态分量,但考虑到 s_3 对于 $s_{1,2}$ 造成的瞬态分量的影响,由式(3.72)和式(3.74)可有

$$t_p = \frac{1}{\omega_n\sqrt{1-\zeta^2}}[\pi + \angle(s_1 - s_3)] = \frac{1}{0.893}\left[\pi + \arctan\frac{0.893}{2.5 - 0.45}\right] = 3.98\text{ s}$$

$$\sigma_p\% = \frac{|s_3|}{|s_1 - s_3|}e^{-\zeta\omega_n t_p} \times 100\% = \frac{2.5}{|2.05 + j0.893|}e^{-1.791} = 20.8\%$$

$$t_s \approx \frac{1}{\zeta\omega_n}\ln\left(\frac{2}{\Delta} \cdot \frac{|s_3|}{|s_1 - s_3|}\right) = \frac{1}{0.45}\ln\left(\frac{2}{0.02} \times \frac{2.5}{|2.05 + j0.893|}\right) = 10.48\text{ s}, \quad \Delta = 2\%$$

将该三阶系统降阶为一个二阶系统则有

$$\frac{C(s)}{R(s)} = \frac{\omega_n^2}{s^2 + 2\zeta\omega_n s + \omega_n^2} = \frac{1}{s^2 + 0.9s + 1}$$

因此,性能指标可以近似估算为

$$\sigma_p\% \approx e^{-\pi\zeta/\sqrt{1-\zeta^2}} \times 100\% = e^{-1.791} = 20.5\%$$

$$t_s \approx \frac{4}{\zeta\omega_n} = 8.89\text{ s}, \quad \Delta = 2\%$$

3.5　线性系统的稳定性

在考虑控制系统的分析和设计时,稳定性是最为重要的问题。从实际应用的观点看,一个不稳定的系统是没有什么用处的。几乎所有的控制系统在实际运行中都会遭受外来或内在扰动的作用,例如负载或电源的波动、系统参数或环境的变化。如果一个不稳定的系统遭受到任何外来或内在扰动的作用,那么它的物理变量将随时间增大而偏离正常的工作点并发散。

在任何情况下,稳定性的概念都是用于区分两类系统:有用的系统和没有用的系统。一个控制系统必须是稳定的才能有用。这样一种稳定和不稳定的特征描述指的是绝对稳定性。具有绝对稳定性的系统称为稳定的系统。一旦确认一个系统是稳定的,我们可以进一步表征稳定性的稳定程度。在时域里,诸如超调量和阻尼比之类的参数用于为线性时不变系统提供相对稳定性的指示。本节主要关注控制系统的绝对稳定性。

许多物理系统开环本来就是不稳定的。引入反馈有助于不稳定的被控对象变为稳定,然后可以借助于合适的控制器调节系统的瞬态性能。对于开环稳定的被控对象,反馈仍然可以用于改善系统的性能。

3.5.1 稳定性的概念和定义

稳定性的概念可以通过考虑如图 3.15 所示的单摆说明,图中 O 为单摆的支点。如果静止在 A 点的单摆被移至 C 点,它会回到它原来的平衡位置 A。这种平衡位置和响应称为是稳定的。另一种情况,如果单摆倒置在另一个平衡位置 B 后释放,它将跌落并且不能回到它原来的平衡位置 B。这种平衡位置和响应称为是不稳定的。

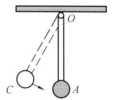
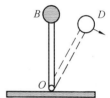

图 3.15 单摆的稳定性

控制系统的稳定性是以类似上面的方式定义的。如果在外部扰动造成的非零初始条件的作用下,线性时不变系统的响应随时间增大而逐渐衰减并且当时间增至无穷大时趋近于零,即系统趋近于原来的平衡位置,则该系统称为是渐近稳定的,或者简称为稳定的,否则该系统称为是不稳定的。

设描述线性时不变系统的微分方程为

$$a_0 \frac{d^n c(t)}{dt^n} + a_1 \frac{d^{n-1} c(t)}{dt^{n-1}} + \cdots + a_{n-1} \frac{dc(t)}{dt} + a_n c(t) =$$
$$b_0 \frac{d^m r(t)}{dt^m} + b_1 \frac{d^{m-1} r(t)}{dt^{m-1}} + \cdots + b_{m-1} \frac{dr(t)}{dt} + b_m r(t) \quad (3.75)$$

式中,$c(t)$ 为输出;$r(t)$ 为输入;系数 $a_0, a_1, \cdots, a_n$ 和 $b_0, b_1, \cdots, b_m$ 为常数,而且 $n \geq m$。在零输入的情况下,条件

$$c(t) = \dot{c}(t) = \cdots = c^{(n-1)}(t) = 0$$

满足齐次方程,并定义为该系统的平衡位置。当时间趋向无穷大时,如果在非零初始条件 $c(0), \dot{c}(0), \cdots, c^{(n-1)}(0)$ 作用下的系统回到它的平衡位置,则称该系统是渐近稳定的,或者简称是稳定的,否则该系统是不稳定的。

按照数学的方式,稳定性的定义可以叙述为:当时间趋向无穷大时,如果由式(3.75)所描述的线性时不变系统在非零初始条件 $c(0), \dot{c}(0), \cdots, c^{(n-1)}(0)$ 的作用下,其零输入响应趋近于零,即

$$\lim_{t \to \infty} c(t) = 0 \quad (3.76)$$

则该系统称为是稳定的,否则该系统是不稳定的。

3.5.2 稳定性的条件

假设在零输入条件下系统的动态方程为

$$a_0 \frac{d^n c(t)}{dt^n} + a_1 \frac{d^{n-1} c(t)}{dt^{n-1}} + \cdots + a_{n-1} \frac{dc(t)}{dt} + a_n c(t) = 0 \quad (3.77)$$

在非零初始条件下取拉普拉斯变换，并整理结果可得

$$(a_0 s^n + a_1 s^{n-1} + \cdots + a_{n-1} s + a_n) C(s) = M_0(s) \quad (3.78)$$

或者

$$C(s) = \frac{M_0(s)}{a_0 s^n + a_1 s^{n-1} + \cdots + a_{n-1} s + a_n} \quad (3.79)$$

式中 $M_0(s)$ 是一个 s 的多项式。由于该系统的特征方程为

$$a_0 s^n + a_1 s^{n-1} + \cdots + a_{n-1} s + a_n = 0 \quad (3.80)$$

式(3.79)意味着 $c(t)$ 是由系统的极点支配的。由此，式(3.76)成立的条件要求系统的特征根必须全都具有负实部，即系统的所有极点都必须位于右半 s 平面。

3.5.3 稳定性的代数判据

尽管线性时不变系统的稳定性可以通过找出特征方程的根来检验，由于求解高阶方程是很麻烦的事情，实际上有时候这一方法难以应用。应当认识到，检验一个系统的稳定性并没有必要确定特征方程所有根的确切位置。事实上，只需要知道是否所有的根都在复平面的左半部分。代数判据只要利用特征方程的系数就可以确定位于右半复平面的根的数目，而并不需要为了特征根本身去求解特征方程。

1. 稳定性的必要条件

考虑系统的特征方程为多项式形式，即

$$\Delta(s) = a_0 s^n + a_1 s^{n-1} + \cdots + a_{n-1} s + a_n = 0 \quad (3.81)$$

式中所有系数均为实数且 $a_0 > 0$。

假设系统的特征根为 $s_1, s_2, \cdots, s_n$，根据基本的代数知识，可以知道存在以下关系

$$\begin{aligned}
\frac{a_1}{a_0} &= (-1) \sum_{i=1}^{n} s_i \\
\frac{a_2}{a_0} &= \sum_{i=1, j=1}^{n} s_i s_j \quad (i \neq j) \\
&\vdots \\
\frac{a_n}{a_0} &= (-1)^n \prod_{i=1}^{n} s_i
\end{aligned} \quad (3.82)$$

显然，所有的比值都必须为非零正实数，否则至少有一个根具有正实部。

因此，由式(3.82)可知，特征方程的所有系数都必须为非零正实数，否则至少有一个根具有正实部。这一必要条件可以通过观察就可以检验。但是，这一条件不是充分条件。经常有这样的情况，一个多项式所有的系数都是非零的正实数，但仍然有右半复平面上的根，因此需要进一步的分析。

2. 赫尔维茨判据

考虑系统的特征方程为式(3.81)中的多项式形式,即
$$\Delta(s) = a_0 s^n + a_1 s^{n-1} + \cdots + a_{n-1} s + a_n = 0$$
系统稳定即其特征方程所有的根全都位于左半 s 平面的充分必要条件为多项式的所有赫尔维茨行列式 $D_i(i=1,2,\cdots,n)$ 都必须为正。式(3.81)多项式的赫尔维茨行列式由下式给出为

$$D_1 = a_1, \quad D_2 = \begin{vmatrix} a_1 & a_3 \\ a_0 & a_2 \end{vmatrix}, \quad D_3 = \begin{vmatrix} a_1 & a_3 & a_5 \\ a_0 & a_2 & a_4 \\ 0 & a_1 & a_3 \end{vmatrix}, \cdots,$$

$$D_n = \begin{vmatrix} a_1 & a_3 & a_5 & a_7 & \cdots & \cdots & a_{2n-1} \\ a_0 & a_2 & a_4 & a_6 & \cdots & \cdots & a_{2n-2} \\ 0 & a_1 & a_3 & a_5 & \cdots & \cdots & a_{2n-3} \\ 0 & a_0 & a_2 & a_4 & \cdots & \cdots & a_{2n-4} \\ 0 & 0 & a_1 & a_3 & \ddots & \cdots & a_{2n-5} \\ \vdots & \vdots & \vdots & \vdots & & \ddots & \vdots \\ 0 & 0 & 0 & 0 & 0 & 0 & a_n \end{vmatrix} \tag{3.83}$$

式中下标大于 n 或者为负的元素用零代替。

3. 林纳德 – 奇帕特判据

林纳德 – 奇帕特判据可以看作是简化的赫尔维茨判据。根据林纳德 – 奇帕特判据,系统稳定的充分必要条件是,① 稳定性的必要条件必须满足;② 特征多项式的所有奇次阶或者所有偶次阶赫尔维茨行列式必须全都为正。

例 3.4 确定具有以下开环传递函数单位反馈系统的稳定性。

(1) $G_1(s) = \dfrac{100}{(s+1)(s+2)(s+3)}$;

(2) $G_2(s) = \dfrac{10}{s(s-1)(s+5)}$;

(3) $G_3(s) = \dfrac{10(s+1)}{s(s-1)(s+5)}$。

解 (1) 闭环特征多项式为
$$\Delta(s) = (s+1)(s+2)(s+3) + 100 = s^3 + 6s^2 + 11s + 106$$
由于该多项式没有缺项且系数全都为正,满足稳定性的必要条件。但是,多项式的二阶赫尔维茨行列式为
$$D_2 = \begin{vmatrix} 6 & 106 \\ 1 & 11 \end{vmatrix} = -40$$
因此,根据林纳德 – 奇帕特判据,闭环系统不稳定。

(2) 闭环特征多项式为
$$\Delta(s) = s(s-1)(s+5) + 10 = s^3 + 4s^2 - 5s + 10$$

由于该多项式具有符号为负的系数,根据稳定性的必要条件,无需应用赫尔维茨判据就可以知道闭环系统不稳定。

(3) 闭环特征多项式为
$$\Delta(s) = s(s-1)(s+5) + 10(s+1) = s^3 + 4s^2 + 5s + 10$$

该多项式没有缺项且系数全都为正。由于
$$D_2 = \begin{vmatrix} 4 & 10 \\ 1 & 5 \end{vmatrix} = 10$$

根据林纳德-奇帕特判据可知,闭环系统稳定。

这些例子表明,开环系统和闭环系统的稳定性之间没有一定的联系。

4. 劳斯判据

由于要计算多项式的赫尔维茨行列式,赫尔维茨判据或者林纳德-奇帕特判据的应用对于高阶系统不太方便。在这种情况下,通过劳斯判据可以把计算工作简化为一张称之为劳斯阵列的表。

这一方法可以按以下步骤应用。

第1步:写出多项式形式的特征方程
$$\Delta(s) = a_0 s^n + a_1 s^{n-1} + \cdots + a_{n-1} s + a_n = 0 \tag{3.84}$$

式中所有系数均为实数。

第2步:如果满足稳定性的必要条件,必须按照以下方式组成劳斯阵列(示例是一个六阶系统的劳斯阵列)

s^6	a_0	a_2	a_4	a_6
s^5	a_1	a_3	a_5	0
s^4	$B_1 = \dfrac{a_1 a_2 - a_0 a_3}{a_1}$	$B_2 = \dfrac{a_1 a_4 - a_0 a_5}{a_1}$	$B_3 = \dfrac{a_1 a_6 - a_0 \cdot 0}{a_1} = a_6$	0
s^3	$C_1 = \dfrac{B_1 a_3 - a_1 B_2}{B_1}$	$C_2 = \dfrac{B_1 a_5 - a_1 B_3}{B_1}$	$C_3 = \dfrac{B_1 \cdot 0 - a_1 \cdot 0}{a_1} = 0$	0
s^2	$D_1 = \dfrac{C_1 B_2 - B_1 C_2}{C_1}$	$D_2 = \dfrac{C_1 a_6 - B_1 \cdot 0}{C_1} = a_6$	0	0
s^1	$E_1 = \dfrac{D_1 C_2 - C_1 D_2}{D_1}$	0	0	0
s^0	$F_1 = \dfrac{E_1 a_6 - D_1 \cdot 0}{E_1} = a_6$	0	0	0

该阵列一次构成一行,每一行在该行的左侧做有标记,即第一行左侧标有特征多项式中出现的 s 的最高幂次,第二行为次高幂次,依此类推。阵列的前两行直接由式(3.84)的系数获得,而后面的每一行则由该行的前面两行推得如上例所示。这一过程一直进行到 s^0 行,即所有行全都完成为止。在为给定多项式制定劳斯阵列时,有些元素可能不存在。在计算后续行的内容时,这些元素作为零处理。

第 3 步：一旦劳斯阵列已经完成，就可以根据劳斯判据确定系统的稳定性。劳斯判据表明，如果劳斯阵列第一列的所有元素符号相同，则特征方程（3.84）的根全在左半 s 平面；而且如果在第一列中符号有变化，则符号变化的次数指示了右半 s 平面上根的数目。这一要求既是必要条件，也是充分条件。

例 3.5　应用劳斯判据确定系统是否稳定，描述系统的特征方程为
$$\Delta(s) = 2s^4 + s^3 + 3s^2 + 5s + 10 = 0$$

解　特征方程指出该方程没有缺项，而且系数全都为正，所以组成劳斯阵列如下

s^4	2	3	10
s^3	1	5	0
s^2	−7	10	
s^1	6.43	0	
s^0	10		

第一列中有两次符号变化，从 1 变到 −7 和从 −7 变到 6.43，因此，该方程有两个根在右半 s 平面，系统不稳定。

劳斯阵列内的任意一行的系数都可以乘以或者除以一个正数而不会改变第一列的符号。这样可以简化系数的估算，在组成劳斯阵列时减少工作量。

例 3.6　给定特征方程
$$\Delta(s) = s^6 + 3s^5 + 2s^4 + 9s^3 + 5s^2 + 12s + 20 = 0$$
用劳斯判据考虑系统是否稳定。

解　特征方程表明该方程没有缺项，而且系数全都为正。劳斯阵列为

s^6	1	2	5	20
s^5	3	9	12	
	1	3	4	（除以 3 后）
s^4	−1	1	20	
s^3	4	24		
	1	6		（除以 4 后）
s^2	7	20		
s^1	22	0		
s^0	20			

注意，s^5 行除以 3 和 s^3 行除以 4 后，数的大小减小了。而且，这样做没有改变问题的结果，即第一列中还是有两次符号改变，因此方程有两个根在右半 s 平面，系统不稳定。

代数判据可以用于确定反馈系统稳定时开环增益或其他参数的取值范围。

例 3.7 确定使负反馈系统稳定时开环增益 K 的取值范围。已知系统的开环传递函数为

$$G(s) = \frac{K}{s(0.1s+1)(0.25s+1)}$$

解 该系统的特征方程为

$$D(s) = 0.025s^3 + 0.35s^2 + s + K$$

根据稳定性的必要条件我们有

$$K > 0$$

而且,根据林纳德－奇帕特判据,若系统稳定则以下条件应当满足

$$D_2 = \begin{vmatrix} 0.35 & K \\ 0.025 & 1 \end{vmatrix} > 0$$

即

$$K < 14$$

因此,闭环系统稳定的条件是开环增益必须满足

$$0 < K < 14$$

在组成劳斯阵列时,其第一列有两种特殊情况,每一种特殊情况都必须单独处理,并需要对阵列计算过程做出适当的修改。

特殊情况 1 非零行中首元为零。

当某一行的首元为零,但其他元不全为零时,第一列中的零元将妨碍阵列的完成,这时可以用一个很小的正数代替这个零元来克服这一问题。

例 3.8 描述系统的特征方程为

$$\Delta(s) = s^4 + s^3 + 2s^2 + 2s + 5 = 0$$

用劳斯判据确定系统的稳定性。

解 特征方程没有缺项,而且系数全都为正。在本例中,劳斯阵列组成如下。

s^4	1	2	5
s^3	1	2	0
s^2	ε	5	0
s^1	$2 - \dfrac{5}{\varepsilon} \approx -\dfrac{5}{\varepsilon}$	0	0
s^0	5	0	0

在第一列中符号改变两次,所以系统不稳定,并且在右半 s 平面有两个特征根。

特殊情况 2 某一行所有元全都为零。

当某一行的所有元全都为零时,完成劳斯阵列的步骤如下:

① 由前一行组成辅助方程。

② 用辅助方程对 s 求导得到的系数替代该全零行,由此完成劳斯阵列。

在这种情况下,辅助方程的根也就是原方程的根。而且,这些根位于复平面上关于原点对称的地方。注意,辅助方程的阶数总是偶数,这一阶数还指示出关于复平面原点对称

的根的个数。

例3.9 某系统具有如下特征方程
$$\Delta(s) = s^3 + 2s^2 + s + 2 = 0$$
用劳斯判据确定系统的稳定性。

解 特征方程没有缺项,而且系数全都为正。劳斯阵列为

s^3	1	1
s^2	2	2
s^1	0	0

可以看到,s^1 行为零行。辅助方程可以由前一行即 s^2 行获得。因此辅助方程为
$$2s^2 + 2 = 0$$
辅助方程对 s 求导得到新的方程为
$$4s = 0$$
这一新方程的系数插入 s^1 行,劳斯判据得以完成为

s^3	1	1
s^2	2	2
s^1	4	0
s^0	2	

由于第一列中没有符号的改变,系统没有正实部的根。由辅助方程可知,系统有一对纯虚根 $s = \pm j1$,系统不稳定。

3.5.4 相对稳定性

利用代数判据验证稳定性仅仅提供了稳定性问题的一部分答案。代数判据通过确定特征方程是否有根位于右半 s 平面揭示一个系统的绝对稳定性。但是,如果该系统满足代数判据是绝对稳定的,那么就会希望确定它的相对稳定性,也就是说该系统距离不稳定有多远。对于主导极点实部为 $-\zeta\omega_n$ 的系统,相对稳定性可以根据阻尼系数 ζ(对超调量的限制)和阻尼因子 $\zeta\omega_n$(对响应速度的限制)加以规定。

如果系统是稳定的,有时候想知道离虚轴最近的极点距离虚轴有多远。通过平移 s 平面上的纵轴得 $\tilde{s}$ 平面,代数判据也可以用来获取这一信息。这样,在特征多项式 $\Delta(s)$ 中用 $\tilde{s} - a$ 取代 s,就可以得到一个新的多项式 $\Delta(\tilde{s})$。对这一多项式应用代数判据将告诉我们 $\Delta(\tilde{s})$ 所有的根是否全部都在左半 $\tilde{s}$ 平面内,也就是 $\Delta(s)$ 所有的根是否全部都在 s 平面上直线 $s = -a$ 的左半边。

例3.10 考虑例3.7中的特征多项式
$$\Delta(s) = 0.025s^3 + 0.35s^2 + s + K$$
现在确定使直线 $s = -1$ 右边没有根时所需 K 的取值。

解 把纵轴向左移动1个单位,即

则有

$$s = \tilde{s} - 1$$

$$\Delta(\tilde{s}) = 0.025(\tilde{s}-1)^3 + 0.35(\tilde{s}-1)^2 + (\tilde{s}-1) + K =$$
$$= \tilde{s}^3 + 11\tilde{s}^2 + 15\tilde{s} + (40K - 27)$$

根据林纳德 - 奇帕特判据,如果 $\Delta(\tilde{s})$ 所有的根全部都在左半 $\tilde{s}$ 平面内,那么以下条件必须满足:

① $40K - 27 > 0$,即 $K > 0.675$;

② $D_2 = \begin{vmatrix} 11 & 40K-27 \\ 1 & 15 \end{vmatrix} > 0$,即 $K < 4.8$。

因此,如果 $0.675 < K < 4.8$ 则 $\Delta(\tilde{s})$ 的根全部都在左半 $\tilde{s}$ 平面,即 $\Delta(s)$ 的根全部都在 s 平面上直线 $s = -1$ 的左半边。

3.6 稳态误差

控制系统的时间响应的特征可以用瞬态响应和稳态响应表示。在一个物理系统中,由于特定系统的本性所致,实际响应的稳态与期望的状态完全一致的情况是很少的。稳态误差是用于衡量特定类型的输入施加于控制系统时系统稳态精度的一种判别标准。由于稳态误差可以是参考输入或者扰动引起的,稳态误差也可以看成系统跟随输入信号和/或抑制扰动信号的能力。

3.6.1 稳态误差的定义

1. 误差的定义

在理论上,把期望输出 $c_d(t)$ 与实际输出 $c(t)$ 之间的差别定义为控制系统的误差函数,即定义误差

$$E(s) = C_d(s) - C(s) \quad (3.85)$$

是合理的。在输入 $r(t) = c_d(t)$ 的情况下,输入信号代表了期望输出,误差函数就是

$$E(s) = R(s) - C(s) \quad (3.86)$$

对于大多数如图 3.16(a) 所示的单位反馈系统,输入通常都可以看成期望输出;对于如图 3.16(b) 所示的系统,通常指令输入正代表了期望的输出。

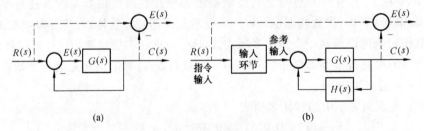

图 3.16 误差的一种定义

但是有时候给定系统的输入仅仅是参考输入,而且要提供一个与输出数量级相同的输入可能是难以做到或者不太方便的,甚至于可能连输出相同的量纲都难以保证。例如,有可能需要用一个低压信号控制高压的输出;在另一种情况下,对于速度控制系统使用电压信号或位置输入控制输出轴的速度更为实际。在这些情况下,误差信号不能简单地定义为参考输入与受控输出之间的差别。因此,如图3.17(a)所示,通常在反馈通路引入一个非单位环节 $H(s)$,构成一个具有典型结构的反馈系统。例如,如果一个10 V参考电压用于调节一个100 V的电源,那么 $H(s) = 0.1$。作为另一个例子,如果一个参考输入电压用于控制一个速度控制系统的输出速度,在反馈通路内就需要有一台测速发电机,而且 $H(s) = K_t s$。在这些情况下,明智的做法是把这样一种典型的非单位反馈系统的误差定义为

$$\varepsilon(s) = R(s) - B(s) = R(s) - H(s)C(s) \tag{3.87}$$

式中 $B(s)$ 是系统的主反馈信号,如图3.17(a)所示。

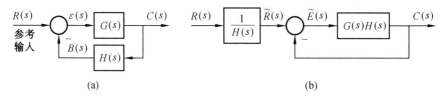

图 3.17 误差的另一种定义

通常,式(3.85)中定义的 $E(s)$ 称为输出端误差,而式(3.87)中定义的 $\varepsilon(s)$ 则称为输入端误差或者偏差。把图3.17(a)变换为图3.17(b)中的等效方框图,可以说明误差这两种定义之间的内在联系。由于在关系上 $\tilde{R}(s)$ 代表期望的输出,所以有

$$\tilde{E}(s) = \tilde{R}(s) - C(s) = C_d(s) - C(s) = E(s) \tag{3.88}$$

和

$$\tilde{E}(s) = \frac{1}{1 + G(s)H(s)}\tilde{R}(s) = \frac{1}{1 + G(s)H(s)} \cdot \frac{1}{H(s)}R(s) = \frac{1}{H(s)}\varepsilon(s) \tag{3.89}$$

比较式(3.88)和式(3.89),可以得到误差两个定义之间的一个简单关系

$$E(s) = \frac{1}{H(s)}\varepsilon(s) \tag{3.90}$$

尤其在单位负反馈系统的情况下有

$$E(s) = \varepsilon(s) \tag{3.91}$$

式(3.90)表明,由 $\varepsilon(t)$ 获取 $e(t)$ 不是困难的事情。

2. 稳态误差

作为时间的函数,误差信号由两部分组成:瞬态分量和稳态分量。由于稳态误差只对稳定系统才有意义,瞬态分量必定随着时间趋向于无穷大而趋近于零。因此,反馈控制系统的稳态误差定义为时间趋向于无穷大时的误差,即

$$e_{ss} = \lim_{t \to \infty} e(t) \tag{3.92.1}$$

$$\varepsilon_{ss} = \lim_{t \to \infty} \varepsilon(t) \tag{3.92.2}$$

3.6.2 用终值定理估算稳态误差

控制系统的误差可以是输入和/或扰动引起的。在这一节首先讨论由输入引起的误差。给定一个系统后,根据误差的定义和系统的动态描述可以获得误差传递函数 $\Phi_e(s)$ 或者 $\Phi_\varepsilon(s)$。由此,误差函数的基本表达式就是

$$E(s) = \Phi_e(s) R(s) \qquad (3.93.1)$$

$$\varepsilon(s) = \Phi_\varepsilon(s) R(s) \qquad (3.93.2)$$

利用终值定理,如果 $sE(s)$ 或者 $s\varepsilon(s)$ 在虚轴上(原点除外)和右半 s 平面内没有极点,则系统的稳态误差为

$$e_{ss} = \lim_{s \to 0} sE(s) = \lim_{s \to 0} s\Phi_e(s) R(s) \qquad (3.94.1)$$

$$\varepsilon_{ss} = \lim_{s \to 0} s\varepsilon(s) = \lim_{s \to 0} s\Phi_\varepsilon(s) R(s) \qquad (3.94.2)$$

该式表明,稳态误差将与系统的结构特征和输入要求两者都有关系。

应当强调,如果系统不稳定,稳态误差就没有意义了。由于 $\Phi_e(s)$ 或者 $\Phi_\varepsilon(s)$ 没有极点位于虚轴上和右半 s 平面内的充分必要条件是闭环系统稳定,所以在计算稳态误差之前应当先检验闭环系统的稳定性。另一方面,如果输入信号是阶跃、斜坡和抛物线函数,则 $R(s)$ 的所有极点全都位于原点。因此,如果系统稳定而且输入是典型测试信号,那么就可以应用终值定理计算稳态误差。

例 3.11 单位负反馈系统的开环传递函数为

$$G(s) = \frac{K(0.5s + 1)}{s(s + 1)(2s + 1)}$$

确定对于单位斜坡输入的稳态误差。

解 闭环系统的特征方程为

$$\Delta(s) = 2s^3 + 3s^2 + (1 + 0.5K)s + K = 0$$

由于二阶赫尔维茨行列式为

$$D_2 = \begin{vmatrix} 3 & K \\ 2 & 1 + 0.5K \end{vmatrix}$$

系统稳定的充分必要条件为 $D_2 > 0$,即 $0 < K < 6$。

由于 $R(s) = 1/s^2$,误差函数为

$$E(s) = \Phi_e(s) R(s) = \frac{s(s + 1)(2s + 1)}{s(s + 1)(2s + 1) + K(0.5s + 1)} \cdot \frac{1}{s^2}$$

因此,在 $0 < K < 6$ 的情况下,稳态误差为

$$e_{ss} = \lim_{s \to 0} sE(s) = \frac{1}{K}$$

例 3.12 图 3.18 为一个速度调节系统。确定该系统对于单位阶跃输入的稳态误差。

解 误差传递函数为

$$\Phi_\varepsilon(s) = \frac{1}{1 + \dfrac{K_1}{0.07s + 1} \cdot \dfrac{K_2}{0.24s + 1} \cdot \alpha K_t} = \frac{(0.07s + 1)(0.24s + 1)}{(0.07s + 1)(0.24s + 1) + \alpha K_1 K_1 K_t}$$

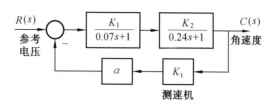

图 3.18 例 3.12 的速度调节系统

由于这个二阶系统必定稳定,根据终值定理可有

$$\varepsilon_{ss} = \lim_{s \to 0} s\Phi_\varepsilon(s)R(s) = \lim_{s \to 0} s \frac{(0.07s+1)(0.24s+1)}{(0.07s+1)(0.24s+1)+\alpha K_1 K_2 K_t} \cdot \frac{1}{s} = \frac{1}{1+\alpha K_1 K_2 K_t}$$

例 3.13 某系统见图 3.19,误差定义为 $E(s)=R(s)-C(s)$。(1)确定以 K_1 和 k 表示的对于单位阶跃输入的稳态误差;(2)选择使稳态误差为零时的 K_1。

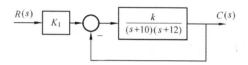

图 3.19 例 3.13 的系统

解 (1)系统总的传递函数为

$$\Phi(s) = \frac{K_1 k}{(s+10)(s+12)+k}$$

按定义,误差传递函数为

$$E(s) = R(s) - C(s) = R(s) - \Phi(s)R(s) = [1-\Phi(s)]R(s) = \frac{(s+10)(s+12)+k-K_1 k}{(s+10)(s+12)+k} \cdot \frac{1}{s}$$

由于该系统总是稳定的,由终值定理可有

$$e_{ss} = \lim_{s \to 0} sE(s) = \lim_{s \to 0} s \frac{(s+10)(s+12)+k-K_1 k}{(s+10)(s+12)+k} \cdot \frac{1}{s} = \frac{120+k-K_1 k}{120+k}$$

(2)令 $e_{ss}=0$ 得到

$$\frac{120+k-K_1 k}{120+k} = 0$$

即稳态误差为零时有

$$K_1 = \frac{120+k}{k} = 1 + \frac{120}{k}$$

3.6.3 应用静态误差系数估算 ε_{ss}

对于许多控制系统而言,通常由阶跃、斜坡和抛物线函数这样一些典型输入引起的稳

态误差表征系统的稳态性能。通过引入系统型别和静态误差系数的概念,可以简化由这些典型输入引起的稳态误差的估算。

1. 系统型别

一般情况下,非单位负反馈系统的开环传递函数可以写成

$$G(s)H(s) = \frac{K \prod_{i=1}^{m}(\tau_i s + 1)}{s^v \prod_{j=v+1}^{n}(T_j s + 1)} \quad (3.95)$$

式中,K 为开环增益;$\tau_i(i=1,2,\cdots,m)$ 和 $T_j(j=v+1,v+2,\cdots,n)$ 均为常数。

一个反馈控制系统的型别指的是开环传递函数中的纯积分因子数,即 $G(s)H(s)$ 在 $s=0$ 处极点的阶数。

注意系统的型别与系统的阶数两者之间的区别。系统的阶数是特征多项式中 s 的最高幂次。例如,系统的开环传递函数为

$$G(s) = \frac{10(s+1)}{s^2(3s+1)}$$

该系统为 2 型系统,而系统的阶数则为三阶。

对于具有图 3.17(a) 中所示典型结构的反馈系统,如果终值定理的条件能够满足,由输入引起的稳态误差的表达式可以改写为

$$\varepsilon_{ss} = \lim_{s \to 0} s \cdot \frac{1}{1 + G(s)H(s)} \cdot R(s) = \lim_{s \to 0} s \cdot \frac{1}{1 + \frac{K}{s^v}} \cdot R(s) =$$

$$\lim_{s \to 0} \frac{s^{v+1}}{s^v + K} R(s) \quad (3.96)$$

该式表明,在这种情况下稳态误差与系统的型别 v 和开环增益 K,以及参考输入 $R(s)$ 有关。

在下面的讨论中,输入将依次考虑几种典型信号(阶跃函数、斜坡函数和抛物线函数)中的一种,而具有图 3.17(a) 所示典型结构反馈系统的稳态误差将作为系统型别的函数进行估算。

2. 静态位置误差系数

如果施加于系统的参考输入是一个幅值为 r_0 的阶跃信号,即 $R(s) = r_0/s$,则

$$\varepsilon_{ss} = \lim_{s \to 0} s \cdot \frac{1}{1 + G(s)H(s)} \cdot \frac{r_0}{s} = \lim_{s \to 0} \frac{r_0}{1 + G(s)H(s)} \quad (3.97)$$

为了方便起见,我们定义

$$K_p = \lim_{s \to 0} G(s)H(s) = \lim_{s \to 0} \frac{K}{s^v} = \begin{cases} K & v = 0 \\ \infty & v \geq 1 \end{cases} \quad (3.98)$$

其中 K_p 称为静态位置误差系数。这样,式(3.97) 就变为

$$\varepsilon_{ss} = \frac{r_0}{1 + K_p} = \begin{cases} \dfrac{r_0}{1 + K} & v = 0 (K_p = K) \\ 0 & v \geq 1 (K_p \to \infty) \end{cases} \quad (3.99)$$

式(3.99)表明,对于0型系统,阶跃输入引起的稳态误差 ε_{ss} 总是大小有限的,但 ε_{ss} 将随着开环增益的增大而减小。另一方面,为了使 ε_{ss} 为零, $G(s)H(s)$ 必须至少有一个纯积分因子。

3. 静态速度误差系数

如果施加于系统的参考输入为 $v_0 t \cdot 1(t)$,即 $R(s) = v_0/s^2$,则有

$$\varepsilon_{ss} = \lim_{s \to 0} s \cdot \frac{1}{1 + G(s)H(s)} \cdot \frac{v_0}{s^2} = \lim_{s \to 0} \frac{v_0}{s[1 + G(s)H(s)]} = \frac{v_0}{\lim_{s \to 0} sG(s)H(s)} \tag{3.100}$$

为了方便起见,定义

$$K_v = \lim_{s \to 0} sG(s)H(s) = \lim_{s \to 0} \frac{K}{s^{v-1}} = \begin{cases} 0 & v = 0 \\ K & v = 1 \\ \infty & v \geq 2 \end{cases} \tag{3.101}$$

其中 K_v 称为静态速度误差系数,这样,式(3.100)就变为

$$\varepsilon_{ss} = \frac{v_0}{K_v} = \begin{cases} \infty & v = 0(K_v = 0) \\ \dfrac{v_0}{K} & v = 1(K_v = K) \\ 0 & v \geq 2(K_v \to \infty) \end{cases} \tag{3.102}$$

式(3.102)表明,在斜坡输入的情况下,为了使 ε_{ss} 为零, $G(s)H(s)$ 必须至少有两个纯积分因子。1型系统能够跟踪斜坡输入,但有一个有限大的误差。

4. 静态加速度误差系数

如果施加于系统的输入是 $\dfrac{a_0}{2}t^2 \cdot 1(t)$,即 $R(s) = \dfrac{a_0}{s^3}$,则

$$\varepsilon_{ss} = \lim_{s \to 0} s \cdot \frac{1}{1 + G(s)H(s)} \cdot \frac{a_0}{s^3} = \lim_{s \to 0} \frac{a_0}{s^2[1 + G(s)H(s)]} = \frac{a_0}{\lim_{s \to 0} s^2 G(s)H(s)} \tag{3.103}$$

为了方便起见,定义

$$K_a = \lim_{s \to 0} s^2 G(s)H(s) = \lim_{s \to 0} \frac{K}{s^{v-1}} = \begin{cases} 0 & v \leq 1 \\ K & v = 2 \\ \infty & v \geq 3 \end{cases} \tag{3.104}$$

其中 K_a 称为静态加速度误差系数。这样,式(3.103)变为

$$\varepsilon_{ss} = \frac{a_0}{K_a} = \begin{cases} \infty & v \leq 1(K_a = 0) \\ \dfrac{a_0}{K} & v = 2(K_a = K) \\ 0 & v \geq 3(K_a \to \infty) \end{cases} \tag{3.105}$$

该式表明,当输入为抛物线函数时,为了使 ε_{ss} 为零, $G(s)H(s)$ 必须至少有三个纯积分因

子。

因此,概要地讲,在具有典型结构的反馈系统中,如果系统稳定,那么由典型输入所引起的稳态误差可以直接由开环传递函数获取。表 3.2 列出了静态误差系数、系统型别以及输入类型之间的关系。

应当指出,如果稳态误差是时间的函数,误差系数仅仅给出无穷大的答案,但不能提供误差如何随时间变化的任何信息。而且,当输入不是这三种所提到的类型时,误差系数不能给出关于稳态误差的信息。

表 3.2 阶跃、斜坡和抛物线输入引起的稳态误差的小结

系统型别	误差系数			输入		
v	K_p	K_v	K_a	阶跃 $\varepsilon_{ss} = \dfrac{r_0}{1+K_p}$	斜坡 $\varepsilon_{ss} = \dfrac{v_0}{K_v}$	抛物线 $\varepsilon_{ss} = \dfrac{a_0}{K_a}$
0	K	0	0	$\varepsilon_{ss} = \dfrac{r_0}{1+K}$	$\varepsilon_{ss} \to \infty$	$\varepsilon_{ss} \to \infty$
1	∞	K	0	$\varepsilon_{ss} = 0$	$\varepsilon_{ss} = \dfrac{v_0}{K}$	$\varepsilon_{ss} \to \infty$
2	∞	∞	K	$\varepsilon_{ss} = 0$	$\varepsilon_{ss} = 0$	$\varepsilon_{ss} = \dfrac{a_0}{K}$

例 3.14 考虑图 3.20 所示控制系统。

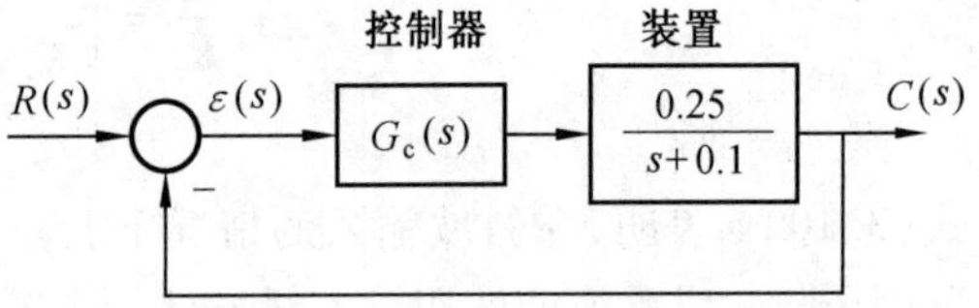

在 $G_c(s) = 10$ 的情况下,开环传递函数为

$$G(s)H(s) = \frac{2.5}{s+0.1}$$

图 3.20 例 3.14 的控制系统

且该系统稳定。我们可以看到,该系统为 0 型,由式(3.98)静态位置误差系数为

$$K_p = \lim_{s \to 0} G(s) = 25$$

如果输入为 $r_0 = 5$,那么稳态误差就是

$$\varepsilon_{ss} = \frac{r_0}{1+K_p} = \frac{5}{26} = 0.192$$

假设 $G_c(s) = \dfrac{s+0.5}{s}$,则开环传递函数变为

$$G(s) = \frac{2.5(s+0.5)}{s(s+0.1)}$$

而且该系统仍然是稳定的。由于系统现在是 1 型的,对于恒值输入的稳态误差为零。在前述方程中,如果装置的参数发生了变化,稳态误差仍然为零。我们还可以注意到,如果输入端施加了单位斜坡信号,稳态误差可以根据静态速度误差系数获得。由于

$$K_v = \lim_{s \to 0} sG(s) = 12.5$$

所以稳态误差为

$$\varepsilon_{ss} = \frac{1}{K_v} = \frac{1}{12.5} = 0.08$$

3.6.4 无差度

无差度是控制系统的一个稳态精度指标,用于以典型测试信号评价控制系统跟踪输入或者抑制扰动作用的能力。如果一个系统即使对于阶跃输入的稳态误差也不为零,则称该系统对于输入具有零阶无差度。如果一个系统对于阶跃输入而且只有对于阶跃输入的稳态误差为零,则称该系统对于输入具有一阶无差度。如果一个系统对于斜坡输入的稳态误差为零,但对于抛物线输入的稳态误差不为零,则称该系统对于输入具有二阶无差度。如果一个系统对于抛物线输入的稳态误差也为零,则称该系统对于输入具有三阶无差度。类似地,可以定义一个系统对于扰动的无差度。

考虑图 3.17(a) 所示系统。假设系统是稳定的,开环传递函数为

$$G(s)H(s) = \frac{KM(s)}{s^v N(s)} \tag{3.106}$$

式中,K 为开环增益;v 为系统型别。典型输入可写为

$$R(s) = \frac{R_l}{s^l} \tag{3.107}$$

式中,$l = 1, 2, \cdots$ 且 R_l 为常数。这样,在指定输入作用下的误差为

$$\varepsilon(s) = \Phi_\varepsilon(s) R(s) = \frac{1}{1 + G(s)H(s)} R(s) = \frac{s^v N(s)}{s^v N(s) + KM(s)} \cdot \frac{R_l}{s^l} \tag{3.108}$$

而稳态误差则为

$$\varepsilon_{ss} = \lim_{s \to 0} s \frac{s^v N(s)}{s^v N(s) + KM(s)} \cdot \frac{R_l}{s^l} = \lim_{s \to 0} \frac{R_l}{s^v + K} s^{v+1-l} \tag{3.109}$$

由该式可见,如果 $v = 0$,对于任何输入 ε_{ss} 都将不为零;如果 $v = 1$,对于阶跃输入 ε_{ss} 将为零,但是对于斜坡和抛物线输入则不为零;如果 $v = 2$,那么对于阶跃和斜坡输入 ε_{ss} 将为零,但是对于抛物线输入则不为零。因此,图 3.17(a) 所示系统对于输入信号具有 v 阶无差度。

在更一般的情况下,如果系统的结构与图 3.17(a) 所示结构不一样,除了按定义外,还可以通过找出整个系统的等效开环传递函数确定该系统对于输入的无差度;而且,如果等效开环传递函数包含 v 个积分因子,则该系统称为等效的 v 型系统。

3.6.5 跟踪参考输入的前馈控制

正如所知,通过提高系统的型别和/或增大开环增益可以改善控制系统的稳态性能。但是,从动态性能的角度看,这两种方法都可能导致稳定性问题,使系统变为不稳定或者相对稳定性下降。为了避免这一缺点,在需要高精度的控制系统中广泛使用复合控制。

图 3.21 中所示复合控制系统用于改善控制系统的稳态性能,该系统在反馈控制的基础上另外引入了一条前馈通路。

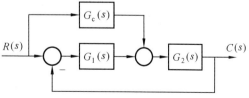

图 3.21 复合控制系统

由图 3.21 中的方框图,整个系统的传递函数为

$$\Phi(s) = \frac{C(s)}{R(s)} = \frac{G_1(s)G_2(s) + G_c(s)G_2(s)}{1 + G_1(s)G_2(s)} \quad (3.110)$$

取

$$G_c(s) = \frac{1}{G_2(s)} \quad (3.111)$$

得到

$$\Phi(s) \equiv 1$$

这意味着该系统将在整个响应过程中完全复现指令输入。

就物理实现而言,由于 $G_c(s)$ 是 $G_2(s)$ 的倒数,实际上实现完全复现几乎是不可能的。另一方面,在很多情况下,近似的复现对于改善控制系统的稳态性能已经足够了。至于前馈信号施加位置的选择问题,应当同时注意降低 $G_c(s)$ 的复杂程度和对于前馈信号功率的要求。此外还应当注意,由于前馈方框的传递函数 $G_c(s)$ 在开环中起着作用,因此它必须是稳定的。

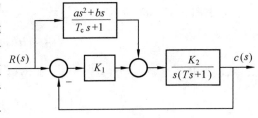

图 3.22 例 3.15 的控制系统

例 3.15 位置跟踪系统的方框图如图 3.22 所示,图中 K_1, K_2, T 以及 T_c 均为正的常数。确定使抛物线输入 $r(t) = t^2/2$ 的稳态误差为零时 a 和 b 的取值。

解 由给定的方框图,整个系统的传递函数为

$$\Phi(s) = \frac{C(s)}{R(s)} = \frac{\dfrac{K_1 K_2}{s(Ts+1)} + \dfrac{K_2(as^2 + bs)}{s(T_c s+1)(Ts+1)}}{1 + \dfrac{K_1 K_2}{s(Ts+1)}} = \frac{K_1 K_2(T_c s+1) + K_2(as^2 + bs)}{(T_c s+1)(Ts^2 + s + K_1 K_2)} = $$

$$\frac{K_2[as^2 + (b + K_1 T_c)s + K_1]}{(T_c s+1)(Ts^2 + s + K_1 K_2)}$$

相应地,误差传递函数为

$$E(s) = R(s) - C(s) = [1 - \Phi(s)]R(s) = \left[1 - \frac{K_1 K_2(T_c s+1) + K_2(as^2 + bs)}{(T_c s+1)(Ts^2 + s + K_1 K_2)}\right] \cdot \frac{1}{s^3} =$$

$$\frac{T_c T s^3 + (T_c + T - K_2 a)s^2 + (1 - K_2 b)s}{(T_c s+1)(Ts^2 + s + K_1 K_2)} \cdot \frac{1}{s^3}$$

通过观察,该系统稳定。令稳态误差

$$e_{ss} = \lim_{s \to 0} sE(s) = \lim_{s \to 0} s \cdot \frac{T_c T s^3 + (T_c + T - K_2 a)s^2 + (1 - K_2 b)s}{(T_c s+1)(Ts^2 + s + K_1 K_2)} \cdot \frac{1}{s^3} = 0$$

结果为

$$a = \frac{T + T_c}{K_2}, \quad b = \frac{1}{K_2}$$

3.7 扰动的抑制

在任何控制系统中都存在一些影响被控对象的输出并且一般不受控制的输入作用,称这些输入作用为扰动,并且通常努力使控制系统设计成这些扰动对于控制系统的影响能够最小。

除了在改善动态响应方面优于开环系统外,利用反馈控制的其他好处之一是使系统对扰动不太灵敏。系统忽视扰动的能力经常称为它的扰动抑制能力。

3.7.1 开环和闭环的扰动抑制

不论扰动是怎么造成的,就其作用而言,它们对控制系统的最终影响与图 3.23 中所示是相同的,图中 $R(s)$ 为参考输入,$D(s)$ 为扰动,而 $C(s)$ 则为输出。在研究中假设系统是稳定的。

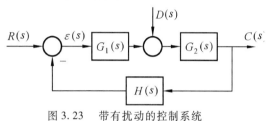

图 3.23　带有扰动的控制系统

为了方便,令 $H(s)=1$,并且

$$G_1(s)=\frac{k_1 M_1(s)}{s^{v_1} N_1(s)},\quad G_2(s)=\frac{k_2 M_2(s)}{s^{v_2} N_2(s)}$$

其中 $M_i(s)$ 和 $N_i(s)(i=1,2)$ 均为 s 的多项式,且

$$\lim_{s\to 0} M_i(s)=1,\quad \lim_{s\to 0} N_i(s)=1,\quad i=1,2$$

开环增益和系统的型别分别为

$$K=\lim_{s\to 0} s^{v_1+v_2} G_1(s) G_2(s)=k_1 k_2$$

$$v=v_1+v_2$$

根据叠加原理,系统的输出为

$$C(s)=\frac{G_1(s)G_2(s)}{1+G_1(s)G_2(s)}R(s)+\frac{G_2(s)}{1+G_1(s)G_2(s)}D(s) \quad (3.112)$$

该表达式表明,由扰动 $D(s)$ 做出的贡献可以看成是由输入 $R(s)$ 建立的稳态输出中的变化部分。增量的大小视为扰动影响的一种度量,如果终值定理的条件满足,增量的大小还可以由终值定理确定

$$\lim_{t\to\infty}\Delta c(t)_{\text{CL}}=\lim_{s\to 0} s\frac{G_2(s)}{1+G_1(s)G_2(s)}D(s) \quad (3.113)$$

类似地,对于开环系统而言,系统的输出为

$$C(s)=G_1(s)G_2(s)R(s)+G_2(s)D(s) \quad (3.114)$$

扰动 $D(s)$ 的影响造成输入 $R(s)$ 建立的稳态输出发生变化,而且变化的大小可以利用终值定理确定

$$\lim_{t\to\infty}\Delta c(t)_{\text{OL}}=\lim_{s\to 0} s G_2(s) D(s) \quad (3.115)$$

由式(3.115)和式(3.113)可以看到,开环系统和闭环系统的扰动抑制比为

$$\frac{\lim_{t\to\infty}\Delta c(t)_{\text{OL}}}{\lim_{t\to\infty}\Delta c(t)_{\text{CL}}} = \lim_{s\to 0}\frac{sG_2(s)D(s)}{s\dfrac{G_2(s)}{1+G_1(s)G_2(s)}D(s)} = \lim_{s\to 0}[1+G_1(s)G_2(s)] =$$

$$1 + \lim_{s\to 0}\frac{k_1 M_1(s)}{s^{v_1}N_1(s)} \cdot \frac{k_2 M_2(s)}{s^{v_2}N_2(s)} = 1 + \lim_{s\to 0}\frac{K}{s^v} > 1 \tag{3.116}$$

式(3.116)表明,闭环系统维持输出的能力比开环系统好。

对于图 3.23 所示系统,定义 $\varepsilon(t) = r(t) - c(t)$,我们有

$$(\varepsilon_{\text{ss,d}})_{\text{CL}} = \lim_{s\to 0} s \frac{-G_2(s)}{1+G_1(s)G_2(s)}D(s) \tag{3.117}$$

$$(\varepsilon_{\text{ss,d}})_{\text{OL}} = \lim_{s\to 0} s[-G_2(s)]D(s) \tag{3.118}$$

$$\frac{(\varepsilon_{\text{ss,d}})_{\text{OL}}}{(\varepsilon_{\text{ss,d}})_{\text{CL}}} = \lim_{s\to 0}[1+G_1(s)G_2(s)] = 1 + \lim_{s\to 0}\frac{K}{s^v} > 1 \tag{3.119}$$

式(3.119)的结果和式(3.116)的结果是一样的,因此,扰动的影响也可以由该扰动所引起的稳态误差 $\varepsilon_{\text{ss,d}}$ 得到反映。

3.7.2 扰动的抑制

1. 扰动对系统的影响

现在分别考虑阶跃和斜坡扰动对于闭环系统的影响。假设

$$H(s) = \frac{k_3 M_3(s)}{N_3(s)}, \quad \lim_{s\to 0}M_3(s) = \lim_{s\to 0}H_3(s) = 1$$

如果系统稳定,得到

$$\varepsilon_{\text{ss,d}} = \lim_{s\to 0} s\frac{-G_2(s)H(s)}{1+G_1(s)G_2(s)H(s)}D(s) = \lim_{s\to 0} s\frac{-\dfrac{k_2 M_2(s)}{s^{v_2}N_2(s)}\cdot\dfrac{k_3 M_3(s)}{N_3(s)}}{1+\dfrac{k_1 M_1(s)}{s^{v_1}N_1(s)}\cdot\dfrac{k_2 M_2(s)}{s^{v_2}N_2(s)}\cdot\dfrac{k_3 M_3(s)}{N_3(s)}}D(s) =$$

$$\lim_{s\to 0} s\frac{-k_2 k_3 s^{v_1}}{s^{v_1+v_2}+k_1 k_2 k_3}D(s) \tag{3.120}$$

(1) 阶跃扰动的影响

假设扰动是一个阶跃函数,即 $D(s) = d_0/s$,由式(3.120)得到

$$\varepsilon_{\text{ss,d}} = \lim_{s\to 0} s \cdot \frac{-k_2 k_3 s^{v_1}}{s^{v_1+v_2}+k_1 k_2 k_3} \cdot \frac{d_0}{s} =$$

$$\begin{cases} \dfrac{-k_2 k_3}{1+k_1 k_2 k_3}d_0 & v_1 = 0, v_2 = 0 \\ \dfrac{-1}{k_1}d_0 & v_1 = 0, v_2 \neq 0 \\ 0 & v_1 = 1, 2, \cdots \end{cases} \tag{3.121}$$

(2) 斜坡扰动的影响

如果扰动是一个斜坡函数,即 $D(s) = d_1/s^2$,有

$$\varepsilon_{\mathrm{ss,d}} = \lim_{s \to 0} s \cdot \frac{-k_2 k_3 s^{v_1}}{s^{v_1+v_2} + k_1 k_2 k_3} \cdot \frac{d_1}{s^2} = \begin{cases} \infty & v_1 = 0 \\ \dfrac{-1}{k_1} d_1 & v_1 = 1 \\ 0 & v_1 = 2,3,\cdots \end{cases} \quad (3.122)$$

2. 抑制扰动的途径

（1）串联补偿

由 3.7.1 节可以看到,扰动对于反馈系统的影响与 v_1 和 k_1,即前向通路在扰动作用点之前部分 $G_1(s)$ 的积分因子数和增益密切相关。因此,如图 3.24 所示,在前向通路扰动作用点之前加入一个包含积分因子的串联补偿装置 $G_c(s)$ 就有可能消除扰动的影响。有时候通过增大 $G_c(s)$ 的增益也有可能抑制扰动。

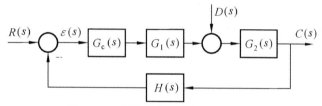

图 3.24　带补偿装置的闭环系统

一定要记住,当系统稳定而且只有当系统稳定时,估算所得到的由输入和/或扰动引起的稳态误差才有意义。因此,当系统中加入补偿装置 $G_c(s)$ 后有必要重新检验系统的稳定性。

例 3.15 考虑图 3.25 中的控制系统。当扰动为单位阶跃函数时,估算采用补偿装置之前由扰动造成的稳态误差,并设计一个能使该扰动引起的稳态误差为零的补偿装置。

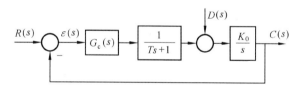

图 3.25　例 3.15 的控制系统

解 采用补偿装置之前,开环传递函数为

$$G_p(s) = \frac{K_0}{s(Ts+1)}$$

该系统是一个二阶系统,系统稳定。由公式(3.121)可有

$$\varepsilon_{\mathrm{ss}} = \varepsilon_{\mathrm{ss,d}} = \lim_{s \to 0} s \cdot \frac{-\dfrac{K_0}{s}}{1 + \dfrac{K_0}{s(Ts+1)}} \cdot \frac{1}{s} = \lim_{s \to 0} s \cdot \frac{-K_0(Ts+1)}{s(Ts+1) + K_0} \cdot \frac{1}{s} = -1$$

仍然由公式(3.121)可以知道,当且仅当 $G_c(s)$ 至少有一个积分因子时,由阶跃扰动引起的稳态误差将为零。假设 $G_c(s) = 1/s$,则可以得到一个新的开环传递函数为

$$G_Xc(s)G_p(s) = \frac{K_0}{s^2(Ts+1)}$$

显然,在采用补偿装置后系统将变为不稳定。因此需要重新设计另一个补偿装置。现在使补偿装置为

$$G_c(s) = \frac{\tau s + 1}{s}$$

这样,开环传递函数就变成

$$G_c(s)G_p(s) = \frac{K_0(\tau s + 1)}{s^2(Ts+1)}$$

根据稳定性的代数判据,该系统当且仅当 $\tau > T$ 时稳定。因此,采用串联补偿装置,即

$$G_c(s) = \frac{\tau s + 1}{s}, \quad \tau > T$$

则由阶跃扰动引起的稳态误差将为零。

(2)前馈补偿

抑制扰动的另一种方法称为前馈补偿,但是只有当扰动可测量时才能应用。这一方法可用图 3.26 说明。

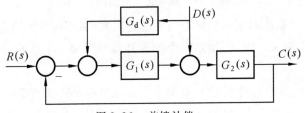

图 3.26　前馈补偿

现在,从扰动到输出的传递函数为

$$\frac{C(s)}{D(s)} = \frac{G_2(s) + G_d(s)G_1(s)G_2(s)}{1 + G_1(s)G_2(s)} \tag{3.123}$$

如果可以使分子为零,即

$$G_d(s) = -\frac{1}{G_1(s)} \tag{3.124}$$

则扰动将被完全抑制。

与跟踪参考输入相类似,这一方法经常用于改善控制系统的稳态性能,而且记住,传递函数仅仅是物理系统近似的模型。因此,如果式(3.124)严格地得到满足,物理系统中扰动抑制的效果将取决于系统模型的精确程度。

习　题

题 3.1　某系统的单位阶跃响应为

$$c(t) = 1 + e^{-t} - e^{-2t}, \quad t \geq 0$$

(a)确定该系统的脉冲响应。

(b) 确定该系统的传递函数 $C(s)/R(s)$。

题 3.2 考虑由题 3.2(a) 图所示方框图描述的系统,为以下题 3.2(b) 图所示各阶跃响应确定两个反馈的极性,图中"0"表示反馈断开。

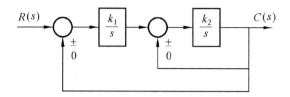

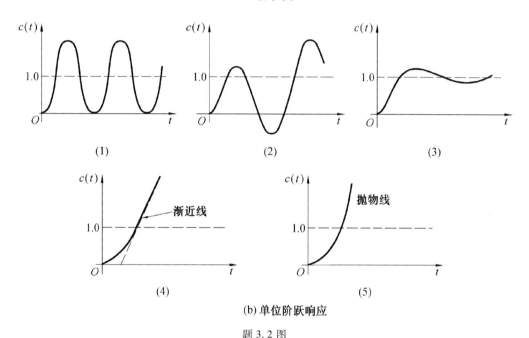

题 3.2 图

题 3.3 考虑以下闭环系统,根据极点在复平面上的位置概略绘制它们的阶跃响应曲线,并解释所得结果。

(a) $\Phi(s) = \dfrac{20}{s^2 + 12s + 20}$ 　　(b) $\Phi(s) = \dfrac{6}{s^3 + 6s^2 + 11s + 6}$

(c) $\Phi(s) = \dfrac{4}{s^2 + 2s + 2}$ 　　(d) $\Phi(s) = \dfrac{12.5}{(s^2 + 2s + 5)(s + 5)}$

题 3.4 某单位负反馈系统的开环传递函数为

$$G(s) = \dfrac{1}{s(s+1)}$$

确定该系统的上升时间、峰值时间、最大(百分比)超调量和调整时间(5% 允许误差带)。

题 3.5 题 3.5 图所示为某二阶系统给出的单位阶跃响应。如果该系统为单位服反馈系统,找出它的开环传递函数。

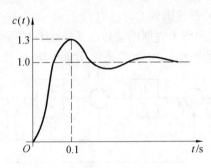

题 3.5 图

题 3.6 题 3.6(a) 图所示为某反馈系统方框图,题 3.6(b) 图所示则为其单位阶跃响应曲线。确定参数 k_1、k_2 和 a 的取值。

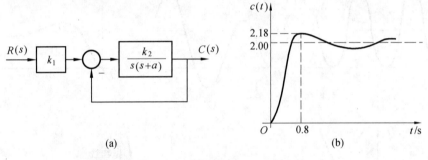

题 3.6 图

题 3.7 某单位负反馈系统的开环传递函数为

$$G(s) = \frac{k}{s(s+\sqrt{2k})}$$

(a) 确定超调量和调整时间(5% 调整标准)。
(b) 调整时间小于 0.75 s 时 k 的取值范围。

题 3.8 对于题 3.6 图所示伺服系统,试确定满足以下闭环系统设计要求时 k 和 a 的取值。
(a) 超调量最多为 40%;
(b) 峰值时间为 4 s。

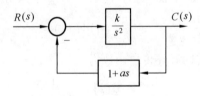

题 3.8 图

题 3.9 某单位反馈系统的开环传递函数为

$$G(s) = \frac{k}{s(s+2)}$$

阶跃响应的指标指定为峰值时间 $t_p = 1.1$ s 和百分比超调量 $\sigma_p = 5\%$。
(a) 确定这两个性能指标能否同时满足。
(b) 如果这两个性能指标不能同时满足，确定一个折中的 k 值，使得幅值时间和百分比超调量的要求放宽相同的百分比。

题 3.10　一个控制系统的传递函数为
$$\frac{C(s)}{R(s)} = \frac{0.33}{(s + 2.56)(s^2 + 0.4s + 0.13)}$$
如果可能，用主导极点法估算峰值时间，最大（百分比）超调量和调节时间（$\Delta = 5\%$）。

题 3.11　系统具有以下特征方程，用代数判据确定系统的稳定性。
(a) $s^3 + 20s^2 + 9s + 20 = 0$
(b) $3s^4 + 10s^3 + 5s^2 + s + 2 = 0$
(c) $s^5 + 2s^4 + 9s^3 + 10s^2 + s + 2 = 0$

题 3.12　一些系统的特征方程给出如下。确定每一种情况下在右半 s 平面内的特征根数和纯虚根数。
(a) $s^3 - 3s + 2 = 0$
(b) $s^3 + 10s^2 + 16s + 160 = 0$
(c) $s^5 + 3s^4 + 12s^3 + 24s^2 + 32s + 48 = 0$
(d) $s^5 + 2s^4 + 3s^3 + 6s^2 - 4s - 8 = -0$

题 3.13　一些系统的特征方程给出如下。对于各系统确定 k 和 T 的取值范围，使得相应的系统稳定。假设 k 和 T 都是正数。
(a) $s^4 + 2s^3 + 10s^2 + 2s + k = 0$
(b) $s^3 + (T + 0.5)s^2 + 4Ts + 50 = 0$

题 3.14　一个单位负反馈控制系统的开环传递函数为
$$G(s) = \frac{K}{s(0.01s^2 + 0.2\zeta s + 1)}$$
确定使闭环系统稳定时 K 和 ζ 的取值范围。

题 3.15　某负反馈系统的开环传递函数由下式给出为
$$G(s)H(s) = \frac{K(s + 1)}{s(Ts + 1)(2s + 1)}$$
参数 K 和 T 可以表示在一个以 K 作为横坐标、以 T 作为纵坐标的平面上。确定闭环系统稳定时的区域。

题 3.16　某单位负反馈系统具有开环传递函数
$$G(s) = \frac{K}{(Ts + 1)(nTs + 1)(n^2Ts + 1)}, \quad 0 \leq n \leq 1$$
(a) 确定使得系统稳定的 K 和 n 的取值范围。
(b) 确定 $n = 1$、0.5、0.1 和 0.01 时信息稳定所需的 K 值。
(c) 讨论当 K 不变时闭环稳定性与 n 的关系。

题 3.17　单位负反馈系统具有开环传递函数

$$G(s) = \frac{K}{s\left(\dfrac{s}{3}+1\right)\left(\dfrac{s}{6}+1\right)}$$

确定使得直线 $s = -1$ 右侧没有闭环极点时所需 K 的取值范围。

题 3.18　某系统具有特征方程

$$s^3 + 10s^2 + 29s + k = 0$$

利用劳斯－赫尔维茨判据确定使复根的实部为 -2 时 k 的取值。

题 3.19　题 3.19 图中系统代表某自动制导飞行器。

(a) 确定系统稳定所需的 τ 值。
(b) 确定当特征方程的一个根为 $s = -5$ 时的 τ 值,以及此时其余根的数值。
(c) 对于在 (b) 中所选的 τ,找出系统对阶跃指令的响应。

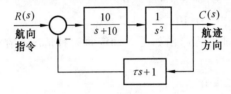

题 3.19 图

题 3.20　一种温度调节器可以用传递函数 $1/(Ts+1)$ 描述。已知测量某容器内的水温时,指示实际水温的 98% 需时一分钟。如果容器被加热且水温以 10 度／分钟的速率线性上升,试估算温度调节器的稳态指示误差。

题 3.21　确定单位负反馈系统在单位阶跃输入、单位斜坡输入以及抛物线输入 $t^2/2$ 作用下的稳态误差。各系统的开环传递函数由下式给出为

(a) $G(s) = \dfrac{50}{(0.1s+1)(2s+1)}$　　(b) $G(s)H(s) = \dfrac{10}{s(s+4)(s+0.5)}$

(c) $G(s) = \dfrac{8(0.5s+1)}{s^2(0.1s+1)}$　　(d) $G(s) = \dfrac{10}{s^2(s+1)(s+5)}$

(e) $G(s) = \dfrac{k}{s(s^2+4s+200)}$

题 3.22　单位负反馈系统的开环传递函数为

$$G(s) = \frac{K}{s(T_1 s+1)(T_2 s+1)}$$

确定输入为 $r(t) = a + bt$ 时使得稳态误差小于 ε_0 的 K、T_1 和 T_2 的取值。假设 K、T_1 和 T_2 都大于零,a 和 b 都是常数。

题 3.23　单位反馈系统的开环传递函数为

$$G(s) = \frac{K}{s(Ts+1)}$$

确定满足以下性能指标时 K 和 T 的取值。

(a) 单位斜坡输入时的稳态误差小于 0.02。
(b) 最大(百分比)超调量小于 30%,调节时间小于 0.3 s。

题 3.24　题 3.24 图所示是某控制系统的方框图,图中误差函数为 $E(s)=R(s)-C(s)$。选择使该系统对于输入具有二阶无差度时 τ 和 b 的取值。

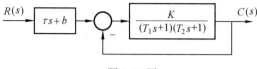

题 3.24 图

题 3.25　题 3.25 图所示是某复合控制系统的方框图。选择使该系统对于输入具有三阶无差度时 a 和 b 的取值。

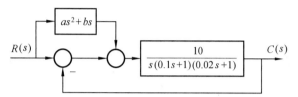

题 3.25 图

题 3.26　一个控制系统的方框图如题 3.26 图所示。在两种情况下分别确定扰动为单位阶跃和单位斜坡时的稳态误差。

(a) $G_1(s)=K_1, G_2(s)=\dfrac{K_2}{s(T_2s+1)}$

(b) $G_1(s)=\dfrac{K_1(T_1s+1)}{s}, G_2(s)=\dfrac{K_2}{s(T_2s+1)}, \quad T_1>T_2$

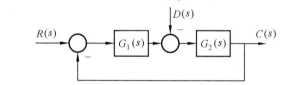

题 3.26 图

题 3.27　题 3.27 图所示是某组合系统的方框图,图中

$$G_1(s)=\dfrac{K_1}{T_1s+1},\quad G_2(s)=\dfrac{K_2}{s(T_2s+1)},\quad G_3(s)=\dfrac{K_3}{K_2}$$

确定使单位阶跃扰动造成的稳态误差为零时前馈方框的传递函数 $G_d(s)$。

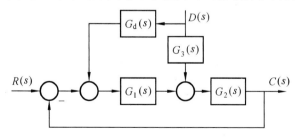

题 3.27 图

第4章 根轨迹法

4.1 反馈系统的根轨迹

4.1.1 引言

在上一章中已经看到,通过改变开环增益或其他参数可以改变闭环传递函数的极点、即控制系统特征方程的根。由于闭环系统的相对稳定性和过渡过程特性与闭环系统特征方程根的位置直接相关,为了获得合适的根的位置常常有必要调节系统的一个或多个参数。因此,当系统的某一个参数变化时,确定特征方程的根在 s 平面上如何移动是非常有用的。

伊凡思(W. R. Evans)于1948年提出的根轨迹法是一种作图的方法,用于在复平面上画出代数方程的一个参数从零变化到无穷大时方程的根的轨迹。根轨迹法提供了直接通过作图"求解"特征方程寻找闭环极点在 s 平面上位置的方法。在有些情况下这些闭环极点的位置可以很精确地寻找到,而在另一些情况下有可能只是大概的位置。但是,无论在哪一种情况下,都可以以合理的精度得到关于系统稳定性和其他性能的信息。

根轨迹法的基本原理是闭环的极点(暂态响应的模式)与开环的零点、极点以及开环增益有关。

4.1.2 二阶系统的根轨迹

通过以下的例子引入根轨迹的概念。

例4.1 考虑图4.1中所示的二阶系统。为了研究增益 k 从零变化到无穷大时对系统的影响,在 s 平面上绘制系统特征方程的根的轨迹。

解 在这种情况下,该系统的闭环传递函数和特征方程分别为

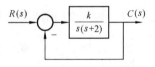

图 4.1 二阶系统的方框图

$$\Phi(s) = \frac{k}{s^2 + 2s + k} \quad (4.1)$$

$$\Delta(s) = s^2 + 2s + k = 0 \quad (4.2)$$

由方程(4.2),特征方程的根为

$$s = -1 \pm \sqrt{1-k} \quad (4.3)$$

首先考虑 $k = 0$ 的情况。这时得到的根为 $s = 0$ 和 $s = -2$,它们也是开环系统的极点,在图4.2中用符号"×"表示。当 k 从0增加到1时,得到的解是实根,但是 $k > 1$ 时得到的解则是一对实部为 -1 的共轭复根。方程(4.2)的根的变化情况如图4.2所示。

第 4 章 根轨迹法

我们可以把轨迹看成从两个开环极点出发。从 A 点出发的轨迹在实轴上向右移动的同时,从 B 点出发的轨迹则向左移动,一直到 $k=1$ 时它们在 C 点相遇。当 k 取值更大时,轨迹则沿直线 CD 和 CE 移动如图 4.2 所示。

一旦已经获得了一个控制系统的根轨迹,就有可能确定当一个参数变化时系统性能变化的情况。例如,由图 4.2 所示的根轨迹图可以得到结论,对于所有的 k 值系统都是稳定的,只有当 $k>1$ 时闭环传递函数的极点才是复数。而且,在特征方程的根为复数的情况下 k 逐渐增大

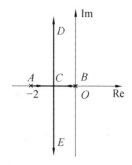

图 4.2 二阶系统的根轨迹

时,尽管系统是稳定的,但阻尼比的数值将逐渐减小,瞬态响应的超调量将逐渐增大。

每一个控制系统的根轨迹都可以按照类似的方式进行分析,以获得由于系统参数变化而导致的系统响应变化的情况。

4.1.3 根轨迹条件

在讨论根轨迹问题时,所考虑的闭环系统可以是图 4.3 中所示的非单位反馈系统。由此,一般情况下闭环传递函数和特征方程分别为

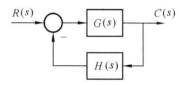

图 4.3 闭环系统方框图

$$\frac{C(s)}{R(s)} = \frac{G(s)}{1+G(s)H(s)} \tag{4.4}$$

$$1 + G(s)H(s) = 0 \tag{4.5}$$

为了绘制根轨迹,开环传递函数通常表示成标准的零极点形式,即伊凡斯形式

$$G(s)H(s) = \frac{K^*(s-z_1)(s-z_2)\cdots(s-z_m)}{(s-p_1)(s-p_2)\cdots(s-p_n)} = \frac{K^*\prod_{i=1}^{m}(s-z_i)}{\prod_{j=1}^{n}(s-p_j)} \tag{4.6}$$

其中 $z_i(i=1,2,\cdots,m)$ 和 $p_j(j=1,2,\cdots,n)$ 分别为开环传递函数 $G(s)H(s)$ 的零点和极点。可变参数 K^* 称为根轨迹增益,以与开环增益相区别。

将式(4.6)代入式(4.5),特征方程经改写后得到

$$\frac{K^*\prod_{i=1}^{m}(s-z_i)}{\prod_{j=1}^{n}(s-p_j)} = -1 \tag{4.7}$$

该式有时称为根轨迹方程。为了满足根轨迹方程或者特征方程,s 平面上任何在根轨迹上的点都必须同时满足以下两个条件

$$|G(s)H(s)| = \frac{K^*\prod_{i=1}^{m}|s-z_i|}{\prod_{j=1}^{n}|s-p_j|} = 1 \tag{4.8}$$

$$\angle G(s)H(s) = \sum_{i=1}^{m} \angle(s-z_i) - \sum_{j=1}^{n} \angle(s-p_j) = (2l+1)\pi,$$
$$l = 0, \pm 1, \pm 2, \cdots \quad (4.9)$$

称式(4.8)为根轨迹的幅值条件,而式(4.9)则称为根轨迹的幅角条件或幅角判据。由于假设K^*是从零变化到无穷大,s平面上的任意一点都能够满足式(4.8)。因此,验证s平面上的一点是在根轨迹上的充分必要条件是满足式(4.9)给出的幅角条件。

原则上,根轨迹法分两步进行:
① 寻找出s平面上满足根轨迹幅角条件的所有的点。
② 寻找出根轨迹上满足根轨迹幅值条件的特定的点。

4.2 绘制根轨迹的法则

通过求解特征方程获取根轨迹、或者用试探的方法搜索s平面上所有满足幅角条件的点来获取根轨迹是非常繁重而乏味的任务。幸运的是借助于根轨迹的性质,通过运用一些构造"法则",不用太费劲就能够获得根轨迹的大致图形。

以下绘制根轨迹的法则是根据闭环控制系统的特征方程或者开环传递函数得到的。这些法则应当仅仅作为构造根轨迹的辅助手段,因为在一般情况下它们并不能给出根轨迹的精确图形。

在下面的讨论中假设开环传递函数以零极点的形式给出为

$$G(s)H(s) = \frac{K^* \prod_{i=1}^{m}(s-z_i)}{\prod_{j=1}^{n}(s-p_j)} \quad (4.10)$$

而具有一个可变参数K^*的根轨迹问题则用以下的特征方程或者根轨迹方程描述

$$\prod_{j=1}^{n}(s-p_j) + K^* \prod_{i=1}^{m}(s-z_i) = 0 \quad (4.11)$$

$$\frac{K^* \prod_{i=1}^{m}(s-z_i)}{\prod_{j=1}^{n}(s-p_j)} = -1 \quad (4.12)$$

相应地,根轨迹的幅角条件就是

$$\sum_{i=1}^{m} \angle(s-z_i) - \sum_{j=1}^{n} \angle(s-p_j) = (2l+1)\pi, \ l=0, \pm 1, \pm 2, \cdots \quad (4.13)$$

4.2.1 根轨迹的连续性、对称性和分支数

法则1 根轨迹可以看到是由一些连续的分支组成的,分支数等于式(4.11)中特征方程的阶数。而且,这些根轨迹分支是关于s平面的实轴对称的。

说明 根轨迹的一条分支就是K^*从零变化到无穷大时特征方程的一个根的轨迹。这一法则的成立是因为根轨迹的分支数必定等于式(4.11)中特征方程的根数。另一方

面,(对于物理系统,)系统已假定采用实系数的有理函数(两个多项式之比)作为模型。因此,如果特征方程有一个复数根,那么该复数根的共轭复数也一定是该方程的一个根、即实系数代数方程的复根总是以成对的共轭复数出现。

4.2.2 根轨迹的起点和终点

法则2 当根轨迹增益 K^* 从零增大到无穷大时,根轨迹起始于 $G(s)H(s)$ 的极点而终止于 $G(s)H(s)$ 的零点。对于通常情况下物理系统的数学模型,开环传递函数的极点将多于零点,即 $n > m$,这时根轨迹的 $(n-m)$ 条分支将随 $K^* \to \infty$ 而趋向无穷远处。

证明 将式(4.12)改写为

$$\frac{\prod_{i=1}^{m}|s-z_i|}{\prod_{j=1}^{n}|s-p_j|} = \frac{1}{K^*} \tag{4.14}$$

当 $K^* \to 0$ 时,式(4.14)的取值趋向无穷大,与此相应的是 s 的取值逼近于开环极点 $p_j(j=1,2,\cdots,n)$。当 $K^* \to \infty$ 时,式(4.14)的取值趋近于零,相应地则有 s 的取值逼近于开环零点 $z_i(i=1,2,\cdots,m)$。由于把无穷远处的零点和极点都包含在内时有理函数的零点和极点的总数必定相等,在 $n > m$ 的情况下,$G(s)H(s)$ 有 $(n-m)$ 个零点位于无穷远处。

4.2.3 根轨迹的渐近线

法则3 当 $n > m$,即开环传递函数在无穷远处有零点时,随着 $K^* \to \infty$ 将有 $(n-m)$ 条分支沿 $(n-m)$ 条渐近线趋向无穷远处。这些渐近线相聚在实轴上的一点,且交点为

$$\sigma_a = \frac{\sum_{j=1}^{n}p_j - \sum_{i=1}^{m}z_i}{n-m} \tag{4.15}$$

这些渐近线与实轴正方向的夹角为

$$\varphi_a = \frac{(2l+1)\pi}{n-m}, \quad l = 0,1,2,\cdots,n-m-1 \tag{4.16}$$

证明 式(4.11)可以改写为

$$\left[s^n - \sum_{j=1}^{n}p_j \cdot s^{n-1} + \cdots + \prod_{j=1}^{n}(-p_j)\right] + K^*\left[s^m - \sum_{i=1}^{m}z_i \cdot s^{n-1} + \cdots + \prod_{i=1}^{m}(-z_i)\right] = 0$$

或者

$$\frac{s^n - \sum_{j=1}^{n}p_j \cdot s^{n-1} + \cdots + \prod_{j=1}^{n}(-p_j)}{s^m - \sum_{i=1}^{m}z_i \cdot s^{n-1} + \cdots + \prod_{i=1}^{m}(-z_i)} = -K^*$$

方程的左侧进行长除法运算,在 s 很大的情况下,除了前两项外均忽略不计,得到

$$s^{n-m} - \left(\sum_{j=1}^{n}p_j - \sum_{i=1}^{m}z_i\right)s^{n-m-1} = -K^*$$

即
$$s^{n-m}\left[1 - \frac{1}{s}\left(\sum_{j=1}^{n} p_j - \sum_{i=1}^{m} z_i\right)\right] = -K^*$$

或者
$$s\left[1 - \frac{1}{s}\left(\sum_{j=1}^{n} p_j - \sum_{i=1}^{m} z_i\right)\right]^{\frac{1}{n-m}} = (-K^*)^{\frac{1}{n-m}}$$

进行两项式展开并在结果所得级数中仅保留前两项则有
$$s\left[1 - \frac{1}{n-m} \cdot \frac{1}{s}\left(\sum_{j=1}^{n} p_j - \sum_{i=1}^{m} z_i\right)\right] = (-K^*)^{\frac{1}{n-m}}$$

即
$$s - \frac{1}{n-m}\left(\sum_{j=1}^{n} p_j - \sum_{i=1}^{m} z_i\right) = (-K^*)^{\frac{1}{n-m}}$$

或者
$$s = \frac{1}{n-m}\left(\sum_{j=1}^{n} p_j - \sum_{i=1}^{m} z_i\right) + (K^*)^{\frac{1}{n-m}} \cdot e^{(2l+1)\pi/(n-m)},\ l = 0,1,2,\cdots,n-m-1 \quad (4.17)$$

参考图4.4,式(4.17)表明有$(n-m)$个闭环极点位于无穷远处,每个闭环极点的位置由两个向量确定:一个共同的实向量为

$$\sigma_a = \frac{\sum_{j=1}^{n} p_j - \sum_{i=1}^{m} z_i}{n-m}$$

和一个各不相同的复向量为

$$(K^*)^{\frac{1}{n-m}} \cdot e^{(2l+1)\pi/(n-m)},\ l = 0,1,2,\cdots,n-m-1$$

因此,$(n-m)$条渐近线相交于式(4.15)给出的点σ_a,而渐近线与实轴正方向的夹角φ_a则由式(4.16)给出。

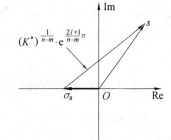

图4.4 渐近线的向量图

例4.2 假设需要画出一个负反馈系统的根轨迹,其开环传递函数为

$$G(s)H(s) = \frac{K^*}{s(s+1)(s+2)}$$

该开环传递函数没有零点,三个极点为$p_1 = 0, p_2 = -1$和$p_3 = -2$。使用选定的符号将它们放置在s平面上。通常,"开环"极点"×"标记,而开环零点用"●"标记,而特定的闭环极点则用"·"标记。

由法则1,根轨迹由三条连续的分支组成,这三条分支关于s平面的实轴是对称的。

由法则2,由于$n=3$和$m=0$,当$K^*=0$时这三条分支起始于三个开环极点,而$K^* \to \infty$时它们则趋向无穷远处。

由法则3,这三条分支沿三条渐近线趋向与无穷远处。这些渐近线的交点为

$$\sigma_a = \frac{0-1-2}{3-0} = -1$$

而它们与实轴的夹角则为

$$\varphi_a = \frac{(2l+1)\pi}{3-0} = \pm\frac{\pi}{3}, \pi\ (l = 0, \pm 1)$$

结果如图 4.5 所示。

4.2.4 实轴上的根轨迹

法则 4 实轴上任一区段当且仅当其右侧的开环零极点总数为奇数时是根轨迹的一部分。

证明 这一法则的证明是基于以下的观察结果：

① 对于实轴上的任一试验点 s_t，如图所示从 $G(s)H(s)$ 的共轭复数极点或零点画出的向量的夹角相加之和为零。例如，图 4.6 中有

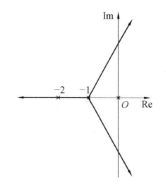

图 4.5 例 4.2 系统的根轨迹渐近线

$$\angle(s_t - p_3) + \angle(s_t - p_4) = 0$$

因此，在式 (4.13) 中对于幅角条件的贡献仅来自 $G(s)H(s)$ 的实数极点和零点。

② 只有位于点 s_t 右侧的 $G(s)H(s)$ 的实数极点和零点才可能对式 (4.13) 中的幅角条件做出贡献，因为位于该点左侧的实数极点和零点的贡献为零。例如，在图 4.6 中有

$$\angle(s_t - p_2) = 0$$

③ $G(s)H(s)$ 在点 s_t 右侧的每个实极点对于式 (4.13) 的贡献为 $-180°$，$G(s)H(s)$ 在点 s_t 右侧的每个实零点对于式 (4.13) 的贡献为 $180°$。

最后一点观察表明，s_t 是根轨迹上的点则其右侧开环极点和零点数之和必须为奇数。

例 4.2 中的系统在实轴上的根轨迹如图 4.7 所示。

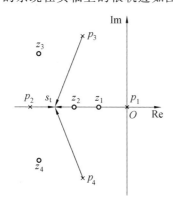

图 4.6 实轴上根轨迹的条件

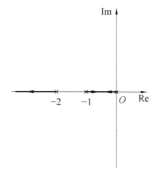

图 4.7 例 4.2 实轴上的根轨迹

4.2.5 根轨迹的起始角和终止角

根轨迹的起始角（或终止角）是根轨迹离开开环极点（或到达开环零点）时的角度，即根轨迹在起始点（或终止点）的切线与实轴正方向的夹角。

法则 5 根轨迹离开开环极点 p_k 时的起始角为

$$\theta_{p_k} = (2l+1)\pi + \sum_{i=1}^{m}\angle(p_k - z_i) - \sum_{j=1,j\neq k}^{n}\angle(p_k - p_j), \quad l = 0, \pm 1, \pm 2, \cdots$$

或者简记为

$$\theta_{p_k} = 180° + \sum_{i=1}^{m}\angle(p_k - z_i) - \sum_{j=1,j\neq k}^{n}\angle(p_k - p_j) \quad (4.18)$$

类似地，根轨迹到达开环零点 z_k 时的终止角为

$$\theta_{z_k} = 180° + \sum_{j=1}^{n}\angle(z_k - p_j) - \sum_{i=1,i\neq k}^{m}\angle(z_k - z_i) \quad (4.19)$$

证明 假设 s_t 是离开开环极点 p_k 的根轨迹分支上的一个点，并且非常靠近该极点。这样 s_t 必定满足式(4.13)，即

$$\sum_{i=1}^{m}\angle(s_t - z_i) - [\sum_{j=1}^{k-1}\angle(s_t - p_j) + \angle(s_t - p_k) + \sum_{j=k+1}^{n}\angle(s_t - p_j)] = (2l+1)\pi,$$

$$l = 0, \pm 1, \pm 2, \cdots$$

或者

$$\angle(s_t - p_k) = -(2l+1)\pi + \sum_{i=1}^{m}\angle(s_t - z_i) - \sum_{j=1,j\neq k}^{n}\angle(s_t - p_j), l = 0, \pm 1, \pm 2, \cdots$$

使 s_t 无限逼近该极点 p_k，不失一般性，取 $l = -1$ 得到

$$\theta_{p_k} = 180° + \sum_{i=1}^{m}\angle(p_k - z_i) - \sum_{j=1,j\neq k}^{n}\angle(p_k - p_j)$$

类似地有

$$\theta_{z_k} = 180° + \sum_{j=1}^{n}\angle(z_k - p_j) - \sum_{i=1,i\neq k}^{m}\angle(z_k - z_i)$$

如果一个开环极点或零点为 q 重，则起始角为

$$\theta_{p_k} = \frac{1}{q}\Big[(2l+1)\pi + \sum_{i=1}^{m}\angle(p_k - z_i) - \sum_{\substack{j=1 \\ j\neq k,\cdots,k+q-1}}\angle(p_k - p_j)\Big], \, l = 0,1,2,\cdots,q-1$$

(4.20)

而终止角由式(4.21)给出

$$\theta_{z_k} = \frac{1}{q}\Big[(2l+1)\pi + \sum_{j=1}^{m}\angle(z_k - p_j) - \sum_{\substack{i=1 \\ i\neq k,\cdots,k+q-1}}^{m}\angle(z_k - z_i)\Big], \, l = 0,1,2,\cdots,q-1$$

(4.21)

根轨迹的起始角（或终止角）对于复极点或复零点的情况特别有意义，因为在完成根轨迹的绘制时有关的信息非常有用。尽管这一法则也适用于实极点或实零点的情况，但在许多情况下，甚至是对于多重实极点或实零点也没有必要使用上述公式确定起始角或终止角。如果 q 个极点或零点重合在一起，那么根轨迹各分支在 q 重极点或零点处的切线将等分 $360°$，即离开该极点或到达该零点的两条相邻分支之间的夹角均为

$$\theta = 2\pi/q \quad (4.22)$$

例 4.3 单回路反馈系统的开环传递函数为

$$G(s)H(s) = \frac{K^*(s+5)}{s(s^2+4s+8)}$$

确定根轨迹的起始角。

解 如图 4.8 所示,该开环传递函数有一个零点和三个极点

$$z_1 = -5, \quad p_1 = 0, \quad p_{2,3} = -2 \pm j2$$

由法则 4 可确定实轴上的根轨迹如图 4.8 所示,由观察得到

$$\theta_{p_1} = 180°$$

然后,由式(4.18) 有

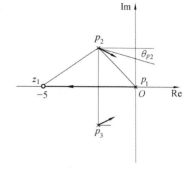

图 4.8 例 4.3 的起始角

$$\theta_{p_2} = 180° + \angle(p_2 - z_1) - [\angle(p_2 - p_1) + \angle(p_2 - p_3)] = \\ 180° + \arctan\frac{2}{3} - \left[\left(180° - \arctan\frac{2}{2}\right) + 90°\right] = -12°$$

由于根轨迹是关于实轴对称的,所以得到

$$\theta_{p_3} = 12°$$

4.2.6 根轨迹的分离点和分离角

随着 K^* 的变化,当闭环系统的特征方程含有重根时,根轨迹的两条(或更多)分支在分离点处相会合而后又相互分离,如图 4.9 所示。

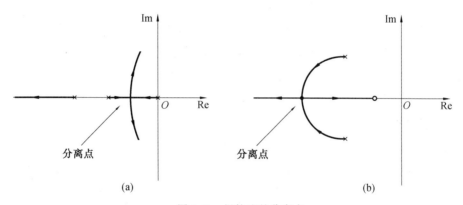

图 4.9 根轨迹的分离点

1. 根轨迹的分离点

法则 6 根轨迹上的分离点必定满足方程

$$\frac{\mathrm{d}}{\mathrm{d}s}\left[\frac{\prod_{i=1}^{m}(s-z_i)}{\prod_{j=1}^{n}(s-p_j)}\right] = 0 \tag{4.23}$$

证明 由式(4.11),特征方程可以改写成

$$D(s) = Q(s) + K^*P(s) = 0 \tag{4.24}$$

其中 $Q(s) = \prod_{j=1}^{n}(s - p_j), P(s) = \prod_{i=1}^{m}(s - z_i)$。

设 s_b 是一个分离点，则特征方程在 s_b 处有重根，即特征方程可以表达成
$$D(s) = (s - s_b)^\gamma D_1(s)$$
其中 γ 是该根的重数。由此，$D(s)$ 关于 s 的导数是
$$D'(s) = \gamma(s - s_b)^{\gamma-1} D_1(s) + (s - s_b)^\gamma D'_1(s) =$$
$$(s - s_b)^{\gamma-1}[\gamma D_1(s) + (s - s_b)D'_1(s)]$$
其中"′"表示关于 s 的导数。由于 $D'(s)$ 在点 s_b 处也有 $(\gamma - 1)$ 重的根，即 $D'(s) = 0$，由式(4.24)得到
$$D'(s) = Q'(s) + K^* P'(s) = Q'(s) - \frac{Q(s)}{P(s)}P'(s) = 0$$
该式可以表达成
$$[Q(s)P'(s) - Q'(s)P(s)]_{s=s_b} = 0$$
即
$$\frac{d}{ds}\left[\frac{\prod_{i=1}^{m}(s - z_i)}{\prod_{j=1}^{n}(s - p_j)}\right] = 0$$

一般而言，关于分离点可有以下结论成立：

① 式(4.23)给出的分离点条件是必要条件，但不是充分条件。换句话说，任一分离点都必须是式(4.23)的解，而且该解还必须是根轨迹上的点。

② 一个分离点关系到的根轨迹分支有可能多于两条。

③ 根轨迹图可以有不止一个分离点。而且，分离点未必总是在实轴上。

2. 根轨迹的分离角

假设 $K^* = K_b^*$ 时存在一个 q 重分离点，而且相应的特征根为 $s_{b1}, s_{b2}, \cdots, s_{bn}$，其中 q 个根为重根。现在以 $s_{b1}, s_{b2}, \cdots, s_{bn}$ 作为开环极点、并以原来系统相同的开环零点构造一个新的系统。那么，原来系统的分离角问题就变为新系统的起始角问题。因此可以做出如下结论：

一般而言，由根轨迹的幅角条件，根轨迹各分支的切线在分离点处均分 360°。如果分离点处根是 q 重的，那么在分离点处根轨迹任意两条相邻分支切线之间的夹角为

$$\theta = \frac{2\pi}{q} \tag{4.25}$$

重新考虑例 4.2 的系统。解分离点方程
$$\frac{d}{ds}\left[\frac{K^*}{s(s+1)(s+2)}\right] = 0$$

得到

$$3s^2 + 6s + 2 = 0$$

该方程的根为

$$s_1 = -0.423, \; s_2 = -1.577$$

由于 $s_2 = -1.577$ 不在根轨迹上，根轨迹的分离点为 $s_b = -0.423$。

根轨迹离开分离点的两条分支之间的夹角为 π。由于实轴上区段 $[0, -1]$ 是根轨迹的一部分，这两条分支垂直离开实轴。图 4.10 中所示为根轨迹的分离点和分离角。

4.2.7 根轨迹与虚轴的交点

法则 7 根轨迹与 s 平面的虚轴相交的交点以及相应的 K^* 值可由以下方程给出

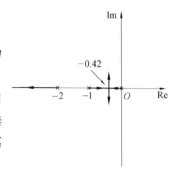

图 4.10 例 4.2 系统的分离点和分离角

$$1 + G(j\omega)H(j\omega) = 0 \qquad (4.26)$$

说明 如果根轨迹与虚轴相交，则特征方程有一对纯虚根，即 $s = \pm j\omega$，它们必定满足特征方程

$$1 + G(j\omega)H(j\omega) = 0$$

在例 4.2 中，特征方程为

$$1 + \frac{K^*}{s(s+1)(s+2)} = 0$$

可以改写为

$$s^3 + 3s^2 + 2s + K^* = 0$$

将 $s = j\omega$ 代入以上方程得到

$$\begin{cases} K^* - 3\omega^2 = 0 \\ \omega(2 - \omega^2) = 0 \end{cases}$$

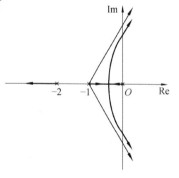

图 4.11 例 4.2 系统的根轨迹

结果为 $\omega = \pm 1.41$ 和 $K^* = 6$，这表明当 $K^* = 6$ 时根轨迹与虚轴相交于 ± 1.41 处。完整的根轨迹画在图 4.11 中。

由该根轨迹可以看到，当 $0 < K^* < 6$ 时闭环系统稳定。

4.2.8 其他有用的结论

1. 根轨迹上增益的估算

有时候需要计算对应于根轨迹上特定点的根轨迹增益或开环增益、或者系统的静态误差系数。

根据根轨迹的幅值条件，在根轨迹上任一点 s_1 处的根轨迹增益 K^* 为

$$K^* = \frac{\prod_{j=1}^{n}|s_1 - p_j|}{\prod_{i=1}^{m}|s_1 - z_i|} \tag{4.27}$$

另一方面,如果根轨迹上特定点处的 K^* 值已经给定,根据定义可以获得开环增益为

$$K = \lim_{s \to 0} s^v G(s)H(s) = \lim_{s \to 0} s^v \frac{K^* \prod_{i=1}^{m}(-z_i)}{\prod_{j=1}^{n}(-p_j)} \tag{4.28}$$

式中,v 为系统的型别。

2. 闭环极点与开环极点之间的关系

闭环特征方程可以写成

$$\prod_{j=1}^{n}(s - p_j) + K^* \prod_{i=1}^{m}(s - z_i) = 0$$

或者

$$\prod_{j=1}^{n}(s - s_j) = 0$$

其中 $p_j(j=1,2,\cdots,n)$ 和 $z_i(i=1,2,\cdots,m)$ 分别为开环极点和零点,而 $s_j(j=1,2,\cdots,n)$ 则为闭环极点。如果 $n - m \geq 2$,则有

$$s^n + \sum_{j=1}^{n}(-p_j)s^{n-1} + \cdots = s^n + \sum_{j=1}^{n}(-s_j)s^{n-1} + \cdots$$

即在 $n - m \geq 2$ 的情况下

$$\sum_{j=1}^{n} s_j = \sum_{j=1}^{n} p_j \tag{4.29}$$

在 $n - m \geq 2$ 的情况下,这一结论有时候对于确定根轨迹的走向是很有用的。

3. 闭环极点与特征方程的系数

假设特征方程为

$$s^n + a_1 s^{n-1} + \cdots + a_{n-1} s + a_n = 0$$

而且该方程的根为 $s_1, s_2, \cdots, s_n$,那么有

$$s^n + a_1 s^{n-1} + \cdots + a_{n-1}s + a_n = (s - s_1)(s - s_2)\cdots(s - s_n)$$

根据代数方程的系数与根的关系,得到

$$\sum_{j=1}^{n} s_j = -a_1 \tag{4.30}$$

$$\prod_{j=1}^{n} s_j = (-1)^n a_n \tag{4.31}$$

4.3 其他形式的根轨迹

4.3.1 0°根轨迹

1. 0°根轨迹和 180°根轨迹

在上一节中,所考虑的根轨迹满足特征方程

$$1 + G(s)H(s) = 0 \tag{4.32}$$

其中开环传递函数的形式为

$$G(s)H(s) = \frac{K^* \prod_{i=1}^{m}(s - z_i)}{\prod_{j=1}^{n}(s - p_j)} \tag{4.33}$$

即绘制根轨迹所基于的幅角条件为

$$\sum_{i=1}^{m} \angle(s - z_i) - \sum_{j=1}^{n} \angle(s - p_j) = 180° + 2l\pi, \quad l = \pm 1, \pm 2, \cdots \tag{4.34}$$

现在考虑具有式(4.33)中开环传递函数的正反馈系统。在这种情况下,特征方程变为

$$1 - G(s)H(s) = 0 \tag{4.35}$$

即根轨迹方程变为

$$\frac{K^* \prod_{i=1}^{m}(s - z_i)}{\prod_{j=1}^{n}(s - p_j)} = 1 \tag{4.36}$$

而根轨迹的幅角条件则变为

$$\sum_{i=1}^{m} \angle(s - z_i) - \sum_{j=1}^{n} \angle(s - p_j) = 0° + 2l\pi, \quad l = \pm 1, \pm 2, \cdots \tag{4.37}$$

显然,绘制方程(4.36)的根轨迹时有些与幅角条件有关的法则必须修改。由于很简单的原因,满足式(4.37)的根轨迹称为0°根轨迹,而满足式(4.34)的根轨迹则称为180°根轨迹或常规根轨迹

2. 绘制 0°根轨迹的基本法则

不难了解,法则3~5和法则7与根轨迹的幅角条件有关。绘制0°根轨迹时,这些法则应当修改如下:

① 修改后的法则3:根轨迹的渐近线。渐近线与实轴之间的夹角为

$$\varphi_a = \frac{2l\pi}{n - m}, \quad l = 0, 1, 2, \cdots, n - m - 1 \tag{4.38}$$

② 修改后的法则4:实轴上的根轨迹。实轴上的任一区段是根轨迹一部分的充分必要条件为该区段右侧的开环极点和零点的总数不是奇数。

③ 修改后的法则 5：根轨迹的起始角和终止角。根轨迹离开复极点 p_k 时的起始角为

$$\theta_{p_k} = 0° + \sum_{i=1}^{m} \angle(p_k - z_i) - \sum_{j=1, j \neq k}^{n} \angle(p_k - p_j) \tag{4.39}$$

根轨迹到达复零点 z_k 时的终止角为

$$\theta_{z_k} = 0° + \sum_{j=1}^{n} \angle(z_k - p_j) - \sum_{i=1, i \neq k}^{m} \angle(z_k - z_i) \tag{4.40}$$

④ 修改后的法则 7：根轨迹与虚轴的交点。根轨迹与 s 平面虚轴相交的交点以及相应的 K^* 值可由以下的特征方程获取

$$1 - G(j\omega)H(j\omega) = 0 \tag{4.41}$$

例 4.4 绘制正反馈系统的根轨迹，其开环传递函数为

$$G(s)H(s) = \frac{1}{s(s+1)(s+2)}$$

解 由于特征方程为 $1 - G(s)H(s) = 0$，根轨迹满足式(4.37)中的幅角条件，该根轨迹是 0° 根轨迹。开环极点为 $p_1 = 0, p_2 = -1$ 和 $p_3 = -2$，没有开环零点。

由法则 1，因为 $n = 3$，根轨迹由三条分支组成。

由法则 2，因为 $n = 3$ 且 $m = 0$，当 $K^* = 0$ 时根轨迹的三条分支起始于三个开环极点，当 $K^* \to \infty$ 时三条分支终止于无穷远处。

由修改后的法则 3，根轨迹的渐近线相交于

$$\sigma_a = \frac{0 - 1 - 2}{3 - 0} = -1$$

渐近线与实轴的夹角则为

$$\varphi_a = \frac{2l\pi}{3 - 0} = 0, \pm\frac{2\pi}{3} \quad (l = 0, \pm 1)$$

由修改后的法则 4，复实轴上的区段 $[-2, -1]$ 和 $[0, \infty]$ 构成根轨迹的一部分。

由法则 6，经观察可知在实轴的区段 $[-2, -1]$ 上有一个分离点。由例 4.2，分离点方程有两个解

$$s_1 = -0.423, \quad s_2 = -1.577$$

但是，在这种情况下分离点为 $s_b = -1.577$。

根轨迹离开分离点的两条分支之间的夹角为 π。而且，由于区段 $[-2, -1]$ 是根轨迹的一部分，这两条分支垂直离开实轴。

由修改后的法则 7，该系统的特征方程为

$$s^3 + 3s^2 + 2s - K^* = 0$$

将 $s = j\omega$ 代入得到

$$\begin{cases} -3\omega^2 - K^* = 0 \\ -\omega^3 + 2\omega = 0 \end{cases}$$

由于 K^* 是从零变化到无穷大，上面方程的唯一解为 $\omega = 0$ rad/s，即除原点外根轨迹与虚轴不相交。

最终的根轨迹图见图 4.12。

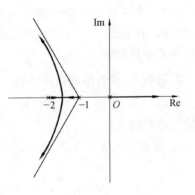

图 4.12 例 4.4 系统的 0° 根轨迹

4.3.2 参数根轨迹

以下将考虑变化参数不是增益的情况。重要的是记住根轨迹是特征方程的根的轨迹。现在用一个例子说明绘制参数根轨迹的过程。

例 4.5 绘制图 4.13 所示系统当参数 T_a 从零变化到无穷大时的根轨迹。

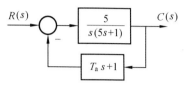

图 4.13 例 4.5 的控制系统

解 首先写出该系统的特征方程

$$s(5s+1) + 5(T_a s + 1) = 0$$

然后将不含 T_a 的项组合在一起,并将上面的式子除以这些项得到

$$1 + \frac{5T_a s}{5s^2 + s + 5} = 0$$

即

$$\frac{T_a s}{s^2 + 0.2s + 1} = -1$$

由于这一表达式的形式与式(4.12)的形式相同,可以得到一个等效的开环传递函数,即

$$G_e(s)H_e(s) = \frac{T_a s}{s^2 + 0.2s + 1}$$

而等效的根轨迹增益则为 $K_e^* = T_a$。

等效开环传递函数有一个零点位于原点,有两个极点为 $p_{1,2} = -0.1 \pm j0.99$。该系统的根轨迹如图 4.14 所示。其中分离点为 $s_b = -1$,起始角为 $\theta_p = \pm 185.7°$。

在一般的情况下,系统的特征方程必须表示成多项式的方程。然后,记变化参数为 α,不含 α 的项

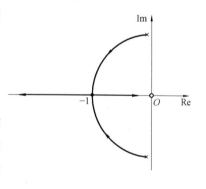

图 4.14 例 4.5 系统的参数根轨迹

组合成函数 $D_e(s)$,剩余的包含 α 的项则组合成函数 $N_e(s)$,使特征方程表示成

$$D_e(s) + \alpha N_e(s) = 0 \tag{4.42}$$

除以 $D_e(s)$ 后得到

$$1 + \alpha \frac{N_e(s)}{D_e(s)} = 0 \tag{4.43}$$

这一方程可以写成

$$1 + K_e^* \frac{s^m + b_1 s^{m-1} + \cdots + b_m}{s^n + a_1 s^{n-1} + \cdots + a_{n-1}s + a_n} = 0 \tag{4.44}$$

或者

$$1 + \frac{K_e^* \prod_{i=1}^{m}(s - z_i)}{\prod_{j=1}^{n}(s - p_j)} = 0 \tag{4.45}$$

然后可以由等效传递函数

$$G_e(s)H_e(s) = \frac{K_e^* \prod_{i=1}^{m}(s - z_i)}{\prod_{j=1}^{n}(s - p_j)}$$

绘制关于等效增益 K_e^* 的根轨迹。

4.4 根轨迹法的应用

根轨迹法提供了图示信息,因此近似的概略绘图可用于获取关于系统稳定性和其他性能的定性的信息,包括以系统型别和静态误差系数表征的稳态性能和以超调量和调整时间等为表征的动态性能。

在本节中将以一些例子说明根轨迹法的应用。

例 4.6 (参数设计)考虑开环传递函数为

$$G(s)H(s) = \frac{K}{s(s+1)(0.5s+1)}$$

的单位负反馈系统,确定使闭环系统具有一对阻尼比为 0.5 的主导极点时的开环增益。

解 将开环传递函数改写为零极点形式

$$G(s)H(s) = \frac{K^*}{s(s+1)(s+2)}$$

可以画出根轨迹如图 4.15 所示(见例 4.2)。

画出 $\zeta = 0.5$ 的等 ζ 线,如图 4.15 所示。等 ζ 线与根轨迹相交于

$$s_{1,2} = -0.33 \pm j0.58$$

由此形成了一对闭环系统的共轭复极点。由于这是一个没有零点的三阶系统,由式(4.29)可有

$$s_3 = -2.34$$

由于

$$\frac{|\text{Re}(s_3)|}{|\text{Re}(s_1)|} = \frac{2.34}{0.33} > 7$$

其中 $s_{1,2}$ 可以认为是一对主导极点。

由于闭环极点必定满足根轨迹的幅值条件,令

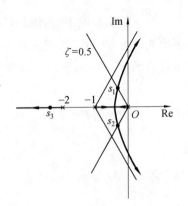

图 4.15 根轨迹与主导极点

$$|G(s)H(s)|_{s=s_3} = \left|\frac{K}{s(s+1)(0.5s+1)}\right|_{s=-2.34} = 1$$

得到

$$K = |s(s+1)(0.5s+1)|_{s=s_3} = 0.525$$

例 4.7 （附加极点的影响）比较图 4.16(a) ~ (c) 的根轨迹。

图 4.16(a) 显示了一个开环传递函数为

$$G(s)H(s) = \frac{k}{s(s+a)}$$

的系统的根轨迹。可以看到,渐近线的角度为 ±90°,该系统总是稳定的。

图 4.16(b) 显示出附加极点 $s = -b$ 引起根轨迹的复数部分向右半 s 平面弯曲。右侧的渐近线的角度变为 ±60°。如果 k 的取值超过临界值,该系统就会变为不稳定。

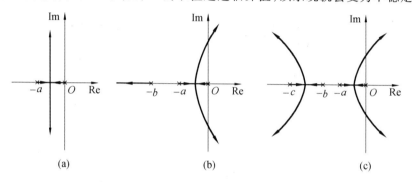

图 4.16　附加极点的影响

如果另一个极点 $s = -c$ 加入 $G(s)H(s)$ 如图 4.16(c) 所示,根轨迹的复平面部分更加向右移动,右侧的渐近线的角度改变为 ±45°。系统的稳定性状况受到更大的限制。

大致上讲,在开环传递函数加入左半 s 平面的极点具有将原来的根轨迹向右半 s 平面推的效果。

例 4.8 （附加零点的影响）比较图 4.17(a) ~ (d) 的根轨迹。

图 4.17(b) 显示出当开环传递函数

$$G(s)H(s) = \frac{k}{s(s+a)}$$

加入一个零点 $s = -b, b > a$ 时的根轨迹。与图 4.17(a) 所示原来的根轨迹相比较,图 4.8(b) 所得的根轨迹向左弯曲,而且系统的相对稳定性得以改善。

比较图 4.17(c) 和 (d),当零点 $s = -c, c > b$,加入开环传递函数

$$G(s)H(s) = \frac{k}{s(s+a)(s+b)}$$

时可以看到类似的影响。

大致上讲,在开环传递函数中加入左半 s 平面的零点具有使原来的根轨迹向左半 s 平面推移的效果。

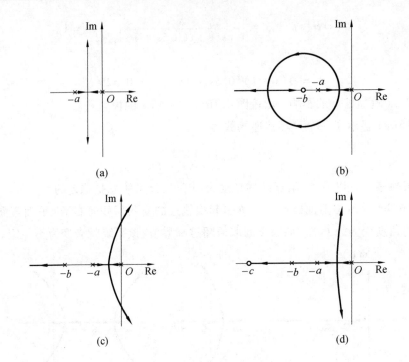

图 4.17 附加零点的影响

习 题

题 4.1 绘制以下开环传递函数 $0 < k < \infty$ 时的根轨迹。

(a) $G(s)H(s) = \dfrac{k(s+5)}{s(s+2)(s+3)}$

(b) $G(s)H(s) = \dfrac{k(s+5)}{s(s+1)(s+3)(s+4)}$

(c) $G(s)H(s) = \dfrac{k(s+3)}{s^2+4s+8}$

(d) $G(s)H(s) = \dfrac{k(s+20)}{s(s^2+20s+200)}$

(e) $G(s)H(s) = \dfrac{k}{s(s+3)^2}$

(f) $G(s)H(s) = \dfrac{K}{s(0.05s^2+0.4s+1)}$

题 4.2 考虑开环传递函数为

$$G(s) = \dfrac{k(s+1)}{s^2+4s+5}$$

的单位反馈系统。(a) 确定根轨迹离开复极点时的起始角。(b) 确定根轨迹进入实轴时的进入点。

题 4.3 一个单位负反馈系统装置的传递函数为

$$G(s) = \dfrac{k(s^2+20)(s+1)}{(s^2-2)(s+10)}$$

绘制 k 变化时的根轨迹。

题 4.4 一个控制系统的开环传递函数为

第4章 根轨迹法

$$G(s)H(s) = \frac{k(s^2 + 4s + 8)}{s^2(s + 4)}$$

若希望主导极点的阻尼比等于 0.5,试用根轨迹证明所需 k 值为 7.35,且主导极点为 $s = -1.3 \pm j2.2$。

题 4.5 一个单位反馈系统具有开环传递函数

$$G(s) = \frac{k}{s(s + 2)(s + 5)}$$

试确定:(a) 实轴上的分离点以及该点对应的增益;(b) 当两个根位于虚轴上时的增益和根;(c) $k = 6$ 时的根;(d) 绘制根轨迹。

题 4.6 一个单位反馈系统的开环传递函数为

$$G(s) = \frac{k(s + a)}{s^2(s + 1)}$$

试确定使根轨迹分别具有零个、一个以及两个分离点时的 a 值,$s = 0$ 处的分离点不计。绘制所有三种情况下 $0 < k < \infty$ 时的根轨迹。

题 4.7 一个负反馈系统的传递函数为

$$G(s) = \frac{k}{s^2(s + 2)(s + 5)}, \quad H(s) = 1$$

(a) 绘制该系统的根轨迹,指出根轨迹在虚轴上的穿越点以及在这些点上相应的 k 值。(b) 若反馈回路的传递函数改为 $H(s) = 2s + 1$,确定修改后系统稳定性与 k 的关系。研究由于 $H(s)$ 的变化给根轨迹带来的影响。

题 4.8 一个反馈控制系统的特征方程为

$$\Delta(s) = s^2 + (k + 5)s + 2k + 4 = 0$$

绘制该系统作为 k 的函数时的根轨迹(仅考虑 k 大于零)。

题 4.9 单位反馈系统的装置为

$$G(s) = \frac{0.25(s + a)}{s^2(s + 1)}$$

绘制该系统作为 a 的函数时的根轨迹(仅考虑 a 大于零)。

题 4.10 一个正反馈控制系统的开环传递函数为

$$G(s)H(s) = \frac{k}{s(s^2 + 4s + 4)}$$

绘制 $0 < k < \infty$ 时该系统的根轨迹。

第5章 频率响应法

5.1 引 言

前面曾经指出过,在实际应用中一个反馈控制系统的性能用它的时域响应特征衡量比较合适。但在系统分析时,尤其是在高阶系统的情况下,确定一个控制系统的时域响应通常比较困难。就系统的设计而言,还没有一种统一的方法可以直接根据给定的诸如最大超调量、上升时间和调节时间之类的性能指标设计系统。根轨迹法看来在给定开环传递函数时足以预测闭环系统的瞬态响应。这一方法的缺点是它依赖于开环传递函数的存在,根轨迹本身正是根据开环传递函数绘制的。在分析和设计线性反馈控制系统时,频率响应法是一种非常实用和重要的方法。最终的目标是利用系统的开环频率响应了解闭环系统的时域行为。

5.1.1 系统对于正弦输入的稳态响应

一个系统的频率响应是由该系统对于正弦输入信号的稳态响应得到的。

假设一个传递函数为 $G(s)$ 的线性系统是稳定的,而且输入是一个正弦信号

$$r(t) = A_r \sin \omega t \tag{5.1}$$

即

$$R(s) = \frac{A_r \omega}{s^2 + \omega^2} \tag{5.2}$$

则该系统的输出为

$$C(s) = G(s)R(s) = G(s) \cdot \frac{A_r \omega}{(s-j\omega)(s+j\omega)} =$$

$$\frac{B_1}{s-j\omega} + \frac{B_2}{s+j\omega} + C_g(s) \tag{5.3}$$

式中 $C_g(s)$ 是部分分式展开式中所有起源于 $G(s)$ 分母的项的总和。因为假设系统是稳定的,$C_g(s)$ 中的这些项将随时间的增大而衰减至零。因此,在式(5.3)中只有前两项对于稳态响应有所贡献。

由式(5.2)和式(5.3)

$$B_1 = \frac{A_r}{j2} G(j\omega) \tag{5.4}$$

$$B_2 = -\frac{A_r}{j2} G(-j\omega) \tag{5.5}$$

对于给定的 ω 值,由于 $G(j\omega)$ 是复数,将 $G(j\omega)$ 表示成以下形式较为方便

$$G(j\omega) = |G(j\omega)| e^{j\varphi} \tag{5.6}$$

式中

$$\varphi = \angle G(j\omega) \tag{5.7}$$

然后,由式(5.3)到式(5.7)可得到输出的稳态值为

$$c_s(t) = B_1 e^{j\omega t} + B_2 e^{-j\omega t} = A_r |G(j\omega)| \frac{e^{j\varphi} e^{j\omega t} - e^{-j\varphi} e^{-j\omega t}}{j2} =$$

$$A_r |G(j\omega)| \sin(\omega t + \varphi) = A_c \sin(\omega t + \varphi) \tag{5.8}$$

现在看到,在稳态下最终得到的输出是一个角频率与输入信号相同的正弦信号,它只是在幅值和相位上与输入波形不同。而且,对于一个指定的频率,稳态时的幅值为

$$A_c = A_r |G(j\omega)| \tag{5.9}$$

而输出正弦量相对于输入正弦量的相位移为

$$\varphi = \angle G(j\omega) \tag{5.10}$$

例 5.1 设系统的传递函数为

$$G(s) = \frac{5}{s+2}$$

输入为 $2\sin 3t$,试求系统的稳态输出。

解 由观察可知该系统稳定。当角频率 $\omega = 3$ 时有

$$|G(j\omega)|_{\omega=3} = 1.387$$

$$\angle G(j\omega)|_{\omega=3} = -56.3°$$

系统的稳态输出为

$$c_s(t) = A_r |G(j\omega)| \sin(\omega t + \varphi) = 2 \times 1.387 \sin(3t - 56.3°) =$$

$$2.774 \sin(3t - 56.3°)$$

顺便指出,由于系统的时间常数是 0.5 s,施加输入信号大约 1.5 s($\Delta = 5\%$)后,输出将到达稳态。

5.1.2 频率响应

已经看到,一个稳定的线性系统对于正弦输入的响应也是一个频率与输入相同的正弦信号。但是,输出的幅值和相位与输入正弦信号的幅值和相位不同,而且它们差别的大小是输入信号频率的函数。这样就可以研究频率变化时系统对于正弦输入的稳态响应。

对于一个线性系统或环节,输出的稳态分量与输入正弦量之比定义为该系统或环节的频率特性或频率响应。由于输出稳态分量的幅值和相位两者都是频率的函数,输出与输入正弦量的幅值比定义为幅值 – 频率特性,或者简称为幅频特性;而输出正弦量相对于输入正弦量的相位移定义为相位 – 频率特性,或者简称为相频特性。

频率响应法的一个优点是有现成的不同频率和幅值范围的正弦测试信号可供使用。因此,用实验的方法确定一个系统的频率特性就很容易完成,对于用实验的方法分析一个系统这也是一种极为可靠而又不太复杂的方法。事实上,可以应用频率响应法设计控制系统而无需建立传递函数。还有,在验证一个通过推导得到的传递函数时,最常用的方法之一是把根据传递函数计算得到的频率响应与物理系统测量得到的频率响应进行比较。

频率响应法的另一优点是用 $j\omega$ 替代 $G(s)$ 中的 s 就可以由一个系统的传递函数 $G(s)$ 直接获取描述该系统正弦稳态行为的频率响应,即

$$G(j\omega) = G(s)|_{s=j\omega} = M(\omega) e^{j\varphi(\omega)} \tag{5.11}$$

式中,$M(\omega)$ 是幅频特性;$\varphi(\omega)$ 是相频特性。

一旦分析和设计工作是在频域中进行,可以根据时域特性和频域特性之间存在的关系解释系统的时域行为。因此可以认为,在频域中进行控制系统分析和设计的主要目的只不过是利用这种方法作为一种使用方便的手段,而要达到的目标和时域法是相同的。

频率响应法用于分析和设计时的主要缺点在于频域和时域之间的联系是间接的。频率响应特性与相应的瞬态响应特性之间直接相关的程度不是很清楚,在实际应用中频率响应特性是通过使用各种不同的判据调节的,这些判据在一般情况下能得到令人满意的瞬态响应。

5.1.3 频率响应图

有好几种用于研究控制系统稳定性和其他性能的分析方法可以算是频率响应法。这些方法都涉及到频率响应的三个参数,幅值、相角和频率之间关系的研究。对于分析和设计控制系统而言,以某种形式绘制频率响应 $G(j\omega)$ 与频率 ω 之间的关系曲线表示频率响应的特征是很有用的。常见的频率响应图包括极坐标图、伯德图和尼柯尔斯图。

1. 极坐标图

在频域中一个系统的传递函数可以用以下关系描述为

$$G(j\omega) = G(s)|_{s=j\omega} = \text{Re}[G(j\omega)] + j\text{Im}[G(j\omega)] \tag{5.12}$$

或者用另一种方法,也可以用幅值和幅角表示即

$$G(j\omega) = |G(j\omega)| e^{j\angle G(j\omega)} \tag{5.13}$$

一种常用的显示频率响应的方法是极坐标图形式,利用式(5.12)或式(5.13)可以获得极坐标图。在这样的图中,当频率 ω 从零变化到无穷大时,频率响应的幅值和相角(或者它的实部和虚部)绘制在极坐标系统中。极坐标图常用于表示开环频率特性,有时也称为开环幅相频率特性曲线。

2. 伯德图

极坐标图的局限性是一目了然的。在一个已有的系统中加入极点或零点时将需要重新计算频率响应。而且,这种形式频率响应的计算相当麻烦,也不能指出加入的各个极点或零点的影响。伯德图的引入简化了频率响应的绘图过程。

伯德图由两张单独的图组成,一张是单位为分贝的幅值与频率的关系曲线,另一张是相位与频率的关系曲线。这两张图要用图 5.1 所示的半对数纸,其中 ω 轴采用对数分度。此外,幅频特性曲线不是简单的幅值 M,而是如图 5.1 所示表示成分贝数

$$L(\omega) = M(\mathrm{db}) = 20\lg M = 20\lg |G(\mathrm{j}\omega)| \tag{5.14}$$

伯德图也常用于表示开环频率特性,并称为开环对数频率特性曲线。

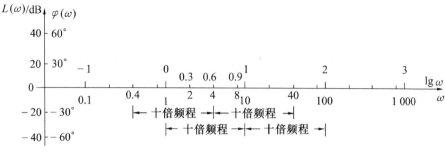

图 5.1 半对数坐标

在频率响应法中,系统的分析主要是基于开环系统的伯德图表示进行的。由于如下原因,伯德图比极坐标图更受到人们的欢迎:第一,借助于直线的近似表示方法可以很快地画出开环频率响应曲线。第二,频率采用对数分度使得关系到系统各种性能的各个频段都可以得到同等的重视。第三,原因是在伯德图上很容易看到改变系统的一个参数时对系统性能的影响。

3. 尼柯尔斯图

尼柯尔斯图有时也称为对数幅相频率响应曲线,将在 5.6 节中介绍。

5.2 基本环节的伯德图

在根轨迹法中,将开环传递函数写成零极点形式特别有用。与此不同,在频率响应法中,开环传递函数通常写成典型因子形式,即伯德形式

$$G(s) = \frac{K \prod (\tau_i s + 1)}{s^v \prod (T_j s + 1) \prod [(s^2/\omega_{nk}^2) + (2\zeta_k/\omega_{nk})s + 1]} \tag{5.15}$$

一般情况下,对应于式(5.15)所示传递函数具有以下一些基本因子:

① 因子 K,即常数,对应于增益环节;
② 因子 s^{-1},即位于原点的极点,对应于积分环节;
③ 因子 s,即位于原点的零点,对应于微分环节;
④ 因子 $(Ts+1)^{-1}$,即负实轴上的极点,对应于惯性环节;
⑤ 因子 $(\tau s+1)$,即负实轴上的零点,对应于一阶微分环节;
⑥ 因子 $[(s^2 + 2\zeta\omega_n s + \omega_n^2)/\omega_n^2]^{-1}$,即一对负实部共轭复极点,对应于振荡环节。

这些因子中除了增益 K 以外都还可以有除 1 之外的整数幂。各种因子的对数幅频特性和相频特性曲线都很容易就可以画出。

采用这样的表示方法时,由于在伯德图中 $G(\mathrm{j}\omega)$ 的幅值是用对数表示的,$G(\mathrm{j}\omega)$ 中各因子的乘法运算变成了加法,相位关系也是以普通的方法相加或相减。因此,这些因子的曲线可以通过作图的方法加在一起得到整个传递函数的频率响应曲线。而且,这些曲线

采用后面所介绍的渐近线近似表示时还可以进一步简化伯德图的绘制过程。

对于频率响应法而言,熟悉各种典型因子即典型环节的对数幅频特性和相频特性是基本的要求。因此是逐个地求取这些因子的对数幅频特性和相频特性。

5.2.1 增益环节

当开环传递函数写成典型因子形式时,常数项正好就是开环增益。因此,这一因子也称为增益环节。由于增益与频率无关,它的对数幅频特性为

$$L(\omega) = 20 \lg K \tag{5.16}$$

是一条水平的直线。只要 K 大于零,则相频特性为

$$\varphi(\omega) = 0° \tag{5.17}$$

就与 ω 轴重合。

图 5.2 所示为一个常数 K 的伯德图。

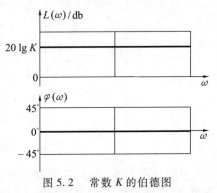

图 5.2 常数 K 的伯德图

5.2.2 积分环节

当传递函数写成典型因子形式时,因子 s^{-1} 意味着纯积分。因此,这一因子有时也称为积分环节。该环节的对数幅频特性为

$$L(\omega) = 20 \lg \left| \frac{1}{j\omega} \right| = -20 \lg \omega \tag{5.18}$$

相频特性则为

$$\varphi(\omega) = -90° \tag{5.19}$$

对数幅频特性曲线是一条斜率为 -20 db/dec 的直线,并与 ω 轴相交于 $\omega = 1$ 处。

对于在原点有多重极点的情况,可以类似地有对数幅频特性,即

$$L(\omega) = 20 \lg \left| \frac{1}{(j\omega)^v} \right| = -20 v \lg \omega \tag{5.20}$$

和相频特性,即

$$\varphi(\omega) = -v \, 90° \tag{5.21}$$

在这种情况下,由于是多重极点的缘故,对数幅频特性曲线的斜率为 $-20 v$ db/dec。

积分环节 s^{-1} 和 s^{-2} 的对数幅频特性曲线图和相频特性曲线图如图 5.3 所示。

5.2.3 微分环节

传递函数在原点的零点,即因子 s 有时称为微分环节。微分环节的对数幅频特性为

$$L(\omega) = 20 \lg |j\omega| = 20 \lg \omega \tag{5.22}$$

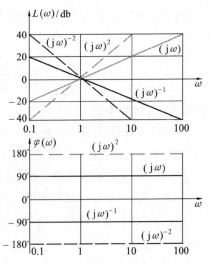

图 5.3 积分环节和微分环节的伯德图

式中的斜率为 +20db/dec,相频特性则为
$$\varphi(\omega) = 90° \tag{5.23}$$
微分环节 s 和 s^2 的对数幅频特性图和相频特性图也如图 5.3 所示。

5.2.4 惯性环节

传递函数的负实极点,即因子为
$$\frac{1}{Ts+1}$$
有时称为惯性环节。该环节的对数幅频特性为
$$L(\omega) = 20\lg\frac{1}{\sqrt{T^2\omega^2+1}} = -20\lg\sqrt{T^2\omega^2+1} \tag{5.24}$$
当 $\omega \ll 1/T$ 时,其渐近线为
$$L(\omega) = -20\lg 1 = 0 \text{ db} \tag{5.25}$$
而 $\omega \gg 1/T$ 时,渐近线则为
$$L(\omega) = -20\lg T\omega = -20\lg\omega - 20\lg T \tag{5.26}$$
这是一条斜率为 -20 db/dec 的直线,并与0db线相交于 $\omega = 1/T$ 处。两条渐近线在角频率 $\omega = 1/T$ 处相会,该角频率称为转折频率。

惯性环节的对数幅频特性的伯德图如图 5.4 所示。为了进行比较,表 5.1 为该环节幅频特性的精确值和使用近似表示所得到的值。可以看到,精确的幅频特性与渐近特性之间最大的差异为 3 db。因此,$\omega > 1/T$ 时用高频渐近线替代精确特性,以及 $\omega < 1/T$ 时用低频渐近线替代精确特性是合理的设想。

惯性环节的相频特性为
$$\varphi(\omega) = -\arctan T\omega \tag{5.27}$$
相频特性曲线及其直线近似特性如图 5.4 所示。这一直线近似特性在转折频率处的相位

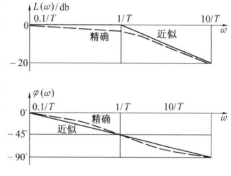

图 5.4 惯性环节的伯德图

值是正确的,而且对于所有的频率与实际的相频特性曲线之间的差异都在 6° 之内。该近似特性将为简便地确定传递函数 $G(s)$ 的相频特性的形态提供一种有用的手段。但是,经常还会需要精确的相频特性曲线,这可通过必要的计算获取。

表 5.1 惯性环节的精确值和近似值

$T\omega$	0.1	0.5	0.76	1	1.31	2	5	10
$L(\omega)$ 的精确值 /dB	-0.04	-1.0	-2.0	-3.0	-4.3	-7.0	-14.2	-20.04
$L(\omega)$ 的近似值 /dB	0	0	0	0	-2.3	-6.0	-14.0	-20.0
$\varphi(\omega)$ 的精确值 /(°)	-5.7	-26.6	-37.4	-45.0	-52.7	-63.4	-78.7	-84.3
$\varphi(\omega)$ 的近似值 /(°)	0	-31.5	-39.5	-45.0	-50.3	-58.3	-76.5	-90.0

5.2.5 一阶微分环节

传递函数的负实零点,即因子 $G(s) = \tau s + 1$ 有时称为一阶微分环节。它的伯德图可以通过与惯性环节相类似的方法获取。该环节的对数渐近幅频特性曲线如图 5.5 所示。小于转折频率时为零,而大于转折频率时则用一条斜率为 +20 db/dec 的直线表示。实际的对数幅频特性曲线位于直线近似特性之上,在转折频率处具有最大的误差为 +3 db。相频特性曲线及其直线近似特性如图 5.5 所示,图中相位角是从 0° 变化到 +90°。

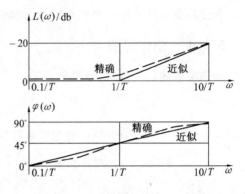

图 5.5 一阶微分环节的伯德图

注意,如果 $\tau = T$,一阶微分环节 $\tau s + 1$ 的对数幅频特性曲线和相频特性曲线是惯性环节 $(Ts + 1)^{-1}$ 的对数幅频特性曲线和相频特性曲线关于频率轴的镜像。

5.2.6 振荡环节

传递函数的共轭复极点即形式为

$$G(s) = \frac{\omega_n^2}{s^2 + 2\zeta\omega_n s + \omega_n^2}$$

的二次因子有时称为振荡环节。这种二次因子经常会发生在反馈系统的传递函数内。当 $\zeta \geq 1$ 时,该二次式可以因式分解为两个包含实极点的一阶因式,它们的伯德图可以按前述方式绘制。但是当 $\zeta < 1$ 时,该二次式包含共轭复极点,绘制伯德图时不能作因式分解。

振荡环节的频率响应为

$$G(j\omega) = \frac{\omega_n^2}{\omega_n^2 - \omega^2 + j2\zeta\omega_n\omega} \tag{5.28}$$

或者

$$G(j\omega) = \frac{1}{\left(1 - \dfrac{\omega^2}{\omega_n^2}\right) + j2\zeta\dfrac{\omega}{\omega_n}} \tag{5.29}$$

对数幅频特性为

$$L(\omega) = -20\lg\sqrt{\left(1 - \frac{\omega^2}{\omega_n^2}\right)^2 + \left(2\zeta\frac{\omega}{\omega_n}\right)^2} \tag{5.30}$$

当 $\omega \ll \omega_n$ 时,渐近特性为

$$L(\omega) = 0 \text{ db} \tag{5.31}$$

特性曲线与 0 db 线重合。

当 $\omega \gg \omega_n$ 时,渐近特性为

$$L(\omega) = -20\lg\frac{\omega^2}{\omega_n^2} = -40\lg\omega + 40\lg\omega_n \tag{5.32}$$

特性曲线如图 5.6 所示,是一条斜率为 -40 db/dec,与 0 db 线相交于转折频率 $\omega = \omega_n$ 处的直线。这两条渐近线在转折频率 $\omega = \omega_n$ 处会合。

在 $\omega = \omega_n$ 处,实际的幅频特性为

$$|G(j\omega_n)| = \frac{1}{2\zeta} \tag{5.33}$$

或者

$$L(\omega_n) = 20\lg\frac{1}{2\zeta} \tag{5.34}$$

因此,在转折频率附近实际的特性曲线将有别于渐近特性,其差别则是阻尼比 ζ 的函数。对式(5.29)的幅频特性关于 ω 求导并置其为零,结果显示精确的幅频特性有峰值时

$$\omega_m = \omega_n\sqrt{1-2\zeta^2} \tag{5.35}$$

这表明只有当 $\zeta < \sqrt{2}/2$ 时才可能发生极值。如果这一条件满足,则得到的极值为

$$M_m = \frac{1}{2\zeta\sqrt{1-\zeta^2}}, \quad \zeta < \frac{\sqrt{2}}{2} \tag{5.36}$$

即

$$L(\omega_m) = 20\lg\frac{1}{2\zeta\sqrt{1-\zeta^2}} \tag{5.37}$$

是以分贝表示的幅频特性的极值。该环节的伯德图如图 5.6 所示。注意,实际的幅频特性曲线可以在直线近似特性之下也可以在其之上。

振荡环节的相位为

$$\varphi(\omega) = -\arctan\frac{2\zeta\omega/\omega_n}{1-(\omega/\omega_n)^2} \tag{5.38}$$

在低频部分,相位有

$$\lim_{\omega\to 0}\varphi(\omega) = -\arctan 0 = 0° \tag{5.39}$$

考虑到适当的象限,在高频部分的相位有

$$\lim_{\omega\to\infty}\varphi(\omega) = -180° \tag{5.40}$$

但是,如图 5.6 所示,相位从 0° 转变为 -180° 的速率也取决于阻尼比 ζ。还应注意,在 $\omega = \omega_n$ 处所有的相频特性曲线的取值均为 -90°。尽管在许多情况下不够精确,相频特性有时也采用直线近似特性。近似特性规定为从低于 ω_n 十倍频程开始到高于 ω_n 十倍频程结束,从 0° 变化到 -180° 的一条直线。

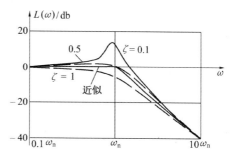

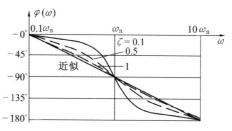

图 5.6　振荡环节的伯德图

5.2.7 非最小相位环节

在前面的讨论中,式(5.15)给出的传递函数的极点和零点全部都限制在左半 s 平面。但是,在构成伯德图时有可能遇到一些不同的情况。

1. 一阶不稳定环节

考虑在右半 s 平面有一个实极点的一阶不稳定环节

$$G(s) = \frac{1}{Ts - 1} \tag{5.41}$$

该环节的对数幅频特性和相频特性分别为

$$L(\omega) = 20 \lg \frac{1}{\sqrt{T^2\omega^2 + 1}} = -20 \lg \sqrt{T^2\omega^2 + 1} \tag{5.42}$$

$$\varphi(\omega) = -\arctan \frac{T\omega}{-1} \tag{5.43}$$

的伯德图如图 5.7 所示。

与频率响应显示于图 5.4 的惯性环节为

$$\frac{1}{Ts + 1}$$

相比较,一阶不稳定环节的幅频特性没有什么变化,这两种环节之间唯一的差别是在相位移上。考虑到适当的象限问题时可以看到,一阶不稳定环节的相角是从 $-180°$ 变化到 $-90°$,它的相角范围大于具有相同幅频特性的惯性环节的相角范围。因此,惯性环节称为是最小相位环节,而一阶不稳定环节则称为是非最小相位环节。

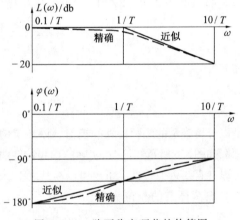

图 5.7 一阶不稳定环节的伯德图

在所有具有相同幅频特性的传递函数中,有着最小相位移范围的传递函数称为最小相位传递函数,否则称为非最小相位传递函数。具有最小相位传递函数的系统或环节称为最小相位系统或环节,否则称为非最小相位系统或环节。当一个系统包含非最小相位元部件或者包含不稳定的内回路时将发生非最小相位移的情况。

2. 理想延时

延时是一种非最小相位特性,通常存在于热力、液压和气动系统内。假设信号施加于一个理想延时,则该理想延时的输出为

$$c(t) = r(t-\tau) \cdot 1(t-\tau) \tag{5.44}$$

式中 τ 是延时时间。对该式作拉氏变换,理想延时的传递函数为

$$G(s) = e^{-\tau s} \tag{5.45}$$

注意,这一传递函数与迄今为止所讨论

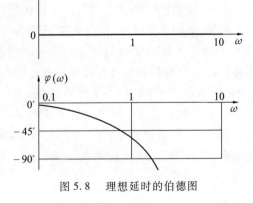

图 5.8 理想延时的伯德图

过的所有其他传递函数不同。它不是 s 的多项式之比,即它不是一个 s 的有理函数。这一事实使包含理想延时的系统的分析变得复杂。例如,稳定性的代数判据不能用于这一类系统,因为系统的特征方程不是 s 的多项式。

式(5.45)给出的传递函数用 $j\omega$ 替代 s 可以得到相应的频率响应,即

$$G(j\omega) = e^{-j\omega\tau} \tag{5.46}$$

因此,如图 5.8 所示,它的对数幅频特性为

$$L(\omega) = 0 \text{ db} \tag{5.47}$$

而相频特性则为

$$\varphi(\omega) = -\tau\omega \text{ rad} = -57.3\ \tau\omega \text{ deg} \tag{5.48}$$

由理想延时带来的相位滞后将随频率增大而无限制地增大。

5.3 开环频率响应

5.3.1 开环传递函数的伯德图

伯德图尤其是开环传递函数的伯德图,是分析和设计控制系统时功能最强的作图工具之一。

正如所知,当一个传递函数写成基本环节的形式时,即

$$G(s) = \frac{K\prod(\tau_i s + 1)}{s^v \prod(T_j s + 1)\prod[(s^2 + 2\zeta_k\omega_{nk}s + \omega_{nk}^2)/\omega_{nk}^2]} \tag{5.49}$$

它的伯德图可以通过各个环节的特性曲线相加而获得。这一方法的简易性可以用一些例子说明。

例 5.2 绘制以下传递函数的伯德图

$$G(s) = \frac{1}{s(Ts+1)}$$

解 由于增益为 1,只需要考虑两个环节。生成对数幅频特性曲线的一种方法是从低频端开始,随着频率向高端移动,用作图的方法把各个环节的幅频特性曲线相加。图 5.9 显示了该传递函数各环节的直线近似表示,以及最终的对数幅频特性曲线。注意直线的斜率是怎样在转折频率处从 −20 db/dec 改变为 −40 db/dec 的。如图 5.9 所示,最终的相频特性曲线可以按相类似的方式通过作图的方法获取。

由于已经知道精确特性与直线近似特性之间在转折频率处有 3db 的幅值误差,若有必要则可以获得较为精确的幅频特性曲线。

例 5.3 绘制以下开环传递函数的伯德图

$$G(s)H(s) = \frac{100(s+2)}{s(s+1)(s+20)}$$

解 首先将传递函数写成典型环节形式,即

$$G(s)H(s) = \frac{10(0.5s+1)}{s(s+1)(0.05s+1)}$$

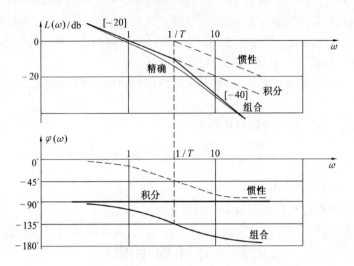

图 5.9 例 5.2 的伯德图

该传递函数有五个串联的环节

$$G_1(s) = 10, \quad G_2(s) = \frac{1}{s}, \quad G_3(s) = \frac{1}{s+1}, \quad G_4(s) = \frac{1}{0.05s+1}, \quad G_5(s) = 0.5s+1$$

绘制各个环节 $G_i(s), i = 1,2,3,4,5$ 的渐近对数幅频特性曲线和相频特性曲线

$$L_i(\omega) = 20 \lg |G_i(j\omega)|$$

$$\varphi_i(\omega) = \angle G_i(j\omega)$$

图 5.10 显示了该传递函数各个环节幅频特性的直线近似表示和相频特性曲线,以及最终的幅频特性曲线和相频特性曲线。还是要注意,直线段的斜率在各转折频率处是如何改变的。

实际上,采用直线近似表示时组合的对数幅频特性可以直接画出。

例 5.4 绘制以下传递函数的伯德图

$$G(s) = \frac{5(0.1s+1)}{s(0.5s+1)[(s^2+30s+50^2)/50^2]}$$

解 该传递函数包含上一节讨论过的所有典型环节。这些环节按照转折频率的顺序为:

(a) 常数增益 $K = 5$;

(b) 积分环节;

(c) 惯性环节,转折频率在 $\omega_1 = 1/T = 2$;

(d) 一阶微分环节,转折频率在 $\omega_2 = 1/\tau = 10$;

(e) 振荡环节,转折频率在 $\omega_3 = \omega_n = 50$。

绘制伯德图可以根据以下步骤按转折频率的顺序进行:

第 1 步:在频率小于第一个转折频率 ω_1 的部分,只有增益环节(a)和积分环节(b)有效,所有其他环节的贡献均为零。因此,$\omega < \omega_1$ 时组合曲线的斜率为 -20 db/dec,而在 $\omega = 1$ 处的高度则为 $20 \lg 5 = 14$ db。

第 2 步:频率大于 ω_1 时,斜率为 -20 db/dec 的惯性环节(c)变为有效,因而必须加入第 1 步所得结果。当斜率相加时,从 ω_1 到 ω_2 的频段内组合曲线总的斜率为 -40 db/dec。

第 5 章 频率响应法

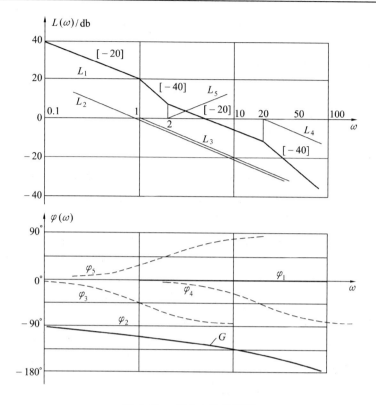

图 5.10 例 5.3 的伯德图

第 3 步:频率大于 ω_2 时,斜率为 $+20$ db/dec 的一阶微分环节(d)变为有效,且必须加入组合曲线。从 ω_2 到 ω_3 的频段内组合曲线总的斜率为 -20 db/dec。

第 4 步:频率大于 ω_3 时,必须加入最后一个环节(e)。振荡环节的斜率为 -40 db/dec,因此组合曲线总的斜率为 -60 db/dec。

最后所得组合的渐近对数幅频特性曲线如图 5.11 所示。

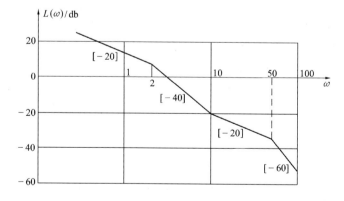

图 5.11 例 5.4 的渐近对数幅频特性曲线图

相频特性可以把各个环节的相频相加,或者直接根据传递函数计算获取。通常,一阶环节的直线近似适用于最初的分析和设计努力。

一般情况下,可以很容易地获取传递函数幅频和相频的近似特性曲线。有必要时,在相对较小的频率范围内,可以利用精确的关系很方便地估算精确的幅频和相频特性。

5.3.2 由伯德图确定传递函数

1. 频率响应的测量

正弦信号可以用于测量控制系统的开环频率响应。通过实验的方法可以获得在许多频率上输出的幅值和相位。这些数据用于获取精确的对数幅频和相频特性曲线。利用幅频特性渐近线的斜率必定为 ±20 db/dec 的整数倍数这一实际情况,在精确的幅频特性图上画出渐近线。对于最小相位传递函数,由渐近幅频特性曲线就可以确认斜率改变处的频率即为相应环节的转折频率,由此可以确定传递函数的各个环节。按照这样的方法,根据这些渐近线就可以确定系统的型别和各个环节大约的时间常数,然后就可以综合成开环传递函数。

在确定传递函数在右半 s 平面是否有任何极点或零点时一定要多加小心。在右半 s 平面的极点或零点的相位变化与在左半 s 平面的极点或零点的相位变化是不同的。例如,一阶不稳定环节具有与惯性环节相同的幅频特性,但它的相位移却发生了变化。

2. 由伯德图确定系统的型别和开环增益

考虑写成典型环节形式的开环传递函数为

$$G(s)H(s) = \frac{K \prod (\tau_i s + 1)}{s^v \prod (T_j s + 1) \prod [(s^2 + 2\zeta_k \omega_{nk} s + \omega_{nk}^2)/\omega_{nk}^2]}$$

显然,在低于最小转折频率的低频部分,对数幅频近似特性的斜率 $-20v$ db/dec 是由开环传递函数中的积分环节数 v 即系统的型别决定的。同时,这一部分的高度取决于常数增益、即开环增益 K 的大小。

(1) 0 型系统

0 型系统的开环积分环节数为 $v = 0$。如图 5.12 所示,它的渐近对数幅频特性曲线在低频部分是一条高度为 $20 \lg K = 20 \lg K_p$ 的水平线,其中 K_p 是静态位置误差系数。

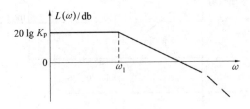

图 5.12 0 型系统的伯德图

(2) 1 型系统

1 型系统的开环积分环节数为 $v = 1$。它的渐近对数幅频特性曲线在低频部分的斜率为 -20 db/dec,而低频部分的渐近线或者它的延长线在 $\omega = 1$ 处的高度则为 $20 \lg K$。而且可以证明,低频部分的渐近线或者它的延长线将与 0 db 线相交于 $\omega = K = K_v$ 处,其中 K_v

是静态速度误差系数。图 5.13 是 1 型系统一些典型的幅频特性曲线。

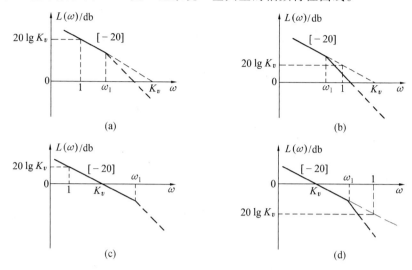

图 5.13 1 型系统的伯德图

(3) 2 型系统

2 型系统的开环积分环节数为 $v=2$。它的渐近对数幅频特性曲线在低频部分的斜率为 $-40\ \text{db/dec}$,而低频部分的渐近线或者它的延长线在 $\omega=1$ 处的高度则为 $20\lg K$。而且可以证明,低频部分的渐近线或者它的延长线将与 0 db 线相交于 $\omega=\sqrt{K}=\sqrt{K_a}$ 处,其中 K_a 是静态加速度误差系数。图 5.13 是 2 型系统一些典型的幅频特性曲线。

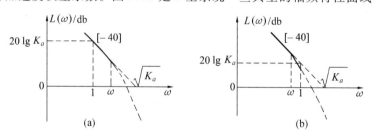

图 5.14 2 型系统的伯德图

3. 确定传递函数的例子

例 5.5 图 5.15 是一直流伺服电机-放大器组合系统(输出为电机轴的角速度,输入为施加的电压)的渐近幅频特性曲线图。确定该组合系统的传递函数。

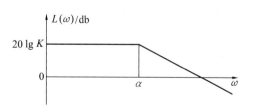

图 5.15 直流伺服电机-放大器组合系统的幅频特性曲线

解 由曲线图很容易得到,组合系统的传递函数为

$$G(s)=\frac{K}{\frac{1}{\alpha}s+1}=\frac{K\alpha}{s+\alpha}$$

例 5.6 由图 5.16 所示渐近对数幅频特性曲线确定一个典型 2 型系统的传递函数,图中转折频率 ω_1 和 ω_2 以及 ω_c 均为已知。假设该传递函数是最小相位的。

解 由渐近幅频特性曲线可见,该传递函数具有以下形式

$$G(s) = \frac{K(\tau s + 1)}{s^2(Ts + 1)} = \frac{K\left(\dfrac{1}{\omega_1}s + 1\right)}{s^2\left(\dfrac{1}{\omega_2}s + 1\right)}$$

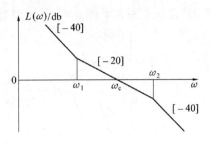

图 5.16 例 5.6 的伯德图

曲线图既没有指明起始部分或其延长线在 $\omega = 1$ 处的高度,也没有指明起始部分或其延长线与 0 db 线的交点。在这种情况下,增益 K 可以利用渐近幅频特性的表达式确定,即

$$L(\omega_c) = 20\lg\left[\frac{K\sqrt{\left(\dfrac{\omega_c}{\omega_1}\right)^2 + 0}}{\omega_c^2\sqrt{0 + 1}}\right] = 0 \quad \text{或者} \quad \frac{K\dfrac{\omega_c}{\omega_1}}{\omega_c^2} = 1$$

求解这一方程得到 $K = \omega_1\omega_c$,因此,传递函数为

$$G(s) = \frac{\omega_1\omega_c[(s/\omega_1) + 1]}{s^2[(s/\omega_2) + 1]}$$

例 5.7 某控制环节的幅频特性曲线如图 5.17 所示,图中实线是渐近特性曲线,而虚线则是精确的特性曲线。如果该环节是最小相位的,试确定它相应的传递函数。

解 由给定的幅频特性曲线,该传递函数有三个串联的环节,并可写成

$$G(s) = \frac{K\omega_n^2}{s(s^2 + 2\zeta\omega_n s + \omega_n^2)}$$

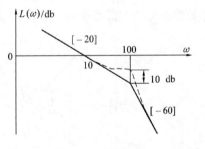

图 5.17 例 5.7 的对数幅频特性曲线

渐近特性曲线与 ω 轴在 $\omega = 10$ rad/s 处相交。由于该传递函数只有一个积分环节,增益环节为

$$K = 10\ s^{-1}$$

很容易看到,振荡环节的转折频率为 $\omega_n = 100$ rad/s。而且,振荡环节在 $\omega = 100$ 处的幅值为 $1/2\zeta$,我们有

$$20\lg\frac{1}{2\zeta} = 10$$

求解此式得到

$$\zeta = 0.158$$

因此,该环节的传递函数为

$$G(s) = \frac{10 \cdot 100^2}{s(s^2 + 2 \times 0.158 \times 100s + 100^2)} = \frac{10^5}{s(s^2 + 31.6s + 10^4)}$$

5.3.3 极坐标图

伯德图的一个缺点是它需要两张单独的图来显示幅值和相位移随频率变化的情况。包含在这两张图内的信息可以组合在极坐标图内。因此常常也需要系统的极坐标图。

1. 绘制极坐标图

极坐标图可由式(5.12)或式(5.13)获取，并用以下的例子说明。

例 5.8 绘制例 2.1 中简单 RC 网络的极坐标图。

解 该网络的传递函数为

$$G(s) = \frac{1}{RCs + 1} = \frac{1}{Ts + 1}$$

而频率响应则为

$$G(j\omega) = \frac{1}{jT\omega + 1}$$

这样，极坐标图可由以下关系获取

$$G(j\omega) = \frac{1}{T^2\omega^2 + 1} - j\frac{T\omega}{T^2\omega^2 + 1}$$

或者

$$G(j\omega) = \frac{1}{T^2\omega^2 + 1} e^{-j\arctan\frac{T\omega}{1}}$$

首先考虑 $\omega = 0$ 和 $\omega = \infty$ 的情况。

$\omega = 0$ 时

$$\text{Re}[G(j\omega)] = 1, \quad \text{Im}[G(j\omega)] = 0$$

$\omega = \infty$ 时

$$\text{Re}[G(j\omega)] = \text{Im}[G(j\omega)] = 0$$

这两个点显示于图 5.18 中。实部和虚部的轨迹很容易证明是一个圆心在点(0.5,0)的圆的一部分如图 5.18 所示。当 $\omega = 1/T$ 时，实部和虚部相等，且相位角为 $-45°$。极坐标图上的箭头指示频率增大的方向。

例 5.9 绘制极坐标图，传递函数为

$$G(s) = \frac{10}{s(s + 1)}$$

解 该传递函数的频率响应为

$$G(j\omega) = \frac{10}{j\omega(j\omega + 1)}$$

幅频特性和相频特性分别为

$$|G(j\omega)| = \frac{10}{\omega\sqrt{\omega^2 + 1}}$$

和

$$\angle G(j\omega) = -90° - \arctan\omega$$

由表 5.2 给出的关系可以得到极坐标图,精确的极坐标图如图 5.19 所示。

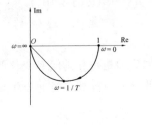
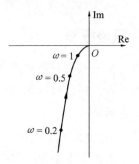

图 5.18　例 5.8 的极坐标图　　　　图 5.19　例 5.9 的极坐标图

表 5.2　例 5.9 给出的传递函数的频率响应

ω	$\mid G(j\omega) \mid$	$\angle G(j\omega)$	ω	$\mid G(j\omega) \mid$	$\angle G(j\omega)$
0	∞	$-90°$	5.0	0.390	$-168.7°$
0.1	99.5	$-95.71°$	7	0.202	$-171.9°$
0.2	49.03	$-101.3°$	10	0.100	$-174.3°$
0.5	17.89	$-116.6°$	20	0.025	$-177.1°$
1.0	7.07	$-135°$	50	0.004	$-178.9°$
2.0	2.24	$-153.4°$	70	0.002	$-179.2°$

2. 概略的极坐标图

在下一节将能看到,在根据开环频率响应确定闭环系统的稳定性时极坐标图是非常有用的。在这种情况下可根据一些关键部分绘制一条概略的曲线就足够了,这些关键部分可以利用以下判据确定。

考虑开环传递函数由典型环节形式描述为

$$G(s)H(s) = \frac{K\prod_{i=1}^{m}(\tau_i s + 1)}{s^v \prod_{j=v+1}^{n}(T_j s + 1)} \quad (5.50)$$

为方便起见,设 $n > m$。

(1) 极坐标图的低频起始部分(当 $\omega \to 0$ 时)

在频率响应中令 $\omega \to 0$,得到

$$\lim_{\omega \to 0} G(j\omega)H(j\omega) = \lim_{\omega \to 0} \frac{K}{(j\omega)^v} \quad (5.51)$$

该式指出开环幅相频率特性曲线的起始部分主要取决于系统的型别,即 v 的取值。不同系统型别的极坐标图低频部分的特点如图 5.20 所示。$\omega = 0$ 时的相位角为 $v(-90°)$。

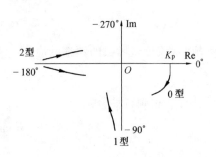

图 5.20　幅相特性曲线的起始部分

(2) 极坐标图的高频终止部分(当 $\omega \to \infty$ 时)

极坐标图的高频终止部分可以确定为

$$\lim_{\omega \to \infty} G(j\omega)H(j\omega) = \lim_{\omega \to \infty} \frac{K(j\omega)^m}{(j\omega)^n} = 0 \cdot e^{j(n-m)\cdot(-90°)} \quad (5.52)$$

注意到式(5.49)中 $n > m$,因此极坐标图将以式(5.52)所确定的相位角,即 $(n-m)(-90°)$ 终止于原点。

(3) 与负实轴的交点

极坐标图与负实轴交点处的频率称为相位穿越频率,记为 ω_g 并确定为

$$\angle G(j\omega_g)H(j\omega_g) = -180° \quad (5.53)$$

然后,交点的位置则由 $G(j\omega_g)$ 或 $|G(j\omega_g)|$ 给出。

例 5.10 传递函数为

$$G(s)H(s) = \frac{5}{s(s+1)(s+2)}$$

绘制概略的极坐标图。

解 根据判据(a),这是一个 1 型系统,当 $\omega \to 0$ 时极坐标图以 $-90°$ 的相位角起始于无穷远处。根据判据(b),当 $\omega \to \infty$ 时极坐标图以 $-270°$ 的相位角趋近于原点。由相位穿越频率的定义有

$$\angle G(j\omega_g)H(j\omega_g) = -90° - \arctan\frac{\omega_g}{1} - \arctan\frac{\omega_g}{2} = -180°$$

求解此式得到

$$\omega_g = \sqrt{2} \text{ rad/s}$$

且

$$|G(j\omega_g)H(j\omega_g)| = \frac{5}{\sqrt{2}\cdot\sqrt{2+1^2}\cdot\sqrt{2+2^2}} = \frac{5}{6}$$

因此,极坐标图与负实轴相交于点 $(-5/6, j0)$。概略的极坐标图如图 5.21 所示。

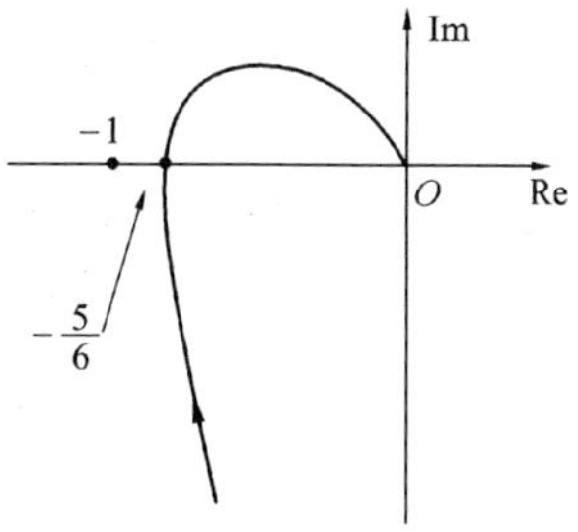

图 5.21 例 5.10 的极坐标图

5.4 奈奎斯特稳定性判据

5.4.1 引 言

如前所述,一个有用的控制系统必须是稳定的。第 3 章中所讨论的代数判据是通过检查对应于闭环传递函数分母的特征多项式确定一个系统是否稳定的。在第 4 章中所介绍的根轨迹法是根据主导极点的阻尼比确定一个系统的相对稳定性的。但是这两种方法都需要知道传递函数,而且传递函数还必须是复变量 s 的有理函数。

在频域内,一个系统的开环频率响应将不仅用于确定闭环系统的稳定性,而且还提供足够多的关于该系统相对稳定性的信息。不仅如此,用正弦输入信号激励系统,用实验的方法能够很容易地获得系统的频率响应,因此即使系统的参数值尚未确定也能应用频率

响应。

频域稳定性判据是由奈奎斯特于1932年提出的,并且依然是研究线性控制系统稳定性的一种基本方法。这一稳定性判据是基于复变理论的一个定理,即柯西的幅角原理。

5.4.2 幅角原理

1. 函数 $F(s)$ 的映射

幅角原理与复变量解析函数的映射理论有关。为了能够有一个全面的了解,先对这一理论作简要的回顾,然后再不加证明地介绍幅角原理。

设 $F(s)$ 是复变量 $s = \sigma + j\omega$ 的一个函数。一般地讲 $F(s)$ 也是复数可以写成

$$F(s) = U(\sigma,\omega) + jV(\sigma,\omega) \tag{5.54}$$

式中 $U(\sigma,\omega)$ 和 $V(\sigma,\omega)$ 是实函数。

定义在 s 平面某一个域内的函数 $F(s)$ 在该域内解析的充分必要条件是它的导数 dF/ds 在该域内连续。可以证明,s 的所有有理函数在 s 平面内除了奇点处外处处解析。因此,所有传递函数在 s 平面内除了在它们的极点处外处处解析。

就像复变量 s 可以表示在一个实轴为 σ 和虚轴为 ω 的平面上一样,$F(s)$ 也可以用一个实轴为 U 和虚轴为 V 的平面表示。前者称为 s 平面,而后者则称为 F 平面。

对于 s 平面内的任意一点 $s = \sigma + j\omega$,根据给定的 σ 和 ω 数值找出相应的 U 和 V 的数值,将 s 平面内的这一点就被"映射"到 F 平面内。例如,考虑函数

$$F(s) = \frac{2s + 3}{s + 5}$$

对于 $s_t = 1 + j2$,得到

$$F(s_t) = \frac{2(1 + j2) + 3}{(1 + j2) + 5} = 0.95 + j0.35$$

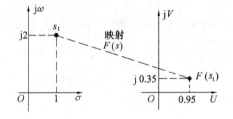

图 5.22 函数 $F(s)$ 的映射

这一"映射"关系如图 5.22 所示。

在这两个平面之间的相应关系称为映射或变换。对于一个解析函数 $F(s)$ 而言,如果 $F(s)$ 在 s 平面内任意给定的某一点是解析的,则该点被映射到 F 平面内唯一的一点。这一概念可以延伸到将 s 平面内一条直线或曲线映射到 F 平面。特别是,如果 $F(s)$ 在 s 平面内的一条光滑曲线上每一点都解析,该曲线将映射为 F 平面内的一条光滑曲线。

2. 幅角原理

现在考虑 s 平面内闭合曲线的映射,这样一条曲线将称为轨线并记为 Γ_s。幅角原理给出了 $F(s)$ 被 Γ_s 所包围的极点和零点数与 Γ_F 将围绕 F 平面原点的周数之间的关系。幅角原理叙述如下:

设 $F(s)$ 是除了在有限个极点处外处处解析的函数。如果 s 平面内的一条闭合曲线 Γ_s 包围了 $F(s)$ 的 Z 个零点和 P 个极点,并且 Γ_s 不经过 $F(s)$ 的任何零点和极点,那么当 s 按顺时针方向沿 Γ_s 绕行一周时,F 平面内相应的闭合曲线 Γ_F 按顺时针方向围绕 F 平面

的原点 N 周,其中 $N = Z - P$。

习惯上认为,沿闭合曲线绕行时右侧的曲线内部区域是被闭合曲线包围的区域。因此,将假设顺时针沿闭合曲线绕行为正,被包围在内的区域是在闭合曲线的右边。

5.4.3 奈奎斯特判据

1. $F(s)$ 的极点和零点

考虑图 5.23 所示闭环系统,它的开环传递函数记为

$$G(s)H(s) = \frac{N(s)}{D(s)} \tag{5.55}$$

式中 $N(s)$ 和 $D(s)$ 均为 s 的多项式。

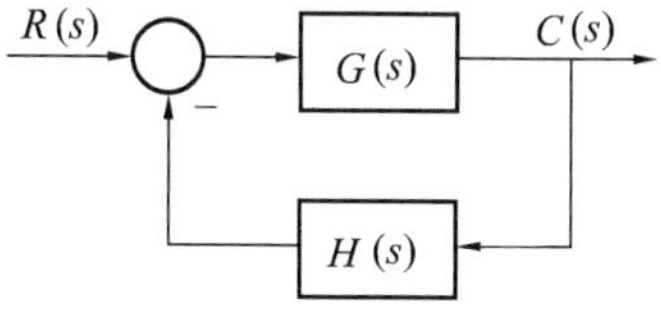

假设规定前面所讨论的函数 $F(s)$ 就是该闭环系统的特征方程,即

图 5.23 闭环系统

$$F(s) = 1 + G(s)H(s) = \frac{D(s) + N(s)}{D(s)} \tag{5.56}$$

注意到这一点是很重要的,开环极点是 $F(s)$ 的极点,而闭合极点则是 $F(s)$ 的零点。

为了确定闭环系统的稳定性,有必要确定是否有任何闭环极点,即 $F(s)$ 的零点在右半 s 平面内。这可以按如下方法解决:

① 在 s 平面内选择一条包围整个右半 s 平面的闭合曲线 Γ_s;

② 在 F 平面内绘制 Γ_F 并确定 Γ_F 围绕原点的周数 N。

2. 奈奎斯特轨线

图 5.24(a) 定义的一条顺时针轨线包围了整个右半 s 平面,并称为奈奎斯特轨线。如果 $F(s)$ 在 s 平面的原点或者 $j\omega$ 轴上某些点有极点,为了满足幅角原理的条件,必须对奈奎斯特轨线进行修改,如图 5.24(b) 所示,沿无穷小半圆绕过这些极点。

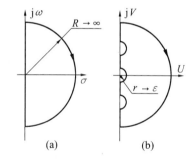

图 5.24 奈奎斯特轨线和修改后的奈奎斯特曲线

3. 奈奎斯特稳定性判据

设 $F(s)$ 在奈奎斯特轨线内有 P 个极点和 Z 个零点。对于一个稳定的系统,必须有 $Z = 0$。即特征方程在奈奎斯特轨线内一定没有根。根据幅角原理,在 F 平面上的 Γ_F 将按顺时针方向围绕原点 N 周,即

$$N = Z - P \tag{5.57}$$

这样,系统稳定的充分必要条件是

$$N = -P \tag{5.58}$$

从而 Z 将为零。进一步可以注意到 F 平面的原点就是 GH 平面内的 $(-1, j0)$ 点。因此,可以得到以下从开环传递函数 $G(s)H(s)$ 出发的奈奎斯特判据。

负反馈控制系统稳定的充分必要条件是,奈奎斯特轨线在 GH 平面内的映射 Γ_{GH} 逆时针方向围绕 $(-1, j0)$ 点的周数等于开环传递函数 $G(s)H(s)$ 在 s 平面奈奎斯特轨线内的

4. 奈奎斯特图

奈奎斯特轨线在 GH 平面内的映射 Γ_{GH} 称为开环传递函数 $G(s)H(s)$ 的奈奎斯特图。$G(s)H(s)$ 的极坐标图是奈奎斯特图的一个主要部分,可以通过实验的方法获取,或者当传递函数已知时通过计算获取。完成奈奎斯特图剩余部分的步骤将用一些例子予以说明。

例 5.11 考虑一个开环传递函数为

$$G(s)H(s) = \frac{6}{(s+1)(0.5s+1)(0.2s+1)}$$

的 0 型系统。绘制奈奎斯特图并确定闭环系统的稳定性。

解 表 5.3 为该开环传递函数的频率响应,可以利用这些数值获得极坐标图。

表 5.3 传递函数的频率响应

ω	幅值	相角	ω	幅值	相角
0	6	0°	4.123	0.476	-180.0°
0.5	5.18	-46.3°	10.0	0.052	-226.4°
1.0	1.76	-130.2°	20.0	0.007	-247.4°

s 平面内的奈奎斯特轨线 Γ_s 可以看成三个分开的部分:AB、BCD 和 DA,如图 5.25(a) 所示。这样可以找到它们在 GH 平面上的映射以获取奈奎斯特图 Γ_{GH},如图 5.25(b) 所示。

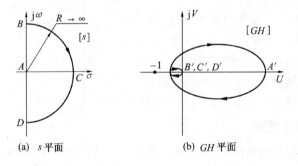

图 5.25 例 5.11 的奈奎斯特轨线和奈奎斯特图

(1) s 平面内的 AB 部分是 $j\omega$ 轴从 $\omega = 0$ 到 $\omega = \infty$。它映射为 GH 平面内的极坐标图 $A'B'$,因为 $s = j\omega$ 且

$$G(s)H(s)|_{s=j\omega} = G(j\omega)H(j\omega)$$

(2) 无穷小半圆 BCD 映射为 GH 平面的原点。

(3) DA 部分是 $j\omega$ 轴的负半轴,它映射为 GH 平面内的曲线 $D'A'$。可以注意到,$D'A'$ 是 $A'B'$ 关于实轴的镜像。这是由于 $G(-j\omega)H(-j\omega)$ 是 $G(j\omega)H(j\omega)$ 的共轭复数。

奈奎斯特图没有包围 GH 平面的 $(-1, j0)$ 点,因此 $N = 0$。而且该传递函数在奈奎斯特轨线内没有任何极点。由于 $Z = N + P = 0$,可见闭环系统稳定。

如果使该系统的开环增益增大 2.5 倍,结果所得的奈奎斯特图将如图 5.26 所示围绕

GH 平面的 (-1, j0) 两周。在这一情况下得到 $N=2$。由于 P 仍然为零,闭环系统在右半 s 平面有两个极点而不稳定。

例 5.12 考虑一个 1 型系统的稳定性,开环传递函数为

$$G(s)H(s) = \frac{K}{s(T_1s+1)(T_2s+1)}$$

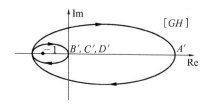

图 5.26 例 5.11 的奈奎斯特图

解 由于这一次开环传递函数在原点有一个极点,在包围整个右半 s 平面的同时,奈奎斯特轨线还必须包括一个绕过该极点的小小的弯路。由此得到无穷小半径 r 的半圆 EFA。结果所得的奈奎斯特轨线 Γ_s 和奈奎斯特图 Γ_{GH} 显示于图 5.27。获取 Γ_{GH} 的过程如下:

① s 平面内的 AB 段是 $j\omega$ 轴从 $\omega=0_+$ 到 $\omega \to \infty$ 部分,并映射为 GH 内的曲线 $A'B'$,即频率响应的极坐标图。

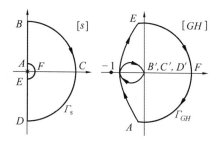

图 5.27 例 5.12 的奈奎斯特图轨线和奈奎斯特图

相位穿越频率可由式(5.53)确定为

$$\angle G(j\omega_g)H(j\omega_g) = -90° - \arctan T_1\omega_g - \arctan T_2\omega_g = -180°$$

因此

$$\omega_g = 1/\sqrt{T_1 T_2}$$

极坐标图与实轴相交的点为

$$|G(j\omega_g)H(j\omega_g)| = \frac{K}{\omega_g\sqrt{(T_1\omega_g)^2+1} \cdot \sqrt{(T_2\omega_g)^2+1}} = \frac{KT_1T_2}{T_1+T_2}$$

即极坐标图与实轴相交于点 $\left(\dfrac{KT_1T_2}{T_1+T_2}, j0\right)$。

② 无穷大半圆 BCD 映射为 GH 平面的原点。

③ DE 段(负 $j\omega$ 轴)映射为频率响应的镜像 $D'E'$。

④ 无穷小半圆 EFA 的表示可以令

$$s = \lim_{r \to 0} re^{j\theta}$$

并使 θ 从 $\omega=0_-$ 时的 $-90°$ 变化到 $\omega=0_+$ 时的 $90°$。由于 r 趋近于零,$G(s)H(s)$ 的映射为

$$\lim_{r \to 0} G(s)H(s) = \lim_{r \to 0} \frac{K}{T_1 T_2} \cdot \frac{1}{re^{j\theta}} = \infty \cdot e^{-j\theta}$$

这一部分映射为一个无穷大的半圆 $E'F'A'$,如图 5.26 所示,幅角从 $\omega=0_-$ 时的 $90°$ 变化到 $\omega=0_+$ 时的 $-90°$,而当 $\omega=0$ 时则通过 $0°$。

利用奈奎斯特判据,当

$$\frac{KT_1T_2}{T_1+T_2} < 1 \quad 即 \quad K < \frac{T_1+T_2}{T_1T_2}$$

时 $Z = N + P = 0$,因此闭环系统稳定。

例 5.13 开环传递函数为

$$G(s)H(s) = \frac{K}{s^2(Ts+1)}$$

确定该 2 型系统的稳定性。

解 令 $s = j\omega$ 可以获得极坐标图。当 ω 趋近于 0_+ 时有

$$\lim_{\omega \to 0_+} G(j\omega)H(j\omega) = \lim_{\omega \to 0_+} \frac{K}{(j\omega)^2} = \infty \cdot e^{-j180°}$$

当 ω 趋近于 $+\infty$ 时有

$$\lim_{\omega \to +\infty} G(j\omega)H(j\omega) = \lim_{\omega \to +\infty} \frac{K}{T(j\omega)^3} = 0 \cdot e^{-j270°}$$

在绕过 s 平面原点的无穷小半圆上 $s = \varepsilon e^{j\theta}$,由此

$$\lim_{\varepsilon \to 0} G(s)H(s) = \lim_{\varepsilon \to 0} \frac{K}{(\varepsilon e^{j\theta})^2} = \infty \cdot e^{-j2\theta}$$

式中 $-90° \leq \theta \leq 90°$。这样这一部分映射为一个整圆,幅角从 $\omega = 0_-$ 时的 $+180°$ 变化到 $\omega = 0_+$ 时的 $-180°$,而当 $\omega = 0$ 时则通过 $0°$。完整的奈奎斯特图如图 5.28 所示。

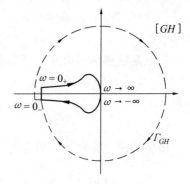

图 5.28 例 5.13 传递函数的奈奎斯特图

由于 $N = 2$ 且 $P = 0$,闭环系统在右半 s 平面有两个根,无论增益 K 怎样,闭环系统都是不稳定的。

例 5.14 考虑两个具有非最小相位开环传递函数的系统的稳定性

(1) $G_1(s) = \dfrac{K}{s(Ts-1)}$;

(2) $G_2(s) = \dfrac{K(\tau s + 1)}{s(Ts-1)}$。

解 两个传递函数都具有一个不稳定环节,因此对于这两个系统有 $P = 1$。

(1) 令 $s = j\omega$ 可以获得频率响应的极坐标图。我们可以注意到,第一个系统的开环相位角从 $-270°$ 变化到 $-180°$,对于极坐标图有

$$\lim_{\omega \to 0_+} |G(j\omega)H(j\omega)| \to \infty \cdot e^{-j270°}$$

和

$$\lim_{\omega \to +\infty} G(j\omega)H(j\omega) = 0 e^{-j180°}$$

奈奎斯特图如图 5.29(a) 所示。

不论开环增益如何增大,总是有 $N = 1$,又由于 $P = 1$,闭环系统有两个根在右半 s 平面,因而闭环系统总是不稳定的。

(2) 第二个系统对于指定的开环增益的奈奎斯特图,如图 5.29(b) 所示。在这一情况下,尽管奈奎斯特图与情况 1 不同,仍然有 $N = 1$ 和 $P = 1$,因此对于指定的开环增益该闭环系统不稳定。

但是，如果增大开环增益，奈奎斯特图将如图 5.29(c)所示。现在由于 $N = -1$ 和 $P = 1$，对于较大的开环增益闭环系统变为稳定了。

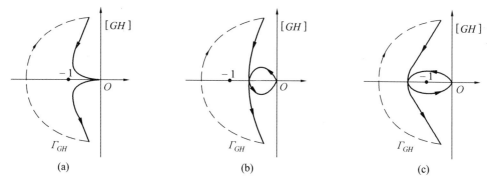

图 5.29 例 5.14 的奈奎斯特图

5. 简化的奈奎斯特判据

(1) 基于一半 Γ_{GH} 的奈奎斯特判据

因为奈奎斯特轨线 Γ_s 在 GH 平面内的映射，即奈奎斯特图 Γ_{GH} 是关于实轴对称的，可以只用一半 Γ_{GH} 就可以确定系统的稳定性。因此，奈奎斯特稳定性判据可以叙述如下：

反馈控制系统稳定的充分必要条件是奈奎斯特图当 ω 从零变化到无穷大时逆时针围绕 $(-1, j0)$ 点的周数等于 $P/2$，其中 P 为 $G(s)H(s)$ 在 s 平面上奈奎斯特轨线内的极点数。

(2) 基于穿越的奈奎斯特判据

由上面的例子可以看到，奈奎斯特图与负实轴上 $(-\infty, -1)$ 段相交的交点对于确定闭环系统的稳定性是很重要的，因为如果奈奎斯特图围绕 $(-1, j0)$ 点则必定与负实轴在这一区段相交。如果奈奎斯特图与负实轴在 $(-1, j0)$ 段相交，则称为在 Γ_{GH} 平面上发生了"穿越"。如果穿越后 Γ_{GH} 相角的增量为正则称为正穿越，否则称为负穿越。例如，在图 5.30 中，Γ_{GH} 与负实轴相交三次，但是只有两次定义如上的"穿越"。在图中正穿越标记为 $(+)$，而负穿越则标记为 $(-)$。

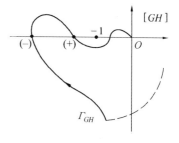

图 5.30 正"穿越"和负"穿越"

给出上述定义后，奈奎斯特稳定性判据可以陈述如下：

反馈控制系统稳定的充分必要条件是当 ω 从零变化到无穷大时

$$N_+ - N_- = P/2 \tag{5.59}$$

其中 N_+ 为正穿越次数，N_- 为负穿越次数，而 P 则为开环传递函数 $G(s)H(s)$ 在右半 s 平面内的极点数。

例如，如果在例 5.11 中开环增益为 6，则当 ω 从零变化到无穷大时有

$$N_+ - N_- = 0 - 0 = 0$$

由于 $P = 0$，系统稳定；如果开环增益为 15，则有

$$N_+ - N_- = 0 - 2 = -2 \neq P/2$$

闭环系统变为不稳定。类似地,例 5.13 中有 $N_+ - N_- = 0 - 1 = -1$ 和 $P = 0$,系统变为不稳定。至于例 5.14,从零变化到无穷大时,在图 5.29(a) 的情况下有 $N_+ = 0, N_- = 1/2$,$N_+ - N_- = -1/2 \neq P/2$,闭环系统不稳定;在情况(b)下,我们仍然有 $N_+ = 0, N_- = 1/2$,对于较小的开环增益闭环系统稳定;但是,在情况(c)下,我们有 $N_+ = 1, N_- = 1/2, N_+ - N_- = 1/2 = P/2$,对于较大的开环增益闭环系统稳定。

5.4.4 应用伯德图时的稳定性

正如所知,可以利用"穿越"确定系统的稳定性。注意,如果在 GH 平面上发生了"穿越",那么如图 5.31 所示,在伯德图上幅频特性曲线大于 0 db 时相频特性曲线将与 $-180°$ 线相交。因此,"穿越"可以定义同上,而且奈奎斯特稳定性判据可以按照伯德图上的 0 db 和 $-180°$ 点规定同上。

由于"穿越"与奈奎斯特图 Γ_{GH} 有关,而 Γ_{GH} 包括了 s 平面上无穷小半圆绕行部分的映射,在伯德图上应用奈奎斯特判据时,相频

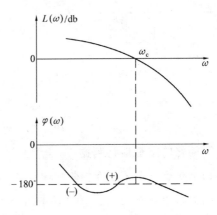

图 5.31 伯德图中的"穿越"

特性曲线应当加以修改,即相频特性曲线的起始部分应当向上移动 $v \cdot 90°$,其中 v 为开环传递函数 $G(s)H(s)$ 中的积分环节数。

例 5.15 考虑以下两种情况下开环传递函数为

$$G(s)H(s) = \frac{K(\tau s + 1)}{s^2(Ts + 1)}$$

的系统的稳定性:(1) $\tau > T$;(2) $\tau < T$。

解 这是一个最小相位传递函数,即 $P = 0$。

(1) $\tau > T$ 的情况,伯德图见图 5.32(a)。由于 $v = 2$,使相频特性曲线的起始部分向上移动 $180°$。当 ω 从零变化到无穷大时,相频特性曲线总是在 $-180°$ 线之上,由此可有

$$N_+ - N_- = 0 - 0 = 0$$

因此,无论开环增益如何增大,闭环系统总是稳定的。

(2) $\tau < T$ 的情况,伯德图如图 5.32(b) 所示。类似地,相频特性曲线的起始部分向上移动 $180°$。但是,在这一情况下由于有一次负穿越而有

$$N_+ - N_- = 0 - 1 = -1$$

因此,无论开环增益降低多少,闭环系统总是不稳定的。

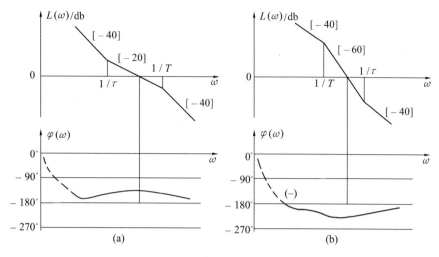

图 5.32　例 5.15 的伯德图

5.5　相对稳定性

在大部分实际情况下,除了要找出闭环系统是否稳定外,还希望确定闭环系统离不稳定有多远。由系统的开环频率响应很容易获取这一信息。在 GH 平面上开环频率响应曲线与 $(-1,j0)$ 点的接近程度是闭环系统相对稳定性的度量。一般情况下度量相对稳定性的两个量是增益裕度和相角裕度。

5.5.1　增益裕度

仍然考虑例 5.12。当 K 增大时,$G(j\omega)H(j\omega)$ 曲线趋近 $(-1,j0)$ 点,并最终围绕 $(-1,j0)$ 点,K 的临界值则为

$$K_c = \frac{T_1 + T_2}{T_1 T_2}$$

当 K 在临界值之下减小时,则稳定程度增加。

如图 5.33(a) 所示,增益裕度定义为正好使系统的频率响应曲线通过临界点所需的增益 K_c 除以系统的实际增益 K,并记为 K_g,即

$$K_g = K_c/K \tag{5.60}$$

临界增益与给定增益之间的裕度是相对稳定性的一种度量。相对稳定性的这一度量称为增益裕度。

由于频率响应对于给定频率的幅值正比于 K 的数值,增益裕度可以用以下在图 5.33(b) 中的长度表示,即

$$K_g = \frac{ON}{OA} = \frac{1}{|G(j\omega_g)H(j\omega_g)|} \tag{5.61}$$

式中 ω_g 为相位穿越频率,且

$$\angle G(j\omega_g)H(j\omega_g) = -180° \tag{5.62}$$

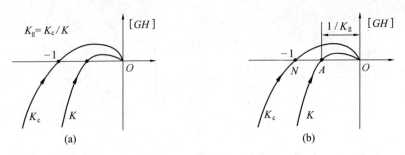

图 5.33　增益裕度的定义

5.5.2　相位裕度

尽管增益裕度是极坐标图接近临界点程度的一种度量,但在有些情况下它不能给出明确的答案甚至给人以误导。图 5.34(a) 所示是一个二阶系统的极坐标图,该极坐标图决不会与负实轴相交,也决不会不稳定。对于所有的增益值,增益裕度均为无穷大。但是,当开环增益增大时极坐标图确实更靠近了临界点。显示于图 5.34(b) 的根轨迹进一步确认了,即使该系统不是不稳定的,但也变为越来越振荡。

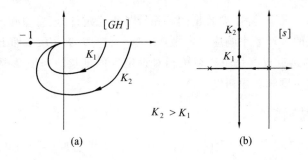

图 5.34　含糊不清的增益裕度

作为另一个例子,图 5.35 显示出两个具有相同增益裕度的系统。但是凭直观可以知道,系统 A 要比系统 B 更接近不稳定。显然,接近临界点的程度还需要另一种度量,这种度量称为相位裕度。

相位裕度定义为稳定的闭环系统为了与 $(-1,j0)$ 点相交而极坐标图必须转过的最小相位角。这一定义仍然用图例说明。在图 5.36 中,相位裕度用角度 γ 指示,在增益穿越频率 ω_c 处极坐标图的幅值 $|G(j\omega)H(j\omega)|$ 等于1,即

$$|G(j\omega_c)H(j\omega_c)| = 1 \tag{5.63}$$

由此,相位裕度为

$$\gamma = 180° + \angle G(j\omega_c)H(j\omega_c) \tag{5.64}$$

第 5 章 频率响应法

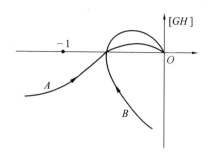

图 5.35 具有相同增益裕度的不同系统

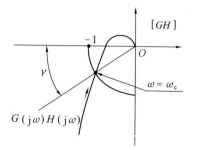

图 5.36 相位裕度

例 5.16 确定欠阻尼二阶系统的相位裕度。

解 设系统的开环传递函数为

$$G(s) = \frac{\omega_n^2}{s(s+2\zeta\omega_n)} \quad (5.65)$$

在增益穿越频率 ω_c 处频率响应的幅值等于 1,因此有

$$\frac{\omega_n^2}{\omega_c\sqrt{\omega_c^2+4\zeta^2\omega_n^2}} = 1$$

整理上式有

$$\omega_c^4 + 4\zeta^2\omega_n^2\omega_c^2 - \omega_n^4 = 0$$

求解 ω_c 可得到

$$\omega_c = \omega_n\sqrt{\sqrt{4\zeta^4+1}-2\zeta^2} \quad (5.66)$$

该系统的相位裕度为

$$\gamma = 180° - 90° - \arctan\frac{\omega_c}{2\zeta\omega_n} = 90° - \arctan\frac{\sqrt{\sqrt{4\zeta^4+1}-2\zeta^2}}{2\zeta} = \arctan\frac{2\zeta}{\sqrt{\sqrt{4\zeta^4+1}-2\zeta^2}} \quad (5.67)$$

5.5.3 伯德图上的相对稳定性

尽管稳定性裕度可以直接由极坐标图获取,它们也经常由伯德图确定,如图 5.37 所示。

用对数(分贝)表示的增益裕度 $K_g(\text{db})$ 是由相位穿越频率 ω_g 处的增益确定的

$$K_g(\text{db}) = 20\lg K_g = -20\lg|G(j\omega_g)H(j\omega_g)| \quad (5.68)$$

类似地,相位裕度 γ 是由增益穿越频

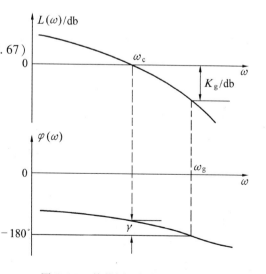

图 5.37 伯德图上的相对稳定性

率 ω_c 处的相位移确定的

$$\gamma = 180° + \angle G(j\omega_c)H(j\omega_c) \qquad (5.69)$$

5.5.4 相位裕度和增益裕度与稳定性的关系

由例 5.16,对于欠阻尼二阶系统有

$$\gamma = \arctan \frac{2\zeta}{\sqrt{\sqrt{4\zeta^4 + 1} - 2\zeta^2}} \qquad (5.70)$$

式中相位裕度 γ 唯一地与阻尼比 ζ 有关。对于行为表现与二阶系统等价的那些系统,可以证明相位裕度与有效阻尼比有关。通常,当相位裕度为 45° 到 60° 时可以获得令人满意的响应。当一个人积累了经验并建立了自己特定的方法时,特定系统所用 γ 的取值就变得更为明显了。

对于一般的非条件稳定系统,系统稳定时用分贝表示时增益裕度必须为正值(数值表示时大于 1),而负的增益裕度则意味着系统不稳定。

系统的阻尼特性也与增益裕度有关。但是与增益裕度相比,相角裕度能给出更好的关于系统阻尼特性以及由此而得的关于系统瞬态响应的估计。

对于如图 5.38 所示的条件稳定系统而言,稳定性与增益裕度的关系要予以修改。该系统对于图中所表示的给定增益是稳定的。随着增益的增大或减小,都可能发生不稳定的情况。这可以用以下的例子说明。

例 5.17 对于图 5.38 中的极坐标图,设增益 $K = 100$ 时有 $a = -5, b = -2$ 和 $c = -0.5$。确定使系统稳定时 K 的取值范围。

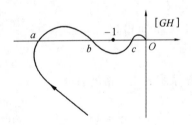

图 5.38 条件稳定系统的极坐标图

解 对于给定系统,由于频率响应的幅值正比与 K 的数值,当 $K_a = 100/5 = 20$、$K_b = 100/2 = 50$ 或者 $K_c = 100/0.5 = 200$ 极坐标图将通过 $(-1, j0)$ 点。

系统稳定的充分必要条件为 $N_+ - N_- = 0$,即极坐标图与负实轴的 $(-\infty, -1)$ 段相交零次或两次。因此当 $K < 20$ 或者 $50 < K < 200$ 时闭环系统稳定,否则不稳定。

5.6 闭环频域分析

在前面几节中,主要关心的是开环系统的频率响应。由开环频率响应能够确定闭环系统的稳定性和相对稳定性。相对稳定性的稳定裕度以一些不太确定的方式与闭环系统的响应特性相联系。但是要记住,反馈回路终究是要闭合的,而且闭环系统的性能需要予以评估。因此,在许多情况下将不得不确定闭环的频率响应。

闭环频率响应可以由闭环传递函数确定,但是如果开环频率响应曲线已经绘制,尤其

5.6.1 由极坐标图确定闭环频率响应

1. 单位反馈系统

首先只考虑单位反馈系统。很容易就可以建立起,开环频率响应和闭环频率响应之间的关系为

$$\Phi(j\omega) = \frac{G(j\omega)}{1 + G(j\omega)} = M(\omega)e^{j\theta(\omega)} \tag{5.71}$$

由开环传递函数的极坐标图,如图 5.39 所示,式 (5.71) 可以给出一个很有意思的图形解释。对于给定的频率 ω_t,向量 $\overrightarrow{OP}$ 表示开环频率响应 $G(j\omega_t)$,向量 $\overrightarrow{AP}$ 则表示 $1 + G(j\omega_t)$。因此,OP 与 AP 之比,即 OP/AP 就表示了 $\omega = \omega_t$ 时闭环频率响应的幅值 $M(\omega_t)$。另一方面,向量 $\overrightarrow{OP}$ 与 $\overrightarrow{AP}$ 之间的夹角,即 $\angle(\overrightarrow{OP} - \overrightarrow{AP})$ 就是闭环频率响应的幅角 $\theta(\omega_t)$。

因此,具体做法是在极坐标图上挑选一些频率已知的点,并使用上面的式子计算闭环的幅值和幅角。

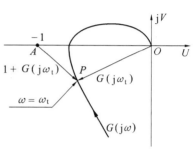

图 5.39 闭环频率响应的计算

如果对足够多的频率重复这一过程,就可以获得闭环系统的频率响应。

换一种方式,考虑 G 平面上所有具有相同幅值和相同幅角的点也可以完成这一任务。这样做法将在 G 平面上生成一些等幅值线和等幅角线。令 G 平面的坐标系统为 U 和 V,则有

$$G(j\omega) = U(\omega) + jV(\omega) \tag{5.72}$$

因此,闭环频率响应的幅值为

$$M = \frac{|U + jV|}{|1 + U + jV|} = \frac{\sqrt{U^2 + V^2}}{\sqrt{(1+U)^2 + V^2}} \tag{5.73}$$

将式 (5.73) 展开得到

$$(1 - M^2)U^2 + (1 - M^2)V^2 - 2M^2 U = M^2 \tag{5.74}$$

这一方程可以重新整理成

$$\left(U - \frac{M^2}{1 - M^2}\right)^2 + V^2 = \left(\frac{M}{1 - M^2}\right)^2 \tag{5.75}$$

对于恒定的 M 值,这是一个圆的方程,圆心在

$$U = M^2/(1 - M^2), \quad V = 0$$

而圆的半径则为 $|M/(1 - M^2)|$。图 5.40 所示为一簇完整的等 M 圆。

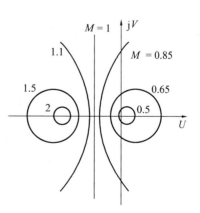

图 5.40 等 M 圆

以相类似的方式，可以获得闭环幅角的等值圆。对于式（5.71），幅角的关系为

$$\theta = \angle(U + jV) - \angle(1 + U + jV)U = \arctan\frac{V}{U} - \arctan\frac{V}{1+U} \quad (5.76)$$

对上式两侧取正切并重新整理可得

$$U^2 + V^2 + U - \frac{V}{N} = 0 \quad (5.77)$$

其中 $N = \tan\theta = $ 常数。在方程的两侧均加上

$$\frac{1}{4} + \frac{1}{4N^2}$$

则得到

$$\left(U + \frac{1}{2}\right)^2 + \left(V - \frac{1}{2N}\right)^2 = \frac{1}{4}\left(1 + \frac{1}{N^2}\right) \quad (5.78)$$

对于恒定的 N 值，这是一个圆的方程，圆心在

$$U = -0.5, \quad V = 1/(2N)$$

而圆的半径则为

$$\frac{1}{2}\sqrt{1 + \frac{1}{N^2}}$$

图 5.41 显示的是一簇完整的等 N 圆。

如图 5.42 所示，把极坐标图画在带有等 M 圆和等 N 圆的复平面上，只要读出特定频率处的幅值和相角，很简单地就能确定闭环的频率响应。

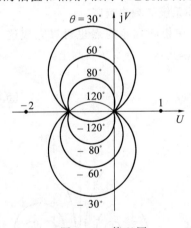

图 5.41　等 N 圆

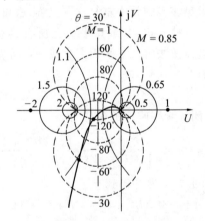

图 5.42　等 M 圆和等 N 圆上的极坐标圆

2. 非单位反馈系统

如果一个系统不是单位反馈系统，那么系统的传递函数需要作些处理。闭环传递函数可以表示成

$$\Phi(s) = \frac{G(s)}{1 + G(s)H(s)} = \frac{G(s)H(s)}{1 + G(s)H(s)} \cdot \frac{1}{H(s)} =$$

$$\frac{\overline{G}(s)}{1 + \overline{G}(s)} \cdot \frac{1}{H(s)} = \overline{\Phi}(s) \cdot \frac{1}{H(s)} \quad (5.79)$$

其中 $\bar{G}(s) = G(s)H(s)$，$\bar{\Phi}(s) = \bar{G}(s)/[1+\bar{G}(s)]$。因此，我们得到

$$M(\omega) = |\bar{\Phi}(j\omega)| \cdot \frac{1}{|H(j\omega)|} = \bar{M}(\omega) \cdot \frac{1}{|H(j\omega)|} \tag{5.80}$$

$$\theta(\omega) = \angle \bar{\Phi}(j\omega) - \angle H(j\omega) = \bar{\theta}(\omega) - \angle H(j\omega) \tag{5.81}$$

其中 $\bar{M}(\omega)$ 和 $\bar{\theta}(\omega)$ 是等效单位反馈系统的幅频特性和相频特性，可以借助等 M 圆和等 N 圆获取。

5.6.2 由伯德图确定闭环频率响应

尽管闭环频率响应可以按照介绍过的方式确定，平常更多的是得自伯德图。将等 M 圆和等 N 圆变换到一个对数幅相特性图上得到的结果是一张称为尼柯尔斯图的新图。在尼柯尔斯图上等 M 圆和等 N 圆呈现为一些等值线，如图 5.43 所示。

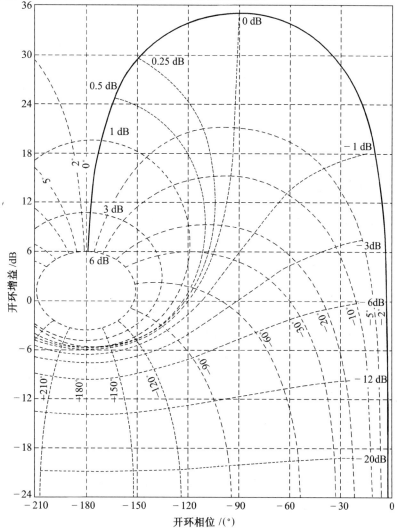

图 5.43 尼科尔斯图

在尼柯尔斯图上开环频率响应是画成以分贝表示的对数幅值与以度数表示的相位角的关系曲线,同时闭环频率响应的等 M 线和等 N 线分别以分贝和度数给出。可以用一个例子说明利用尼柯尔斯图确定闭环频率响应。

例 5.18 单位反馈系统的开环传递函数为

$$G(s) = \frac{10(0.5s + 1)}{s(s+1)(0.05s+1)}$$

确定其闭环频率响应。

解 对于不同的 ω 值,开环对数幅频特性 $L(\omega)$ 和相频特性 $\varphi(\omega)$ 可以从图 5.10 所示的伯德图中找到(见例 5.3)。然后,把 $G(j\omega)$ 的极坐标图画在尼科尔斯图上就可以获得闭环的幅频特性 $M(\omega)$ 和相频特性 $\theta(\omega)$。有关的数据在表 5.4 中给出。最后,由表 5.4 可画出闭环的幅频特性曲线,如图 5.44 所示。

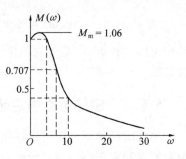

图 5.44 闭环幅频特性曲线

表 5.4 例 5.18 的相关数据

$\omega(\text{rad}\cdot\text{s}^{-1})$	0.4	1	2	4.4	6	10	20	30
$L(\omega)/\text{db}$	28	20	8	0	-2.5	-9.5	-14	-22
$\varphi(\omega)/\text{deg}$	-102	-111.4	-114.7	-120.4	-121.5	-129.4	-140.6	-152
M/db	0	0.25	0.5	0	-1.5	-8	-12.5	-22
M	1	1.03	1.06	1	0.84	0.4	0.24	0.08

5.6.3 频域性能指标

就像时域性能指标允许控制系统的不同设计进行比较一样,也可以定义一些频域的参考标准并将其用于评估整个系统的性能。一个令人感兴趣的问题是,系统的频率响应是如何与该系统预期的瞬态响应相联系的。

1. 二阶系统的性能

由于在实际应用中大量的控制系统满足用主导极点近似表示的二阶系统,值得关注简单的典型二阶系统,它们阐明了为什么这些参考标准是重要的,以及关于系统的性能这些参考标准能告诉我们些什么。

考虑典型二阶系统的传递函数为

$$\Phi(s) = \frac{\omega_n^2}{s^2 + 2\zeta\omega_n s + \omega_n^2} \quad (5.82)$$

其典型的频率响应如图 5.45 所示。

正如所知,当阻尼比 $\zeta < 1/\sqrt{2}$ 时,频率响应在谐振频率 $\omega_m = \omega_n\sqrt{1-2\zeta^2}$ 处的幅值具有一个峰值

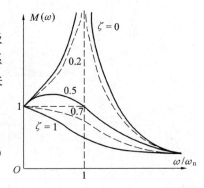

图 5.45 典型二阶系统的频率响应

$$M_{\mathrm{m}} = \frac{1}{2\zeta\sqrt{1-\zeta^2}} \quad (5.83)$$

对于阶跃输入的振荡特性可以通过阻尼比 ζ 与谐振峰值 M_{m} 相联系。由式(5.83)可有

$$\zeta = \frac{\sqrt{2}}{2}\sqrt{1-\sqrt{1-\frac{1}{M_{\mathrm{m}}^2}}} \quad (5.84)$$

和

$$\sigma_{\mathrm{p}} = \mathrm{e}^{-\pi\zeta/\sqrt{1-\zeta^2}} \times 100\% = \mathrm{e}^{-\pi\sqrt{(M_{\mathrm{m}}-\sqrt{M_{\mathrm{m}}^2-1})/(M_{\mathrm{m}}+\sqrt{M_{\mathrm{m}}^2+1})}} \times 100\% \quad (5.85)$$

图 5.46 显示了超调量 σ_{p} 与谐振峰值 M_{m} 的对应关系。可以发现,阶跃输入的超调量随着 M_{m} 增大而增大。一般地讲,谐振峰值 M_{m} 指示了系统的相对稳定性。

在闭环频率响应中,频率响应衰减到零频值的 $\sqrt{2}/2$ 时的频率,即带宽 ω_{b} 是系统如实复制输入信号的能力的量度标准。根据定义有

$$|\varPhi(\mathrm{j}\omega_{\mathrm{b}})| = \left|\frac{\omega_{\mathrm{n}}^2}{(\mathrm{j}\omega_{\mathrm{b}})^2 + 2\zeta\omega_{\mathrm{n}}(\mathrm{j}\omega_{\mathrm{b}}) + \omega_{\mathrm{n}}^2}\right| = \frac{\sqrt{2}}{2}|\varPhi(\mathrm{j}0)| = \frac{\sqrt{2}}{2} \quad (5.86)$$

求解这一方程得到

$$\omega_{\mathrm{b}} = \omega_{\mathrm{n}}\sqrt{1-2\zeta^2+\sqrt{2-4\zeta^2+4\zeta^4}} \quad (5.87)$$

在式子的两侧分别乘以上升时间 t_{p} 和调节时间 t_{s},结果为

$$\omega_{\mathrm{b}} t_{\mathrm{p}} = \pi\sqrt{\frac{1-2\zeta^2+\sqrt{2-4\zeta^2+4\zeta^4}}{1-\zeta^2}} \quad (5.88)$$

和

$$\omega_{\mathrm{b}} t_{\mathrm{s}} = \frac{3}{\zeta}\sqrt{1-2\zeta^2+\sqrt{2-4\zeta^2+4\zeta^4}}, \quad \Delta = 5\% \quad (5.89)$$

当 ζ 保持不变时,随着 ω_{n} 的增高,阶跃响应的上升时间和调节时间都将减小。

正如在频率响应曲线所指示的,系统的带宽 ω_{b} 可以近似地与系统的固有频率 ω_{n} 相关。正如所知,当 ζ 保持不变时,ω_{n} 越大则响应趋近于期望稳态值的速度越快。

这样所希望的频域性能指标是相对较小的谐振峰值和相对较大的带宽。

2. 高阶系统的性能指标

与二阶系统相类似,高阶系统的重要频域性能指标是:

① 谐振峰值 M_{m},或者相对谐振峰值 $M_{\mathrm{r}} = M_{\mathrm{m}}/M(0)$;

② 带宽 ω_{b}。

如果一个高阶系统不能近似为具有主导极点的二阶系统,则其超调量和调整时间可以由闭环频率响应曲线用下列经验公式估算

$$\sigma\% = \left\{41 \cdot \ln\left[\frac{M_{\mathrm{m}} \cdot M(\omega_1/4)}{M^2(0)} \cdot \frac{\omega_{\mathrm{b}}}{\omega_{0.5}}\right] + 17\right\}\% \quad (5.90)$$

$$t_{\mathrm{s}} = \left(13.57\frac{M_{\mathrm{m}}\omega_{\mathrm{b}}}{M(0)\omega_{0.5}} - 2.51\right) \cdot \frac{1}{\omega_{0.5}}, \quad \Delta = 5\% \quad (5.91)$$

其中 M_m, ω_r 和 ω_b 定义如上。如图 5.47 所示,ω_1 是 $M(\omega)$ 衰减到 $M(0)$ 时的频率,$\omega_{0.5}$ 是 $M(\omega)$ 衰减到 $M(0)$ 一半时的频率。

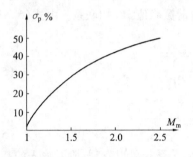
图 5.46 σ_p 与 M_m 的关系

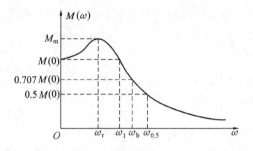
图 5.47 闭环系统幅值与频率的关系

5.7 开环频域分析

对于一个反馈控制系统,闭环频率响应主要取决于它的开环频率响应。尤其是对于单位反馈系统,闭环频率响应更是唯一地取决于它的开环频率响应。一个最使人感兴趣的问题是关于闭环系统的性能,开环频率响应告诉我们些什么。事实上,如果把全部感兴趣的频率分为所谓的低频段、中频段和高频段,那么开环频率响应的各个频段都有一些足以代表系统某些性能的特征,这种方法在反馈系统的分析和设计中是非常有用的。

5.7.1 低频段与稳态性能

对于特定输入测试信号的稳态误差与系统的型别(即开环积分环节数)以及开环增益有关。一般情况下,如果反馈系统的开环传递函数写成

$$G(s)H(s) = \frac{K\prod_{i=1}^{m}(\tau_i s + 1)}{s^v \prod_{j=v+1}^{n}(T_j s + 1)} \tag{5.92}$$

则系统的型别为 v,开环增益为 K。

通常低频段是指开环渐近对数幅频特性曲线第一个转折频率之前的频率范围。渐近对数幅频特性曲线的低频段将具有 $-20v$ db/dec 的斜率,低频段或其延长线在 $\omega = 1$ 处的高度为 $20\lg K$,或者在 $\omega = \sqrt[v]{K}$ 处通过 0 db 线。因此,系统的稳态性能可以由开环对数幅频特性图估算。

5.7.2 中频段与动态性能

中频段是指截至频率(即增益穿越频率)ω_c 附近的区域,它与控制系统的相对稳定性和响应速度密切相关。

1. 二阶系统

考虑单位反馈系统,开环传递函数为

$$G(s) = \frac{\omega_n^2}{s(s+2\zeta\omega_n)}, \quad 0 < \zeta < 1 \quad (5.93)$$

由例 5.16 截止频率为

$$\omega_c = \omega_n \sqrt{\sqrt{4\zeta^2+1} - 2\zeta^2} \quad (5.94)$$

且该系统的相角裕度为

$$\gamma = \arctan \frac{2\zeta}{\sqrt{\sqrt{4\zeta^2+1} - 2\zeta^2}} \quad (5.95)$$

由于系统对阶跃输入信号的超调为

$$\sigma_p = e^{-\pi\zeta/\sqrt{1-\zeta^2}} \times 100\% \quad (5.96)$$

根据表5.5的数据，超调量 σ_p 与相角裕度的关系画在图5.48中。可以看到，当相角裕度 γ 增大时，超调量 σ_p 将减小。

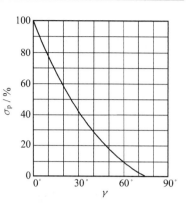

图 5.48 $\sigma_p - \gamma$ 关系图

表 5.5 $\sigma_p - r$

ζ	0	0.2	0.5	0.707	1
$\sigma_p/\%$	100	52.7	16.3	4.3	0
$\gamma/(°)$	0	22.6	51.8	65.5	76.3

继续考虑式(5.93)开环传递函数的欠阻尼二阶系统。在式(5.94)两侧同乘 t_s 得到

$$t_s \omega_c = \frac{3}{\zeta\omega_n} \cdot \omega_n \sqrt{\sqrt{1+4\zeta^4} - 2\zeta^2} = \frac{3\sqrt{\sqrt{1+4\zeta^4} - 2\zeta^2}}{\zeta} \quad (5.97)$$

该式表明，如果阻尼比 ζ 保持不变，即相角裕度 γ 不变，则阶跃响应的调整时间 t_s 将随 ω_c 增大而减小。

2. 高阶系统

要找出 σ_p 与 γ 以及 t_s 与 ω_c 的精确关系很困难。但是，二阶系统显示出的相对稳定性和响应速度关系的趋势与高阶系统的表现是一致的。通常，为了确保足够的相角裕度以满足关于超调的要求，中频段具有 -20 db/dec 的斜率以及足够的宽度是比较合适的。保持所需的相角裕度，通过选择合适的截止频率可以调节响应速度。

而且，如果单位反馈高阶系统的相角裕度为 $34° < \gamma < 90°$，对于阶跃输入信号的超调量和调整时间可以通过以下经验公式估算

$$\sigma\% = 0.16 + 0.4\left(\frac{1}{\sin\gamma} - 1\right) \quad (5.98)$$

$$t_s = \frac{\pi}{\omega_c}\left(2 + 1.5\left(\frac{1}{\sin\gamma} - 1\right) + 2.5\left(\frac{1}{\sin\gamma} - 1\right)^2\right), \quad \Delta = 5\% \quad (5.99)$$

5.7.3 高频段与扰动抑制

高频段是指截止频率 ω_c 十倍频程以上的频率范围。

通常，在较高频率部分系统频率响应的幅值比较小。因此，对于单位反馈高阶系统有

$$|\Phi(j\omega)| = \frac{|G(j\omega)|}{|1+G(j\omega)|} \approx |G(j\omega)| \quad (5.100)$$

该式指出，闭环系统抑制噪声的能力差不多等同于开环系统。

由于作用于控制系统的扰动通常幅值较小而频率较高,如果开环幅频特性在高频段提供较大的衰减,则闭环系统将具有较好的抑制扰动的能力。

习 题

题 5.1 某单位反馈系统的开环传递函数为

$$G(s) = \frac{5}{2s + 1}$$

确定闭环系统由以下输入信号引起的稳态输出:

(a) $r(t) = \sin(t + 30°)$ (b) $r(t) = 2\cos(2t - 45°)$

题 5.2 一个系统的单位阶跃响应为

$$c(t) = 1 - 1.8e^{-4t} + 0.8e^{-9t}, \quad t \geq 0$$

确定该系统的频率响应。

题 5.3 绘制以下传递函数的渐近对数幅频特性曲线和相频特性曲线:

(a) $G(s)H(s) = \dfrac{1}{(2s + 1)(0.5s + 1)}$ (b) $G(s)H(s) = \dfrac{0.5s + 1}{s^2}$

(c) $G(s)H(s) = \dfrac{10(s + 0.2)}{s^2(s + 0.1)}$ (d) $G(s)H(s) = \dfrac{10(s + 2)}{s^2(s + 5)}$

(e) $G(s)H(s) = \dfrac{8(s + 0.1)}{s(s^2 + s + 1)(s^2 + 4s + 25)}$

题 5.4 题 5.4 图给出了一些系统的渐近对数幅频特性曲线。确定各系统的传递函数,并绘制相应的渐近相频特性曲线。假设这些系统具有最小相位传递函数。

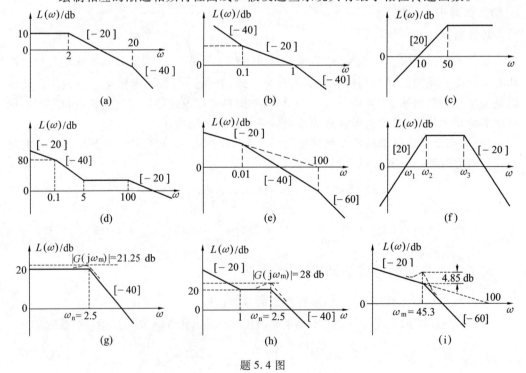

题 5.4 图

题 5.5　题 5.5 图显示了一些系统开环传递函数的极坐标图。确定闭环系统是否稳定。在各图中 p 是位于右半 s 平面内的开环极点数，v 是开环传递函数中的积分环节数。

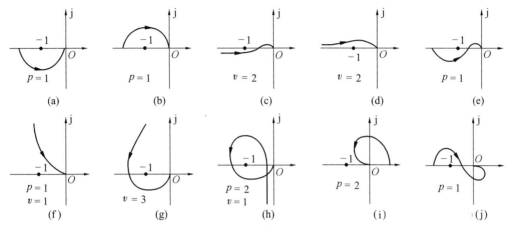

题 5.5 图

题 5.6　绘制以下开环传递函数的极坐标图。只要绘制确定闭环系统稳定性所必须的部分。应用奈奎斯特判据确定闭环系统的稳定性。

(a) $G(s)H(s) = \dfrac{10}{(0.2s+1)(0.5s+2)(s+1)}$

(b) $G(s)H(s) = \dfrac{100}{s(0.1s+2)(0.4s+1)}$

(c) $G(s)H(s) = \dfrac{100}{s(s^2+2s+2)(s+1)}$

(d) $G(s)H(s) = \dfrac{50}{s(s+2)(s^2+4)}$

(e) $G(s)H(s) = \dfrac{s}{1-0.2s}$

题 5.7　绘制以下开环传递函数的极坐标图，并应用奈奎斯特判据使系统稳定时开环增益的最大取值。

(a) $G(s)H(s) = \dfrac{k}{s(s^2+s+4)}$

(b) $G(s)H(s) = \dfrac{k(s+2)}{s^2(s+4)}$

题 5.8　一个负反馈系统具有开环传递函数

$$G(s) = \dfrac{k(s+0.5)}{s^2(s+2)(s+10)}$$

绘制具有渐近特性曲线的伯德图，并应用奈奎斯特判据确定 $k = 10$ 和 $k = 100$ 时系统是否稳定。

题 5.9　某单位反馈系统的装置为

$$G(s) = \frac{11.7}{s(0.05s+1)(0.1s+1)}$$

确定该系统的穿越频率和相角裕度。

题 5.10 某闭环系统具有的开环传递函数为

$$G(s) = \frac{Ke^{-\tau s}}{s}$$

(a) 确定当 $\tau = 0.2$ s 时使相角裕度为 $60°$ 的增益 K。

(b) 绘制 K 为(a)中取值时相角裕度与延迟时间 τ 的关系曲线。

题 5.11 某延时系统具有的开环传递函数为

$$G(s) = \frac{e^{-\tau s}}{s(s+1)}$$

确定保持系统稳定时延迟时间 τ 的取值。

题 5.12 某条件稳定系统对于特定的增益 $K = 50$ 时的极坐标图,如题 5.12 图所示。

(a) 确定该系统是否稳定,如果在右半 s 平面有闭环极点则确定在右半 s 平面有闭环极点数。假定开环系统具有最小相位。

(b) 确定使系统稳定时 K 的取值范围。

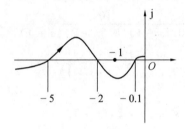

题 5.12 图

题 5.13 考虑某单位反馈系统,其开环传递函数为

$$G(s) = \frac{\tau s + 1}{s^2}$$

确定使系统的相角裕度为 $45°$ 时 τ 的取值范围。

题 5.14 单位反馈系统的开环传递函数为

$$G(s) = \frac{K}{(0.01s+1)^3}$$

(a) 确定使系统的相角裕度为 $45°$ 时 K 的取值范围。

(b) 确定对应于(a)中所得增益的增益裕度。

题 5.15 单位反馈系统的开环传递函数为

$$G(s) = \frac{k}{s(s^2+s+100)}$$

(a) 确定使系统的增益裕度为 20 db 时开环增益的取值。

(b) 确定对应于(a)中所得增益的相角裕度。

题 5.16 某最小相位系统得到渐近对数幅频特性曲线如题 5.16 图所示。估算该系统的相角裕度和增益裕度。

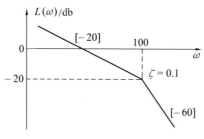

题 5.16 图

题 5.17 单位反馈系统的开环传递函数为

$$G(s) = \frac{1\,000}{s + 100}$$

确定开环系统和闭环系统的带宽,并将结果进行比较。

题 5.18 某单位反馈系统的开环传递函数为

$$G(s) = \frac{16}{s(s + 2)}$$

(a) 确定穿越频率 ω_c 和相角裕度 γ。
(b) 确定闭环系统的谐振频率 ω_r 和相对谐振峰值 M_r。

题 5.19 某单位反馈系统的渐近对数幅频特性曲线如题 5.19 图所示。假设系统具有最小相位。
(a) 确定开环传递函数。
(b) 确定该系统是否稳定。
(c) 如果幅频特性曲线向右平移十倍频程,对于性能指标 σ_p, t_s 和 e_{ss} 有何影响。

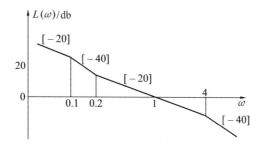

题 5.19 图

第6章　控制系统的校正

6.1　引　言

6.1.1　多重设计约束

反馈控制系统的性能是头等重要的问题,已经看到一个合适的控制系统应当具有以下性质:

① 系统应当是稳定的,并且对于输入指令呈现出可以接受的响应,即被控量应当能以适当的速度跟随输入的变化,并且没有过大的振荡或超调。

② 系统应当以尽可能小的误差运行。

③ 系统应当有能力减小不良扰动的影响。

事实上,一个不需要进行任何调整就能提供优良性能的反馈控制系统确实是很少见的。在期望的最佳性能指标不可能悉数获得的情况下,为了提供既恰当又可以接受的性能,通常有必要在众多要求苛刻且又相互冲突的性能指标之间做出折中的处理,并对系统的参数进行调整。

前面几章已经表明,为了提供期望的系统响应,经常有可能要调整系统的参数。当一个简单的性能要求可以通过选择特定的增益值得到满足时,这一过程称为增益补偿。但是经常发现仅仅调整系统的参数不足以获得所需的性能。为了获得一个合适的系统,更为经常的情况是需要重新考虑系统的结构并对系统重新进行设计。也就是说,必须检查系统的方案,并努力得到一个结果能产生合适系统的新设计。这样,一个控制系统的设计就涉及到系统结构的安排和合适的元部件及其参数的选择。当无法使几个性能要求得以协调时,就必须以某种方式改变系统。为了提供合适的性能而对控制系统进行的变动或调整称为校正。

在重新设计系统以改变系统的响应时,在反馈系统的内部放入附加的元器件或装置校正性能的缺陷。校正用的装置可以是电的、机械的、液压的、气动的,或者其他类型的装置或网络,并称为校正装置。一般情况下,在许多控制系统中电网络用作校正装置。

如果可能,在实际应用中改善控制系统性能的最好也是最简单的方法常常是改变被控对象本身。也就是说,如果系统的设计者能够指定和变动被控对象的设计方案,那么就有可能很容易地改善系统的性能。例如,要改善一个伺服机构位置控制装置的瞬态行为时,我们常常可以为该系统选择一台比较好的电机。因此,控制系统的设计人员应当认识到,变动被控对象的结果可能就是性能改善了的系统。当然,被控对象常常是不能变动的,或者已尽其可能作了变动但结果性能仍然不能令人满意。这时,加入校正装置对于改善系统的性能就变得非常有用了。

6.1.2 校正方式

图 6.1 为校正方式。校正装置可以有好几种方法放置在系统内适当的位置上。附加的元器件可以放置在图 6.1(a) 所示的前向通道内,这种方式称为串联校正。校正装置的传递函数记为 $G_c(s)$,而原有的装置或过程的传递函数则记为 $G_p(s)$。另一种可选择的情况如图 6.1(b) 所示,附加的元器件可以放置在反馈通道内,这种方式称为反馈校正。这两种方案的组合,即复合校正如图 6.1(c) 所示。校正方案的选择取决于性能指标的要求、系统内不同信号点上功率的大小,以及可供使用的校正装置。

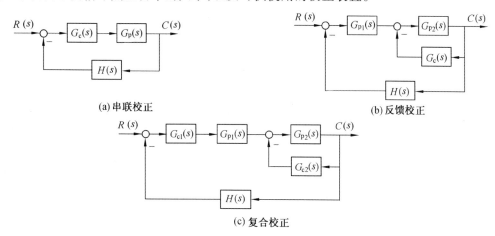

图 6.1 校正方式

6.1.3 串联校正

虽然有许多不同类型的校正装置可以使用,其中最简单的则是串联相位超前、串联相位滞后和串联滞后 - 超前网络。这些串联校正装置都可以用运算放大器电路实现。与其他的频率特性图相比,伯德图更适合用于确定合适的串联校正装置。在进行校正时,是把串联校正装置的频率特性加到未校正系统的频率特性上。

在以下的讨论中,假设图 6.1(a) 所示的校正装置 $G_c(s)$ 和未校正系统一起考虑,使得总的开环增益的设置可以满足稳态性能的要求。然后可以比较顺利地用 $G_c(s)$ 调节系统的动态特性而不影响稳态误差。为了方便起见,未校正系统的开环传递函数 $G_p(s)H(s)$ 记为 $G_o(s)$。

首先考虑一个系统,描述该系统的开环传递函数为

$$G_o(s) = \frac{K}{s(0.2s+1)} \tag{6.1}$$

假设希望闭环系统满足以下的性能要求:

① 单位斜坡引起的稳态误差不大于 0.003 16。
② 相角裕度不小于 45°。

对于第一个要求,静态速度误差系数可由以下公式算出

$$\varepsilon_{ss} = 1/K_v = 1/K \leq 0.003\ 16$$

由此,所需的开环增益为 $K = K_v = 316$。图 6.2 所示是 $K = 316$ 时的伯德图,稳态误差满足要求。由图可见,原系统的增益穿越频率 $\omega_{co} = 40$ rad/s。因此,经过估算相角裕度为 $\gamma_o \approx 7°$。

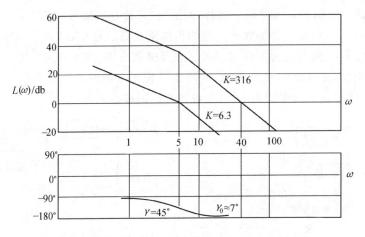

图 6.2 式(6.1)的伯德图

由图还可以看到,如果穿越频率在 $\omega = 5$ rad/s 处则相角裕度将是 45°。因此,为了满足第二个要求,在该频率处的幅值必须为零分贝。对于本例,这将发生在 $K = 6.3$ 的时候,图 6.2 中也显示了这一增益取值时的幅频特性曲线。显然,只靠单一的增益取值是不可能使系统的这两个性能要求都得到满足的。该系统需要以某种方法加以改进,即伯德图的形状不得不以某种方式加以变动,才能使伯德图同时达到这两个性能要求。

本例所述的系统性能要求在许多设计工作中都具有代表意义:稳态误差确定了一个增益值,同时期望的瞬态响应则需要另一个增益值。注意在伯德图中各个要求是怎样与频率的不同区间相联系的:

① 稳态误差与低频段的斜率和幅值有关。

② 相角裕度与通常发生在较高频率处的增益穿越频率有关。

为了使两个要求都能满足,可以做以下两件事情中的一件:

① 保持 $K = 316$ 使稳态误差要求得以满足,但引入另一个在 $\omega = 40$ rad/s 所在区域贡献正相角的环节。

② 保持 $K = 316$ 并保持相频特性曲线不变,但引入另一个使较高频率处幅值衰减的环节,使得修改后的幅频特性曲线在 $\omega = 5$ rad/s 处通过 0 分贝线。

在这两种情况下,所用的方法都涉及如图 6.1(a)中所示的在被校正系统的前向通道中加入校正装置,它们都属于串联校正。在第一种情况下,所采用的方法称为相位超前校正。第二种方法称为相位滞后校正。滞后 – 超前校正是这两种方法的结合。由于改变了系统的开环频率响应,获得所需性能的校正装置有时也称为滤波器。

6.1.4 系统设计的方法

一个控制系统的性能可以按照对于阶跃输入时最大超调量和调节时间的一定要求指

定。而且通常还需要指定对于一些测试信号输入和扰动可允许的最大稳态误差。这些性能指标与闭环传递函数的极点和零点位置有关。由此可以确定闭环传递函数的极点和零点。正如在第4章所看到的,可以很方便地获得系统的一个参数变化时闭环系统的根轨迹。但是,当根轨迹不能形成适当的特征根分布时,必须加入校正装置改变参数变化时的根轨迹。因此可以使用根轨迹法确定合适的校正装置的传递函数,使得最终的根轨迹具有期望的闭环特征根分布。

作为另一钟方法,一个控制系统的性能也可以按照闭环频率响应的相对谐振峰值、谐振频率和带宽,或者开环频率响应的相角裕度、幅值裕度和增益穿越频率指定。为了满足系统性能的要求,如果有必要可以加入适当的校正装置。校正装置的设计根据绘制在极坐标图、伯德图或者尼柯尔斯图上的频率响应曲线确定。由于串联的传递函数在伯德图上很容易解决加入校正装置频率响应的问题,在频率响应法中通常更愿意使用伯德图。

6.2 相位超前校正

6.2.1 相位超前校正装置

相位超前校正装置传递函数的形式为

$$G_c(s) = \frac{\alpha Ts + 1}{Ts + 1} \quad (6.2)$$

其中 $\alpha > 1$。该传递函数的伯德图见图6.3,图中

$$\omega_z = \frac{1}{\alpha T}, \quad \omega_p = \frac{1}{T}$$

由于频率轴上零点出现在先,渐近对数幅频特性曲线在 ω_z 和 ω_p 之间的斜率为 +20 db/dec。

显而易见,这一校正装置在输出和输入之间提供了超前相位,即

$$\varphi(\omega) = \angle G_c(j\omega) = \arctan \alpha T\omega - \arctan T\omega \quad (6.3)$$

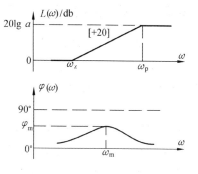

图6.3 相位超前校正装置的伯德图

因此,这种校正装置称为相位超前校正装置。

相位超前校正装置是一种高通滤波器,相对于低频信号而言,高频信号通过相位超前校正装置时得到放大。该装置在高频部分引入了增益,通常不利于稳定。但是,它的正相位角却是有利于稳定的。因此,我们必须小心地选择两个转折频率,使得正相位角的稳定效果起主导作用。

由式(6.3)很容易证明,发生最大相位超前时的频率为

$$\omega_m = \frac{1}{\sqrt{\alpha} T} \quad (6.4)$$

在伯德图中,该频率位于两个转折频率的几何中点。为了确定参数 α 以确保最大的相位超前角,将式(6.3)改写为

$$\varphi = \arctan\frac{\alpha T\omega - T\omega}{1 + \alpha T^2 \omega^2} \tag{6.5}$$

将频率 ω_m 代入上式,结果为

$$\tan\varphi_m = \frac{\alpha - 1}{2\sqrt{\alpha}} \tag{6.6}$$

或者

$$\sin\varphi_m = \frac{\alpha - 1}{\alpha + 1} \tag{6.7}$$

由此,相位角的最大值 φ_m 与参数 α 有关,而且参数 α 为

$$\alpha = \frac{1 + \sin\varphi_m}{1 - \sin\varphi_m} \tag{6.8}$$

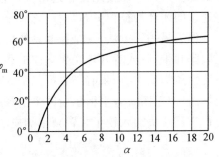

图 6.4 相位超前角与参数 α 的关系

式(6.4)和式(6.8)是用于校正过程时相位超前校正环节的基本方程。需要快速估计相位超前所需 α 的大小时,可以用到图 6.4 所示的关系。

6.2.2 相位超前校正的步骤

首先绘制 $G_o(j\omega)$ 的伯德图。绘制未校正系统伯德图所用的增益是能使稳态误差可以接受所需的增益。然后检验相角裕度,看它是否满足性能指标的要求。如果相角裕度不够,把 $G_c(j\omega)$ 放置在适当的位置上,就可以使相位超前增加到系统的相频特性曲线中。

为了获得最大的附加相位超前,希望放置校正装置时频率 ω_m 正好位于校正后幅频特性曲线穿越 0 分贝轴的频率处,即位于新的截至频率处。由式(6.8),可以根据所需新增相位超前的大小确定必需的 α 的数值。注意到最大的相位超前应当产生于两个转折频率中间位置的 ω_m 处,就可以确定转折频率 $\omega_z = 1/(\alpha T)$ 的位置。因为该校正装置总的幅值为 $20\lg\alpha(db)$,可以预期在 ω_m 处的幅值为 $10\lg\alpha(db)$。确定校正装置传递函数的步骤将通过一些例子详细说明。

例 6.1 仍然考虑系统的开环传递函数为

$$G_o(s) = \frac{K}{s(0.2s + 1)}$$

两个性能指标是:(a) 对于单位斜坡输入的稳态误差为 $\varepsilon_{ss} \le 0.136\%$;(b) 相角裕度为 $\gamma \ge 45°$。

解 正如已计算过的,如果增益 $K = K_v = 316s^{-1}$,则稳态性能要求可以得到满足。其余的步骤如下:

① 绘制未校正系统的伯德图。$K = 316s^{-1}$ 时绘制的 $G_o(j\omega)$ 的伯德图,如图 6.5 所示。

② 估算未校正系统的相角裕度 γ_o。由幅频特性图,未校正系统的穿越频率 ω_{co} 为

40 rad/s。因此未校正系统的相角裕度为

$$\gamma_o = 180° + \angle G_o(j\omega_{co}) = 180° - 90° - \arctan(0.2 \times 40) = 7°$$

一般情况下,确定相角裕度时用公式估算常常比绘制完整的相频特性曲线容易一些。

③ 确定必须增加的相位超前 φ_m。需要加入一个相位超前校正装置,使得新的穿越频率处相角裕度能够增加到 45°。由于新的穿越频率大于未校正系统的穿越频率,未校正系统在新的穿越频率处的相位也将滞后得更多一些。我们选择最大的相位超前为 45°-7°=38° 再加上考虑到附加相位滞后所需的一个相位超前增量(大约5°)。因此,必须增加的相位超前大约为

$$\varphi_m = 38° + 5° = 43°$$

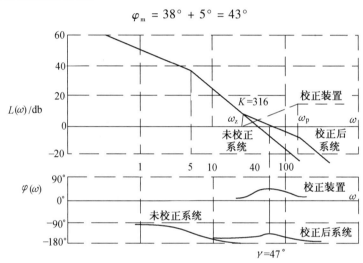

图 6.5 例 6.1 的伯德图

④ 估算参数 α。由式(6.8),有

$$\alpha = \frac{1 + \sin\varphi_m}{1 - \sin\varphi_m} = \frac{1 + \sin 43°}{1 - \sin 43°} = 5.3$$

⑤ 确定期望的穿越频率 ω_c。最大的相位超前发生在 ω_m,而且这一频率将选作新的穿越频率。超前校正装置在 ω_m 处的幅值为 $10\lg\alpha = 10\lg 5.3 = 7.2$ db。由此,$G_o(j\omega)$ 在校正后的穿越频率处幅值为

$$-10\lg\alpha = -7.2 \text{ db}$$

因此

$$\omega_m = \omega_c = 62 \text{ rad/s}$$

⑥ 确定 $G_c(s)$ 的传递函数。由式(6.4)可有

$$\omega_p = \frac{1}{T} = 143, \quad \omega_z = \frac{1}{\alpha T} = 27$$

因此,相位超前校正装置的传递函数为

$$G_c(s) = \frac{0.037s + 1}{0.007s + 1}$$

⑦ 绘制校正后的频率响应曲线。$G_c(j\omega)$ 和 $G_c(j\omega)G_o(j\omega)$ 的伯德图显示在图 6.5

中。

⑧ 检验结果所得的相角裕度 γ。最终的相角裕度为

$$\gamma = 180° + \angle G_o(j\omega_c) + \angle G_c(j\omega_c) = 180° - 175.4° + 42.9° = 47.5°$$

现在两个性能要求都得到了满足。

例 6.2 考虑开环传递函数为

$$G_o(s) = \frac{K}{s(0.1s+1)(0.01s+1)}$$

的系统,希望稳态误差系数 $K_v \geq 100s^{-1}$,相角裕度 $\gamma \geq 30°$,而且穿越频率 $\omega_c \geq 45$ rad/s。

解 由于该系统为 1 型,因此选择

$$K = K_v = 100s^{-1}$$

其余步骤如下:

① 如图 6.6 所示,以 $K = 100s^{-1}$ 绘制未校正系统的伯德图。

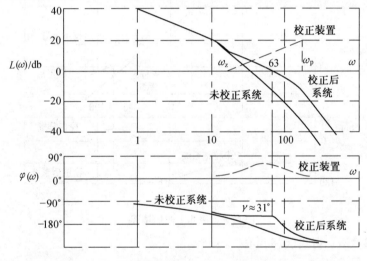

图 6.6 例 6.2 的伯德图

② 在本例中,校正后的穿越频率应当满足响应速度的要求。如果选择 $\omega_c = 50$ rad/s,则 $G_o(j\omega)$ 在该频率出的相角为

$$\angle G_o(j\omega)|_{\omega=50} = -90° - \arctan(0.1 \times 50) - \arctan(0.01 \times 50) = -195°$$

③ 因为希望有相角裕度 $\gamma \geq 30°$,所需的超前相角至少为

$$30° + (180° - 195°) = 45°$$

在留有余地的情况下,选择由校正装置提供的超前相角为 $\varphi_m = 55°$。

④ 由式(6.8)可有

$$\alpha = \frac{1 + \sin 55°}{1 - \sin 55°} = 10$$

由式(6.4)可得,两个转折频率为

$$\omega_z = \frac{1}{\alpha T} = 15.8$$

$$\omega_p = \frac{1}{T} = 158$$

因此,校正装置的传递函数为

$$G_c(s) = \frac{\alpha Ts + 1}{Ts + 1} = \frac{0.063s + 1}{0.0063s + 1}$$

⑤ 校正装置和校正后系统的伯德图如图 6.6 所示。
⑥ 由图 6.3 有

$$20\lg|G_c(\omega_m)| = 10\lg\alpha = 10 \text{ db}$$

而由图 6.6 则有

$$L_o(\omega_m) = 20\lg G_o|(j\omega)|_{\omega=\omega_m} = -8 \text{ db}$$

这样校正后实际的穿越频率大于 $\omega_m = 50$,由图 6.5 可看到合适的校正后穿越频率为

$$\omega'_c = 63 \text{ rad/s}$$

因此,最终的相角裕度为

$$\gamma = 180° + \angle G_o(j\omega'_c) + \angle G_c(j\omega'_c) = 180° - 203.2° + 54.3° = 31.1°$$

最终的相角裕度大于相角裕度的期望值。

6.2.3 评　　述

相位超前校正具有一些其他形式校正所不具备的优点,同时使用时也可能有些麻烦。通过刚刚分析过的例子的观察,对相位超前校正可以归纳出一些结论：
① 相位超前校正提供了超前相位,能使系统的超调量限制在希望的范围内。
② 开环(以及一般情况下闭环)带宽增大。通常这是有利的,因为系统的响应包括较高的频率导致较快的响应速度。但是,如果在较高频率处存在噪声,这也可能带来麻烦。
③ 当 φ_m 附近相频特性曲线比较陡峭时有可能发生麻烦。这是由于新的增益穿越点向右移动时,要求校正装置提供越来越大的相位超前,因此需要很大的 α 值。当校正装置用物理元部件实现时就很难做到。由于这一原因,应当避免 α 的取值大于 15,并可以考虑采用如相位滞后之类其他的校正方法对系统进行校正。

6.3　相位滞后校正

6.3.1　相位滞后校正装置

加入一个具有相位滞后特性的串联校正装置是一种常用的校正方法。相位滞后校正装置的传递函数为

$$G_c(s) = \frac{\beta Ts + 1}{Ts + 1}, \quad \beta < 1 \quad (6.9)$$

这一传递函数的伯德图见图 6.7,图中

$$\omega_p = \frac{1}{T}, \quad \omega_z = \frac{1}{\beta T}$$

图 6.7 表明该校正装置在输出和输入之间

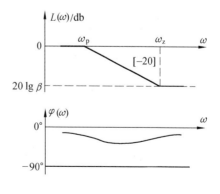

图 6.7　相位滞后校正装置的伯德图

具有相位滞后的特性。因此称为相位滞后校正装置。

但是,认识到这一点上很重要的,相位滞后并不是校正装置有用的作用,而且伯德图相频部分具体的特性在校正装置的全部设计中也没有起重要作用。起重要作用的是校正装置提供的幅频特性衰减量。当 $\omega > \omega_z$ 时,衰减的大小为

$$L_c(\omega_z) = 20\lg \beta \text{ db} \tag{6.10}$$

6.3.2 相位滞后校正的过程

在相位滞后校正中,未校正系统伯德图幅频特性被衰减以使增益穿越频率降低,从而使未校正系统的相频特性图上能产生必需的相位裕度。相位滞后校正装置用于提供幅频特性的衰减,并由此使系统的穿越频率降低。而且通常可以发现,在较低的穿越频率处系统的相位裕度增加,性能指标也就能够得到满足。

当然,由校正装置引起的相位滞后的影响也应当予以考虑。一般情况下,如果对应于校正装置零点的转折频率为 $\omega_z \approx (0.1 \sim 0.2)\omega_c$ 时,相位的滞后大约为 $5° \sim 12°$。

校正的步骤将通过以下的例子详细说明。

例6.3 重新考虑例6.2的系统,设计一个相位滞后校正装置。未校正系统的传递函数为

$$G_o(s) = \frac{K}{s(0.2s + 1)}$$

要求稳态误差系数 $K_v = 316$,同时获得 $45°$ 的相位裕度。

解 ① 绘制未校正系统的伯德图。增益调整到要求的误差系数 $K_v = 316$ 时,未校正系统的伯德图,如图6.8所示。

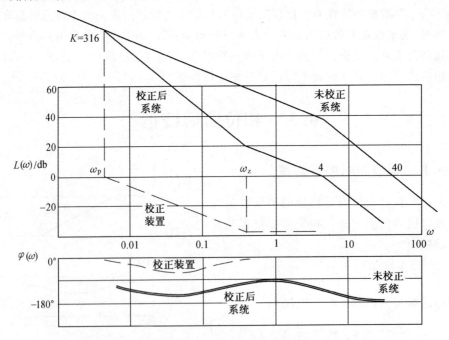

图6.8 加入相位滞后校正的伯德图

② 确定未校正系统的相位裕度。由例6.1 的伯德图,未校正系统的穿越频率为

ω_{co} =40 rad/s,而未校正系统的相位裕度则为 $\gamma_o \approx 7°$。

③ 确定新的穿越频率 ω_c。若允许校正装置有 5° 的相位滞后,在新的穿越频率处有

$$\angle G_o(j\omega) = -180° + 45° + 5° = -130°$$

由相频特性曲线可以发现,如果幅频特性曲线在 $\omega_c = 4$ rad/s 处与 0 db 线相交,则相位裕度的要求可以得到满足。

④ 确定参数 β。由幅频特性图,使 $\omega_c = 4$ rad/s 成为新的穿越频率所必需的衰减等于 38 db。因此可有 $20\lg\beta = -38$ db,即 $\beta = 0.0125$。

⑤ 确定 $G_c(s)$。由图 6.8 可见,转折频率 ω_z 低于新的穿越频率十倍频程时可以确保在 ω_c 处由相位滞后校正装置引起的附加相位滞后只有 5°,因此有

$$\omega_z = 0.1\omega_c = 0.4 \text{ rad/s}$$
$$\omega_p = \beta\omega_z = 0.005 \text{ rad/s}$$

校正装置的传递函数则为

$$G_c(s) = \frac{2.5s + 1}{200s + 1}$$

⑥ 绘制校正后系统的频率响应曲线。

⑦ 检验结果所得的相位裕度。结果所得的相位裕度为

$$\gamma = 180° + \angle G_o(j\omega_c) + \angle G_c(j\omega_c) = 180° - 128.7° - 5.7° = 45.6°$$

这是一个令人满意的结果。

例 6.4 描述系统的开环传递函数是

$$G_o(s) = \frac{K}{s(s+1)(0.5s+1)}$$

为该系统设计一个串联校正装置,使得静态误差系数 $K_v \geq 5s^{-1}$,相位裕度 $\gamma \geq 40°$,而且增益裕度 $L(K_g) \geq 10$ db。

解 为了满足稳态性能的要求,选择 $K = 5$,其余的步骤如下:

① 绘制未校正系统的伯德图。未校正系统的伯德图,如图 6.9 所示。

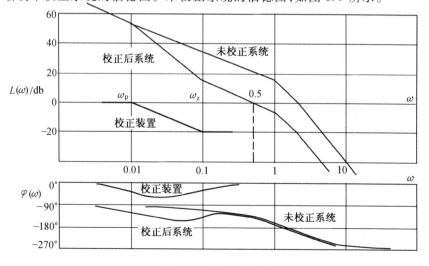

图 6.9 例 6.4 的伯德图

② 分析未校正系统。伯德图表明，系统的穿越频率为 $\omega_{co} \approx 2 \text{ rad/s}$，相位裕度为 $\gamma_o \approx -18°$，而增益裕度则为 $L(K_{go}) \approx -7 \text{ db}$。未校正系统不稳定。由于相频特性曲线在 ω_{co} 附近具有较陡的斜率，可能相位滞后校正较为合适。

③ 确定新的穿越频率 ω_c。考虑到 βT 取值的物理实现以及允许校正装置 12° 的相位滞后，放置滞后校正装置时使新的穿越频率处有

$$\angle G_o(j\omega) = -180° + 40° + 12° = -128°$$

由相频特性图可有

$$\omega_c = 0.5 \text{ rad/s}$$

④ 确定转折频率。由于已取 12° 的安全裕度，转折频率

$$\omega_z = \frac{1}{\beta T} = 0.2\omega_c = 0.1 \text{ rad/s}$$

由幅频特性图，使 $\omega_c = 0.5$ 成为新的穿越频率所需的衰减等于 20 db。因而有 $20 \lg \beta = -20 \text{ db}$，即 $\beta = 0.1$。

这样，另一个转折频率为

$$\omega_p = \frac{1}{T} = 0.01 \text{ rad/s}$$

⑤ 确定 $G_c(s)$。校正装置的传递函数为

$$G_c(s) = \frac{10s + 1}{100s + 1}$$

⑥ 绘制校正后系统的频率特性曲线。

⑦ 检验校正后系统的特性。由伯德图可知，对于校正后系统有 $K_v \geq 5s^{-1}$，$\omega_c = 0.5 \text{ rad/s}$，$\gamma = 40°$ 和 $L(K_g) = 11 \text{db}$。所有设计指标都已满足。

6.3.3 评述

① 相位滞后校正提供必要的阻尼比以使超调限制在要求的范围内。

② 与相位超前校正相比，转折频率的选择不是十分苛刻，因此校正过程比较简单。

③ 由校正后系统可见，相位滞后法减小了开环系统的带宽，并因此也减小了闭环系统的带宽，从而导致响应速度变慢。

④ 与相位超前校正不同，理论上相位滞后校正可以使相位裕度的变化超过 90°。

6.4 滞后-超前校正

在有些情况下，与单独使用超前校正或滞后校正相比较，滞后-超前组合在一起的校正方法可以允许更多的性能要求得到满足。

6.4.1 滞后-超前校正装置

滞后-超前校正装置的传递函数为

$$G_c(s) = \frac{(\beta T_1 s + 1)(\alpha T_2 s + 1)}{(T_1 s + 1)(T_2 s + 1)}, \quad \beta < 1, \quad \alpha > 1 \tag{6.11}$$

在校正装置的设计中,通常认为滞后部分的两个转折频率比超前部分的两个转折频率低。滞后 – 超前校正装置的伯德图如图 6.10 所示。注意图中校正装置的两个部分是如何分开的。伯德图还包括以下的其他特点:

① 低频部分的幅值为 0 db,而高频部分的幅值则为 $20\lg(\alpha\beta)$ db。

② 相位是滞后在先超前在后,但高频和低频部分的相位均为零。

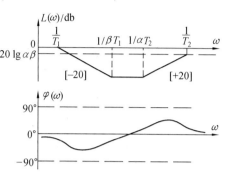

图 6.10 滞后 – 超前校正装置的伯德图

③ 最大的相位滞后和最大的相位超前发生在它们各自的两个转折频率之间。

滞后 – 超前校正装置利用滞后和超前部分各自的长处,通常还没有它们的缺点。例如,滞后 – 超前校正允许引入相位超前使系统稳定,同时在高频提供幅值衰减以滤除噪声。

6.4.2 滞后 – 超前校正的步骤

滞后 – 超前校正的步骤用以下的例子说明。

例 6.5 系统的开环传递函数为

$$G_o(s) = \frac{K}{s(0.1s + 1)(0.01s + 1)}$$

设计一个串联校正装置,使得闭环系统的静态速度误差系数 $K_v \geq 100 s^{-1}$,相位穿越频率 $\omega_c = 20 s^{-1}$,相角裕度 $\gamma \geq 40°$。

解 满足稳态误差要求的开环增益为 $K = K_v = 100$,这时绘制的未校正系统伯德图如图 6.11 所示。

由伯德图可知,未校正系统的穿越频率为 $\omega_{co} = 30$ rad/s,而未校正系统的相角裕度则为 $\gamma_o \approx 2°$。这告诉我们需要某种形式的校正。由于期望的穿越频率低于未校正系统的穿越频率,因此需要相位滞后校正。在 $\omega = 20$ rad/s 处可有

$$\angle G_o(j\omega) = -90° - 63.4° - 11.3° \approx -165°$$

显然,为了确保相角裕度的要求,还需要相位超前校正。

滞后部分考虑 5° 的补偿,在 $\omega_m = \omega_c = 20$ rad/s 处要增加的最大相位超前至少为

$$\varphi_m = 40° - (180° - 165°) + 5° \approx 30°$$

由式(6.8)

$$\alpha \geq \frac{1 + \sin 30°}{1 - \sin 30°} = 3$$

超前部分的两个转折频率应当满足

$$\omega_3 = \frac{1}{\alpha T_2} \leqslant 11.5 \text{ rad/s}$$

$$\omega_4 = \frac{1}{T_2} \geqslant 35 \text{ rad/s}$$

考虑到未校正系统在 $w = 10$ rad/s 处的开环极点,取 $\omega_3 = 10$ rad/s 和 $\omega_4 = 35$ rad/s。在 $\omega_c = 20$ rad/s 处进行估测,可以看到需要 8 db 的衰减,即 $20 \lg |G_c(j\omega)|_{\omega=20} = -8$ db。由此,超前部分的伯德图如图 6.11 所示,在 ω_3 和 ω_4 之间幅频特性斜率为 + 20 db/dec。

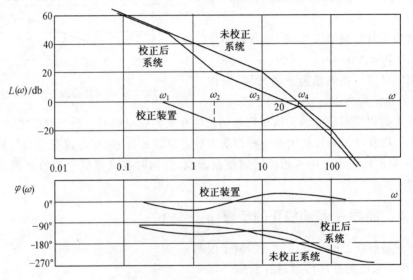

图 6.11 例 6.5 的伯德图

至于滞后部分,也考虑取 5° 的补偿,转折频率 ω_2 定位于

$$\omega_2 = \frac{1}{\beta T_1} = 0.1\omega_c = 2 \text{ rad/s}$$

通过绘制超前部分的幅频特性图可有

$$\omega_1 = \frac{1}{T_1} = 0.35 \text{ rad/s}$$

因此,滞后 - 超前校正装置的传递函数为

$$G_c(s) = \frac{0.5s + 1}{2.86s + 1} \cdot \frac{0.1s + 1}{0.029s + 1}$$

校正后系统的伯德图见图 6.11。最终的相角裕度为

$$\gamma = 180° - 165° + \angle G_c(j\omega_c) = 15° - 4.7° + 33.3° = 43.6°$$

要求的所有性能指标都得到了满足。

由于滞后 - 超前校正法通常允许较多的性能要求,在校正完成之前并不知道系统是否一定能校正。在有些情况下,为了应用系统的设计步骤,为规定的性能要求分配优先顺序并确保设计满足最重要的一些性能要求是很有用的。

6.5 比例-积分-微分(PID)调节器

上一节研究了滞后-超前校正。这种校正在滞后校正后面紧接着超前校正,因此它是二阶的。正如在上一节看到的,比起单独的滞后校正或者超前校正,滞后-超前校正的灵活性要高的多。到目前为止的讨论中,要进行控制的装置都是完全已知的。在实际应用中情况并不总是这样。但是通过引入如图 6.12 所示的比例-积分-微分(PID)调节器,即 PID 校正装置,闭环系统仍然可以获得良好的性能。事实上,比例-积分-微分(PID)调节器有可能是反馈系统中使用最广泛的校正装置。

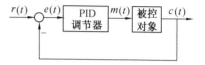

图 6.12 带有 PID 控制器的系统

以误差信号 $e(t)$ 作为调节器的输入,$m(t)$ 作为输出,则比例-积分-微分(PID)调节器可定义为

$$m(t) = K_P e(t) + K_I \int_0^t e(\tau)\mathrm{d}\tau + K_D \frac{\mathrm{d}e(t)}{\mathrm{d}t} \tag{6.12}$$

调节器的传递函数则为

$$G_c(s) = \frac{M(s)}{E(s)} = K_P + \frac{K_I}{s} + K_D s \tag{6.13}$$

被控对象的输入包含三个分量:
① $K_P E$—— 正比于误差信号;
② $K_I E/s$—— 与误差的积分成正比;
③ $K_D s E$—— 与误差的微分成正比。

对于特定的控制系统,要满足设计指标常常并不需要实现所有的三个分量。

如果被控对象不能很精确地知道,经常可由通过试探的方法确定参数 K_P、K_I 和 K_D 的数值。如果被控对象的参数有比较大的变化,可以调节 PID 调节器的参数改善系统的性能。

6.5.1 比例(P)调节器

对于比例调节器,式(6.13)中只有比例增益参数 K_P 不是零。因此,调节器的传递函数为

$$G_c(s) = K_P \tag{6.14}$$

调节器是一个纯增益。

在无需动态校正而只要调节系统内的增益就能获得令人满意的响应的场合下可以考虑采用这种校正装置

6.5.2 比例-积分(PI)调节器

比例-积分(PI)调节器的传递函数为

$$G_c(s) = K_P + \frac{K_I}{s} = \frac{K_P s + K_I}{s} \tag{6.15}$$

它能使系统的型别增加1,用于改善系统的稳态性能。

这种调节器具有一个在原点的极点和一个在负实轴上的零点,传递函数可以表达成

$$G_c(s) = \frac{K_I(\tau s + 1)}{s} \tag{6.16}$$

式中,$\tau = K_P/K_I$。可以看到零点位于 $s = -K_I/K_P$。比例 - 积分(PI)调节器的伯德图见图6.13。

显然,由相频特性图可以看到比例 - 积分(PI)控制器是相位滞后的。如果图6.7中相位滞后校正装置的高频部分增益和零点保持不变,而极点则移到平面的原点(伯德图上 $\lg\omega$ 轴的 $-\infty$),结果就是图6.13的比例 - 积分(PI)特性。

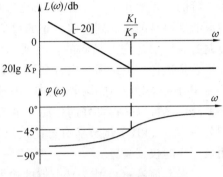

图6.13 PI调节器的伯德图

6.5.3 比例 - 微分(PD)调节器

比例 - 微分(PD)调节器的传递函数为

$$G_c(s) = K_P + K_D s = K_P\left(\frac{K_D}{K_P}s + 1\right) \tag{6.17}$$

因此,$G_c(s)$ 具有一个位于 $s = -K_P/K_D$ 的零点。比例 - 微分(PD)调节器的伯德图见图6.14。

可以看到,这种调节器具有正的相位。因此,比例 - 微分(PD)调节器是一种相位超前校正装置,可用于改善系统的瞬态响应。设计可以应用6.2节中给出的步骤。

在实际应用中,有时候会遇到在所有物理系统中总是存在的噪声问题。如果一个信号随衰减快速变化,那么它的微分就很大。显然,通过比例 - 微分(PD)调节器的高频噪声将被放

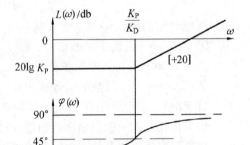

图6.14 PD调节器的伯德图

大。而且,频率越高放大作用越明显。为了防止高频噪声带来的问题,通常有必要在比例 - 微分(PD)调节器的传递函数中加入一个极点,使得传递函数变成

$$G_c(s) = \frac{K_P + K_D s}{s - p_{add}} = K_P\frac{(K_D/K_P)s + 1}{s - p_{add}} \tag{6.18}$$

这样高频部分的增益就限制为 K_D 的数值。校正装置中的极点选择得大于零点的模值,以使校正装置仍然是相位超前的。现在这一校正装置就是一个6.2节中的那种相位超前校正装置。

6.5.4 比例 - 积分 - 微分(PID)调节器

比例 - 积分 - 微分(PID)调节器用于瞬态响应和稳态响应都需要改善的控制系统。

它的传递函数为

$$G_c(s) = K_P + \frac{K_I}{s} + K_D s = \frac{K_D s^2 + K_P s + K_I}{s} \tag{6.19}$$

如前所述,积分分量是相位滞后的,而微分分量则是相位超前的。因此,积分分量在低频部分做出贡献,而微分分量则在高频部分做出贡献。比例-积分-微分(PID)调节器的伯德图见图6.15。

注意图6.15中比例-积分-微分(PID)调节器的伯德图与图6.10中滞后-超前校正装置的伯德图之间的相似之处。可以看到,比例-积分-微分(PID)调节器是一种滞后-超前校正装置。但是比例-积分-微分(PID)调节器要选择的参数只有三个。如前所述,微分部分没有限制的高频增益会造成麻烦,因此通常在微分通道加入一个极点以限制这一增益。在这种情况下,调节器的传递函数为

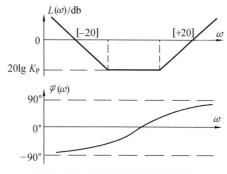

图 6.15 PID 调节器的伯德图

$$G_c(s) = K_P + \frac{K_I}{s} + \frac{K_D s}{s - s_{add}} \tag{6.20}$$

而且,在设计过程中有四个参数必须要确定。

6.6 反馈校正

为了改善系统的性能,除了串联校正外反馈校正也经常用作另一种解决稳态的方案。利用局部反馈校正,可以获得与串联校正差不多相同的效果。而且,还可以获得其他改善系统性能的特殊作用。

6.6.1 反馈校正的作用

局部回路反馈校正的应用如图6.16所示,图中 $G_{p1}(s)$ 是装置的一个部分, $G_{p2}(s)$ 是装置的另一部分, $G_c(s)$ 则是反馈校正装置。包括 $G_{p2}(s)$ 和 $G_c(s)$ 在内的回路称为局部回路,由 $G_{p1}(s)$ 和局部回路串联而组成的闭合回路则称为主回路或外回路。

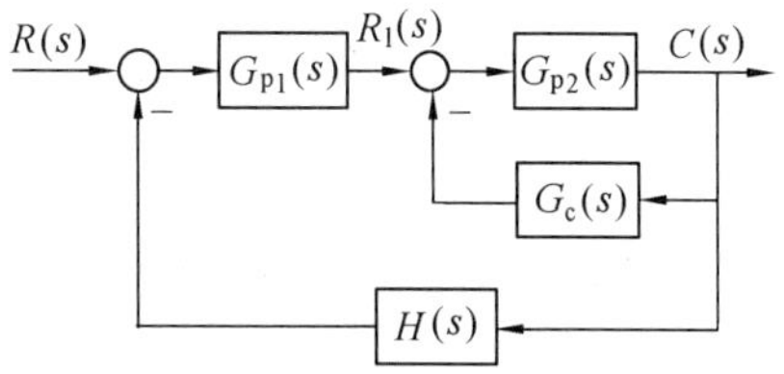

1. 位置反馈

设图6.16中 $G_{p2}(s)$ 的传递函数为

图 6.16 反馈校正

$$G_{p2}(s) = \frac{K_2}{Ts + 1} \tag{6.21}$$

而且局部回路由一个具有比例特性的校正装置组成,即

$$G_c(s) = K_P \tag{6.22}$$

由此,局部闭环的传递函数为

$$\frac{C(s)}{R_1(s)} = \frac{\dfrac{K_2}{Ts+1}}{1+\dfrac{K_2 K_\mathrm{P}}{Ts+1}} = \frac{K'_2}{T's+1} \tag{6.23}$$

其中

$$K' = \frac{K_2}{1+K_2 K_\mathrm{P}}, \quad T' = \frac{T}{1+K_2 K_\mathrm{P}}$$

可以看到,时间常数和增益两者都减小了。而且比例增益越大,负反馈的作用就越强。时间常数的减小有利于加快系统的响应速度。至于增益的减小,通常是不能接受的,但可以在 $G_{\mathrm{p1}}(s)$ 中加入一个适当的放大器予以补偿。

2. 速度反馈

设图 6.16 中的 $G_{\mathrm{p2}}(s)$ 由功率放大器和伺服机构的执行电机组成,$C(s)$ 是一个角位移信号,由此可得

$$G_{\mathrm{p2}}(s) = \frac{K_2}{s(T_\mathrm{m} s+1)} \tag{6.24}$$

考虑 $G_\mathrm{c}(s)$ 是一个测速发电机,传递函数为

$$G_\mathrm{c}(s) = K_\mathrm{t} s \tag{6.25}$$

因此局部闭环的传递函数为

$$\frac{C(s)}{R_1(s)} = \frac{K_2}{1+K_2 K_\mathrm{t}} \cdot \frac{1}{s\left(\dfrac{T_\mathrm{m}}{1+K_2 K_\mathrm{t}}+1\right)} \tag{6.26}$$

使 $K_2 K_\mathrm{t} \gg 1$,则结果可有得到一个简化的方程

$$\frac{C(s)}{R_1(s)} = \frac{1}{K_\mathrm{t} s\left(\dfrac{T_\mathrm{m}}{1+K_2 K_\mathrm{t}}+1\right)} \tag{6.27}$$

与式(6.24)相比可以看到,局部反馈的效果等价于在 $G_{\mathrm{p2}}(s)$ 之前串联加入传递函数

$$G_{\mathrm{eq}} = \frac{1}{K_2 K_\mathrm{t}} \cdot \frac{T_\mathrm{m} s+1}{\dfrac{T_\mathrm{m}}{K_2 K_\mathrm{t}} s+1} = \frac{1}{\alpha} \cdot \frac{\alpha T s+1}{T s+1} \tag{6.28}$$

速度反馈的效果等价于使用相位超前校正装置,同时开环增益减小为原来的 $K_2 K_\mathrm{t}$ 倍。

3. 速度微分反馈

仍然考虑上面的例子。引入一个与测速发电机输出串联的微分网络形成速度微分反馈。假设微分网络的传递函数为 $\dfrac{T_\mathrm{d} s}{T_\mathrm{d} s+1}$,则有

$$G_\mathrm{c}(s) = \frac{K_\mathrm{t} T_\mathrm{d} s^2}{T_\mathrm{d} s+1} \tag{6.29}$$

通过简单的推导,可以获得局部闭环的传递函数为

$$\frac{C(s)}{R_1(s)} = \frac{K_2(T_d s + 1)}{s[(T_m T_d s^2 + (T_m + K_2 K_t T_d + T_d)s + 1)]} \tag{6.30}$$

与式(6.24)相比可以看到,局部反馈的效果等价于在 $G_{p2}(s)$ 之前串联加入传递函数

$$G_{eq} = \frac{(T_m s + 1)(T_d s + 1)}{T_m T_d s^2 + (T_m + K_2 K_t T_d + T_d)s + 1} \tag{6.31}$$

使 $K_2 K_t \gg 1$ 且 $K_2 K_t T_d \gg T_m$,式(6.31)右侧的分母可以近似为

$$T_m T_d s^2 + \left(\frac{T_m}{K_2 K_t} + K_2 K_t T_d\right)s + 1$$

或者

$$\left(\frac{T_m}{K_2 K_t}s + 1\right)(K_2 K_t T_d s + 1)$$

因此,式(6.31)可以近似为

$$G_{eq} = \frac{(T_m s + 1)(T_d s + 1)}{\left(\frac{T_m}{K_2 K_t}s + 1\right)(K_2 K_t T_d s + 1)} \tag{6.32}$$

这是一个滞后 - 超前校正装置。

再把式(6.30)与式(6.24)比较很容易可以看到,内回路的回路增益与采用速度微分反馈之前是相同的。可以证明,如果反馈校正装置中的纯微分因子数大于被包围部分 $G_{p2}(s)$ 中的纯微分因子数,引入反馈校正装置时原系统的开环增益将不会改变。

4. 对系统参数变化灵敏度的影响

反馈系统的一个重要优点是装置参数变化的影响减小了。

考虑图 6.16 所示系统,并设

$$G_{p2}(s) = \frac{K_2}{Ts + 1}, \quad G_c(s) = K_P$$

在没有位置反馈的情况下,如果传递系数从 K_2 变化到 $K_2 + \Delta K_2$,相对增量为 $\Delta K_2 / K_2$。在有位置反馈的情况下,传递系数为

$$K'_2 = \frac{K_2}{1 + K_2 K_P}$$

由于参数变化而导致的增量为

$$\Delta K'_2 = \frac{\partial K'_2}{\partial K_2} \Delta K_2 = \frac{\Delta K_2}{(1 + K_2 K_P)^2}$$

因此,相对增量变为

$$\frac{\Delta K'_2}{K'_2} = \frac{1}{1 + K_2 K_P} \cdot \frac{\Delta K_2}{K_2} \tag{6.33}$$

该式表明,通过加入位置反馈传递系数的相对增量减小为原来的 $(1 + K_2 K_P)$ 之一。

6.6.2 反馈校正

当采用如图 6.16 所示反馈校正时,方框图可以改画为如图 6.17 所示,则

$$G_d(s) = \frac{G_{p2}(s)}{1 + G_{p2}(s)G_c(s)} \quad (6.34)$$

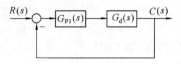

图 6.17 图 6.16 方框图的等效表示

为了应用反馈校正,必须研究新的校正方法。首先需要一些对数幅频特性曲线的直线近似表示,然后研究精确的校正过程。当 $|G_{p2}(j\omega)G_c(j\omega)| \ll 1$ 时,式(6.34)的传递函数可以近似为

$$G_d(j\omega) = \frac{G_{p2}(j\omega)}{1 + G_{p2}(j\omega)G_c(j\omega)} \approx G_{p2}(j\omega) \quad (6.35)$$

当 $|G_{p2}(j\omega)G_c(j\omega)| \gg 1$ 时,则有

$$G_d(j\omega) = \frac{G_{p2}(j\omega)}{1 + G_{p2}(j\omega)G_c(j\omega)} \approx \frac{1}{G_c(j\omega)} \quad (6.36)$$

$|G_{p2}(j\omega)G_c(j\omega)| \approx 1$ 的情况没有规定,这时无论式(6.35)还是式(6.36)都不适用。在近似处理过程中,这种情况不予考虑,式(6.35)和式(6.36)分别用于 $|G_{p2}(j\omega)G_c(j\omega)| < 1$ 和 $|G_{p2}(j\omega)G_c(j\omega)| > 1$ 的情况。

这些近似表示可以用一个简单的例子予以说明。假设 $G_{p2}(s)$ 代表一台电机,其数学描述为

$$G_{p2}(s) = \frac{K_m}{s(T_m s + 1)}$$

令反馈校正为 $G_c(s) = 1$。这一问题用代数方法精确求解是相当简单的事情。不过在这里使用对数幅频特性曲线,并考虑近似处理的情况。近似的对数幅频特性曲线绘制于图 6.18。

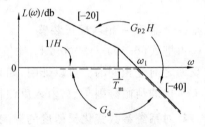

图 6.18 电机的对数幅频特性图

由图 6.18 可见,对于所有低于 ω_1 的频率有 $|G_{p2}(j\omega)G_c(j\omega)| > 1$。根据式(6.36)的近似关系,对于低于 ω_1 的频率,$G_d(j\omega)$ 可以表示为 $1/G_c(j\omega)$。而且,对于所有高于 ω_1 的频率有 $|G_{p2}(j\omega)G_c(j\omega)| < 1$。因此,如图 6.18 所示,对于一直到 ω_1 的频率,$G_d(j\omega)$ 可以用一条斜率为零、高度为 0 db 的直线表示,而对于高于 ω_1 的频率,则用一条斜率为 -40 db/dec 的直线表示。这样,$G_d(s)$ 的传递函数在分母上就有一个 $\omega_n = \omega_1$ 的二次式,即

$$G_d(s) = \frac{\omega_1^2}{s^2 + 2\zeta\omega_1 s + \omega_1^2}$$

当然,对于这一个简单的问题可以由式(6.34)用代数方法获取 $G_d(j\omega)$,其结果为

$$G_d(s) = \frac{K_m}{T_m s^2 + s + K_m} = \frac{\dfrac{K_m}{T_m}}{s^2 + \dfrac{1}{T_m}s + \dfrac{K_m}{T_m}}$$

因此可以有 $\omega_1 = \omega_n = \sqrt{K_m/T_m}$ 和 $\zeta = 1/(2\sqrt{K_m T_m})$。注意,除了一些具体的数据外,近似处理的结果基本上是正确的,在本例中就没有给出 ζ 的数值。根据 $G_d(j\omega)$ 近似的对数幅频特性曲线可以画出相应的近似相位特性曲线。

习 题

题 6.1　某单位反馈系统有

$$G(s) = \frac{1350}{s(s+2)(s+30)}$$

（a）绘制渐近对数幅频特性曲线，并确定增益穿越频率和相角裕度。

（b）相位超前校正装置的传递函数为

$$G_c(s) = \frac{0.25s + 1}{0.025s + 1}$$

对于校正后系统重复（a），并使用经验公式估算最大超调量和调整时间。

题 6.2　某单位反馈系统的开环传递函数为

$$G(s) = \frac{K}{s(s+1)(0.25s+1)}$$

（a）设计一个串联校正装置，使得速度误差系数 $K_v \geq 5$ rad/s，相角裕度 $\gamma \geq 45°$。

（b）如果还要求增益穿越频率 $\omega_c \geq 2$ rad/s，重新设计一个串联校正装置。

题 6.3　某单位反馈系统的开环传递函数为

$$G(s) = \frac{K}{s(0.2s+1)(0.05s+1)}$$

设计一个串联校正装置，使得速度误差系数 $K_v \geq 12$ rad/s，最大超调量 $\sigma_p \leq 25\%$，调整时间 $t_s \leq 1$ s。

题 6.4　某单位反馈系统的开环传递函数为

$$G(s) = \frac{K}{s(0.25s+1)(0.05s+1)}$$

其中 K 设定等于 10 以得到指定的 $K_v = 3.33$ rad/s。设计一个串联校正装置，使得相对谐振峰值 $M_r \leq 1.4$，谐振频率 $\omega_r \geq 10$ rad/s。

题 6.5　单位反馈系统的被控对象为

$$G(s) = \frac{20}{s(0.1s+1)(0.05s+1)}$$

选择一个校正装置 $G_c(s)$ 使得相角裕度至少为 75°。使用一个两级相位超前校正装置

$$G_c(s) = \frac{k(\alpha_1 T_1 s + 1)(\alpha_2 T_2 s + 1)}{(T_1 s + 1)(T_2 s + 1)}$$

要求对于斜坡输入的稳态误差为斜坡输入幅值的 0.5%。

题 6.6　某单位反馈系统的开环传递函数为

$$G(s) = \frac{40}{s(0.2s+1)(0.0625s+1)}$$

（a）设计一相位超前校正装置，使得相角裕度 γ 为 30°，增益裕度为 10～12 db。

（b）设计一相位滞后校正装置，要求相角裕度 γ 为 50°，增益裕度为 30～40 db。

题 6.7 一个位置控制系统的开环传递函数为

$$G(s) = \frac{3}{s(s+1)(0.5s+1)}$$

保持开环增益不变时,确定一个能够提供相角裕度为 45° 的相位滞后校正装置。

题 6.8 单位反馈系统被控对象的传递函数为

$$G(s) = \frac{40}{s(s+2)}$$

要求对于斜坡输入 $r(t) = At$ 的稳态误差小于 $0.05A$,相角裕度为 30°,还要求穿越频率 ω_c 为 10 rad/s。确定是否需要一个超前或滞后校正装置。

题 6.9 题 6.9 图所示是一个控制系统。确定前置放大器的增益并设计一个相位滞后校正装置,使得速度误差系数 $K_v \geq 4s^{-1}$,相对谐振幅值 $M_r = 1.4$。

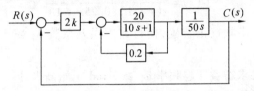

题 6.9 图

题 6.10 某单位反馈系统的被控对象为

$$G(s) = \frac{1}{s(s+1)(0.5s+1)}$$

选择的校正装置是一个 PI 调节器,使得系统对于阶跃输入的稳态误差为零。由此有

$$G_c(s) = k_1 + \frac{k_2}{s} = \frac{k_1 s + k_2}{s}$$

确定一个合适的 $G_c(s)$ 使得:

(a) 最大超调量等于或小于 5%;
(b) 调整时间小于 6 s ($\Delta = 2\%$);
(c) 速度误差系数 K_v 大于 0.9;
(d) 阶跃输入时的峰值时间最小。

题 6.11 考虑单位反馈系统,开环传递函数为

$$G(s) = \frac{20}{s(s+1)(s+3)}$$

希望加入一个滞后 – 超前校正装置

$$G_c(s) = \frac{(s+0.15)(s+0.7)}{(s+0.015)(s+7)}$$

证明校正后系统的相角裕度为 75° 且增益裕度为 24 db。

题 6.12 某单位反馈系统的被控对象为

$$G(s) = \frac{160}{s^2}$$

选择一个滞后 – 超前校正装置,使得最大超调量小于 5%,调整时间($\Delta = 2\%$)小于 1 s,加速度误差系数大于 7 500。

题 6.13 反馈系统的开环传递函数为

$$G(s) = \frac{10(0.316s + 1)}{(0.1s + 1)(0.01s + 1)}$$

确定使系统满足题 6.13 图所示期望开环频率响应时所需要的串联校正装置 $G_c(s)$。

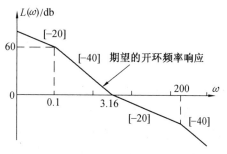

题 6.13 图

题 6.14 考虑题 6.14 图中所示带有反馈校正的控制系统

$$G(s) = \frac{100}{s(1.1s + 1)(0.025s + 1)}, \quad G_c(s) = 0.25s$$

(a)绘制校正前和校正后系统的渐近对数幅频特性曲线;
(b)找出等效的开环传递函数;
(c)估算校正后系统的相角裕度。

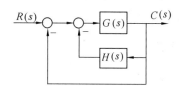

题 6.14 图

题 6.15 一个最小相位控制系统具有题 6.15 图所示的开环对数幅频特性曲线。要求通过题 6.15 图所示反馈校正消除转折频率 20 rad/s 处的谐振峰值。确定反馈校正装置的传递函数。

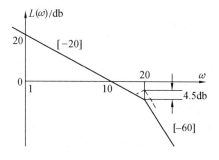

题 6.15 图

第7章 非线性系统分析

7.1 引 言

需要再次指出的是,所有物理系统在本质上都是非线性的。一个基本上算是线性的系统只是在一定的运行范围内线性而已。正如前面已经讲到过的,不能够精确地为一个物理系统建立模型。在通常的情况下,通过增加线性模型的阶数可以提高模型的精度。但是,到了一定程度的时候,阶数的增加将不再会明显地改善系统的模型。在模型的精度仍然不足以满足要求的情况下,就有必要加入非线性特性。

给定一个线性系统后,总是能够确定该系统的稳定性,例如使用劳斯－赫尔维斯判据、奈奎斯特判据或者其他的方法。关于非线性系统则不能做出这样的结论。在考虑给定非线性系统的时候,不存在通用的、总能确定其稳定性的非线性稳定性分析方法。取而代之的是研究出一些专门的方法,没有一种方法适用于所有的非线性系统。对于许多包含若干个非线性特性的系统,可能只有通过仿真才能确定它们的稳定性。

7.1.1 非线性系统的定义

一个非线性系统是不能应用叠加原理的系统。因此,非线性系统的输入输出关系用非线性微分方程描述。下面是一些非线性系统的例子

$$\dot{c}(t) + c^2(t) = r(t)$$
$$\dot{c}(t) + \sin[c(t)] = r(t)$$
$$\ddot{c}(t) + 3\dot{c}(t) + 2\dot{c}(t)c(t) = r(t)$$

第一个方程是非线性的,因为有变量的平方;第二个方程是因为有变量的正弦,而第三个方程则是因为有变量与其导数的乘积。一般地讲,拉普拉斯变换不能用于解任何形式的非线性微分方程。

尽管线性控制系统的分析和设计已经有了很好的研究,但是对于非线性系统而言这方面的问题通常是很复杂的。在很多情况下,有可能难以写出描述非线性系统输入输出关系的微分方程。因此,方框图常常用于非线性系统的研究,非线性特性则可以通过实验的方法获取。而且,使用方框图简化的方法可以使线性部分和非线性部分分别得到简化。如果一个非线性系统的线性部分和非线性部分可以分离成如图7.1所示,则称之为具有基本形式的非线性系统。非线性系统的有些重要方法只适用于那些具有基本形式的非线性系统。在描述非线性系统的方框图中,线性部分通常用传递函数表示。

图 7.1 基本形式的非线性系统

非线性特性可以分类为动态和静态非线性特性。输入

和输出可以通过微分方程联系起来的非线性储能器件称为动态非线性特性。另一方面，输入输出不涉及微分方程的非线性器件则称为静态非线性特性。为了方便，静态非线性特性常用特性曲线表示。图 7.2 所示是一个非线性系统的方框图，图中一个非线性放大器显示为表示饱和特性的曲线。

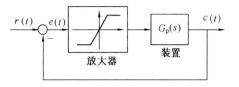

图 7.2　带有静态非线性特性的系统

7.1.2　非线性系统的一些性质

由于非线性系统不遵循叠加原理，它们有许多与线性系统截然不同的性质：

① 非线性系统的响应将不是像线性系统那样的指数或者时间加权指数的情况。对于同一类型输入，响应曲线的形状有可能显著地不同，在有些情况下响应会收敛到一个平衡点，而在另一些情况下响应会变为无界发散；在有些情况下响应会以衰减的振荡收敛到平衡点，而在另一些情况下则没有振荡；不仅如此，响应还可能以不同的频率收敛到平衡点。图 7.3 显示了一个非线性系统对于不同输入幅值的阶跃响应。

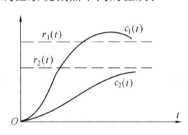

图 7.3　一个非线性系统对于不同输入幅值的阶跃响应

② 有些非线性系统对于一定的初始条件可以是稳定的，但是对于不同的初始条件也有可能是不稳定的。

例如，考虑一个一阶非线性系统，微分方程为

$$\dot{x}(t) = -x(1-x)$$

初始条件为 $x(0) = x_0$ 时，方程的解为

$$x(t) = \frac{x_0 \mathrm{e}^{-t}}{1 - x_0 + x_0 \mathrm{e}^{-t}}$$

图 7.4 画出了在不同初始条件下的响应曲线。从图中可以看到，系统的稳定性与初始条件有关。

③ 在没有输入的情况下有可能存在多个平衡点，因此，对于所有的初始状态而言，到达稳态时的情况有可能是不一样的。如图 7.4 所示，该系统有两个平衡点，$x = 0$ 是一个稳定的平衡点，而 $x = 1$ 则是一个不稳定的平衡点。

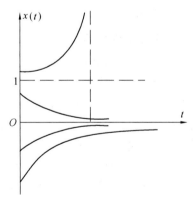

图 7.4　一个非线性系统对于不同初始条件的响应

④ 非线性系统的响应有可能收敛到一个周期振荡。在一个线性时不变系统中，如图 7.5(a) 所示，周期振荡是正弦曲线，其幅值是系统激励的幅值和初始条件两者的函数。在一个非线性系统中，如图 7.5(b) 所示，由许多不同初始条件可以导致相同的周期振荡。

⑤ 周期性输入的非线性系统有可能呈现周期性的输出，其频率是输入频率的分频或

者谐振频率。

⑥ 在有些情况下,对于一个幅值和频率给定的输入正弦信号,非线性系统可能有不止一个稳定的周期状态,因此这种系统称为多模式系统。输入幅值和频率的稍许变化就会导致系统改变状态,这种现象称为跃变响应。图 7.6 说明了一个跃变现象。

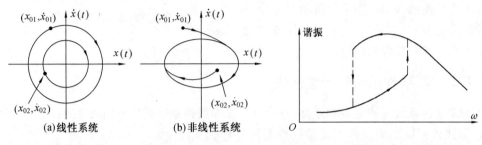

图 7.5　周期振荡　　　　　图 7.6　频率响应中的跳跃谐振

7.1.3　常见的非线性特性

在大部分控制系统中,无法避免某些类型非线性特性的存在。系统中的非线性特性经常分类为固有非线性特性和人为非线性特性。固有非线性特性存在于为了执行某一功能而选择的"估计可能线性"的元部件中。如果不存在固有非线性特性,通常设计人员日子将比较好过。另一方面,人为非线性特性是有意引入系统以执行特殊的非线性功能。鉴于它们的经济性、可靠性、运行特性或者其他优点,人为非线性特性在设计期间已经做了选择。

1. 饱和

饱和是最常见的固有非线性特性之一。例子有饱和了的电子放大器,诸如飞机和船舶上舵的平移和旋转运动系统的机械限位,等等。

饱和的静态特性见图 7.7(a)。只有对于输入的一个有限范围,输出 x 与输入 e 才成比例关系。当输入的幅值超出这一范围时,输出接近一个恒量。尽管对于实际的饱和非线性而言,从一个范围到另一个范围的变化通常是逐渐发生的,在大部分情况下用如图所示一组直线近似表示曲线的精度常常是绰绰有余的。

考虑已饱和环节的增益时,从图 7.7(b) 可以看到,当输入逐渐增大时等效增益逐渐减小。因此,整个控制系统的等效开环增益逐渐减小。

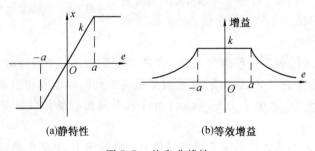

图 7.7　饱和非线性

2. 死区

在许多物理器件中,一直到输入的幅值超过某一确定值时输出才不为零。例如,在为一台直流伺服电机建立数学模型时,假设如果励磁电流保持不变则任一施加于电枢绕组的电压都将导致电枢旋转。实际上,只有当电机产生的力矩足以克服静摩擦时才会造成旋转。结果就是,如果画出稳态角速度和施加电压之间的关系,可以得到图 7.8 所示的特性,该特性表现出的是死区现象。许多其他器件也表现出类似的特性。

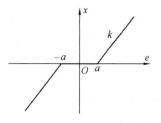

图 7.8　死区的静特性

一般地讲,死区非线性特性将降低控制系统的灵敏度,尤其是降低了误差检测器件的灵敏度。因此,死区也称为不灵敏区。

3. 继电器

继电器经常用于控制系统,因为它在提供大功率放大时的代价相当低。理想继电器的特性见图 7.9(a),图中输入正负号的改变导致输出的突然变化。

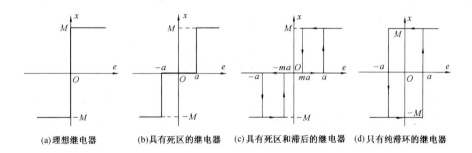

(a)理想继电器　　(b)具有死区的继电器　　(c)具有死区和滞后的继电器　　(d)只有纯滞环的继电器

图 7.9　继电器的静特性

在一个实际的继电器中,铁心线圈中的电流产生磁力使衔铁移动,从而使得触点在一个方向或另一个方向接通。实际上,由于电流必须超过某一确定值时衔铁才能移动,一个真实的继电器将呈现出图 7.9(b) 所示的死区。而且,由于磁滞现象,继电器闭合需要的电流要大于触点断开时的电流。因此,实际上大部分继电器呈现出图 7.9(c) 所示的死区特性和滞环。另一种情况是图 7.9(d) 所示的具有纯滞环的继电器。

4. 间隙

一种建立的模型非常复杂的非线性特性是存在于齿轮中的间隙,因为齿轮不能精确地啮合。当主动齿轮改变方向后,只有消除了齿轮中的间隙时负载齿轮才能重新开始运动。

这一非线性特性造成时间上的滞后,以及最终的磁滞现象。图 7.10 显示的是间隙的静特性,其中间隙的宽度为 2ε。从动力特性来看,间隙的影响等价于引入一个延时环节。另一方面,从稳态性能来看,间隙的影响则等价于引入一个死区。

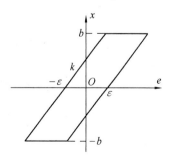

图 7.10　间隙的静特性

7.1.4 研究非线性系统的方法

非线性系统的性能可以通过硬件、仿真、和／或分析的方法进行研究。严格地讲,后两种情况考虑的是系统模型的性能。确实无疑,实际系统的测量和测试能够提供该系统将如何运行的最可靠信息,但这通常是设计的最后阶段。没有经过分析或仿真就建造一个系统并进行测试的做法有可能既是危险的也是代价昂贵的。

研究非线性控制系统的一个重要手段是仿真。利用计算机人们可以承担复杂系统详尽仿真的重任。不过,没有经过预先分析的仿真经常会是徒劳无益的,因为计算机的每一次运行只能提供对于给定参数和输入的解答。为了了解发生了什么并说明可以做出什么样的变动以改善性能,则需要理论提供参考。这样的话,由于是建立在预先分析的基础上,就能够通过计算机仿真以一种深思熟虑的方式完成研究。

尽管已经有许多方法应用于非线性系统的研究,在此考虑两种众所周知的方法。第一种方法是描述函数法,另一种流行的方法是相平面法。相平面法提供一种图解法,可以获得二阶系统非线性微分方程的解。尽管后者不适用于高阶系统,但高阶系统常常可以用近似的二阶系统表示。

7.2 描述函数法

描述函数法是把所熟悉的频率响应法延伸到非线性系统的一种尝试。尽管这一方法是基于简化的近似而不太精确,它的主要优点是它的简易性。描述函数法常常能为设计者给出系统行为的一个正确评价,并能指出要使得系统具有令人满意的性能时应当做些什么改进。

7.2.1 描述函数的定义

这里所考虑的描述函数适用于图 7.11 所示的具有基本结构的系统。通常线性部分 $G(s)$ 还可以包括校正装置和传感器的传递函数。

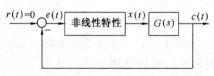

图 7.11 基本结构的非线性系统

为了研究描述函数法,考虑图 7.11 中的情况,系统的输入为零,而非线性特性的输入为正弦信号,即

$$e(t) = A\sin \omega t \tag{7.1}$$

一个具有零输入的系统称为自由系统。如果再加上系统是时不变的,则该系统称为自治系统。这样,在稳态下非线性特性的输出 $x(t)$ 是周期信号,但一般情况下不是正弦信号。由非正弦波形的傅里叶分析可以得到一个基波分量和一些谐波,即

$$x(t) = A_0 + \sum_{n=1}^{\infty}(A_n\cos n\omega t + B_n\sin n\omega t) = A_0 + \sum_{n=1}^{\infty}X_n\sin(n\omega t + \varphi_n) \tag{7.2}$$

式中

$$A_0 = \frac{1}{2\pi}\int_0^{2\pi} x(t)\,\mathrm{d}(\omega t) \tag{7.3}$$

$$A_n = \frac{1}{\pi}\int_0^{2\pi} x(t)\cos n\omega t\,\mathrm{d}(\omega t) \tag{7.4}$$

$$B_n = \frac{1}{\pi}\int_0^{2\pi} x(t)\sin n\omega t\,\mathrm{d}(\omega t) \tag{7.5}$$

$$X_n = \sqrt{A_n^2 + B_n^2}, \quad \varphi_n = \arctan\frac{A_n}{B_n} \tag{7.6}$$

本章仅限于考虑具有奇对称静特性的非线性特性,即我们将仅考虑 $A_0 = 0$ 的情况。

如果进一步假设图 7.11 中的 $G(s)$ 对于谐波具有低通的特性,即与 $|G(\mathrm{j}\omega)|$ 对于 $x(t)$ 的基波分量的值相比,它对于 $x(t)$ 所有其他分量的值都要小。因此,输出 $c(t)$ 可以表达为

$$c(t) = A_c \sin(\omega t + \theta) \tag{7.7}$$

这一假设是描述函数分析的基础。这样,$x(t)$ 中的谐波就不太重要了,因为它们对于 $c(t)$ 只有很小的影响。反馈回非线性特性输入端的谐波可以忽略不计,$x(t)$ 可以近似为

$$x(t) \approx x_1(t) = A_1\cos\omega t + B_1\sin\omega t = X_1\sin(\omega t + \varphi_1) \tag{7.8}$$

为了使如上所描述的非线性系统分析起来比较容易,非线性环节的描述函数定义为非线性环节输出的基波分量与正弦输入的复数比。由式(7.8)看到,$x(t)$ 可以近似为与 $e(t)$ 频率相同但幅值和相位不一样的正弦量。由此,非线性特性可以用一个复数增益替代为

$$N(A,\omega) = \frac{B_1 + \mathrm{j}A_1}{A} \tag{7.9}$$

这一等效增益称为描述函数。描述函数通常是输入正弦信号幅值和相位两者的函数。因为本章仅仅考虑非线性特性,在这种情况下,描述函数仅与输入幅值有关,并可以记为 $N(A)$。而且,对于单值奇对称非线性特性有 $A_1 = 0$,$\varphi_1 = 0$ 和

$$N(A) = \frac{B_1}{A} \tag{7.10}$$

注意描述函数定义的假设条件,只有当以下条件成立时这一定义才有意义:

① 非线性特性的输入为正弦信号;

② 非线性特性具有奇对称静特性;

③ 在非线性特性之后的线性部分具有足够好的低通特性,使得所有谐波都衰减到其大小可以忽略不计。

7.2.2 常见非线性特性的描述函数

1. 饱和

理想饱和特性及其输出波形见图 7.12,其中饱和的静特性为

$$x(t) = \begin{cases} kA\sin\omega t & 0 \leqslant \omega t \leqslant \alpha_1 \\ ka = b & \alpha_1 < \omega t < (\pi - \alpha_1) \\ kA\sin\omega t & (\pi - \alpha_1) \leqslant \omega t \leqslant \pi \end{cases} \tag{7.11}$$

式中

$$\alpha_1 = \arcsin\frac{a}{A}$$

由于理想饱和是单值奇对称非线性特性有

$$A_1 = 0$$

$$B_1 = \frac{1}{\pi}\int_0^{2\pi} x(t)\sin\omega t\,\mathrm{d}(\omega t) = \frac{4}{\pi}\int_0^{\pi/2} x(t)\sin\omega t\,\mathrm{d}(\omega t) =$$

$$\frac{4}{\pi}\Big[\int_0^{\alpha_1} kA\sin\omega t\sin\omega t\,\mathrm{d}(\omega t) + \int_{\alpha_1}^{\pi/2} ka\sin\omega t\,\mathrm{d}(\omega t)\Big] =$$

$$\frac{4kA}{\pi}\Big\{\Big[\frac{1}{2}\omega t - \frac{1}{4}\sin 2\omega t\Big]_0^{\alpha_1} + \frac{a}{A}\Big[-\cos\omega t\Big]_{\alpha_1}^{\pi/2}\Big\} =$$

$$\frac{4kA}{\pi}\Big(\frac{1}{2}\alpha_1 - \frac{1}{4}\sin 2\alpha_1 + \frac{a}{A}\cos\alpha_1\Big) =$$

$$\frac{2kA}{\pi}\Big(\arcsin\frac{a}{A} + \frac{a}{A}\sqrt{1 - \Big(\frac{a}{A}\Big)^2}\Big)$$

由此,饱和特性的描述函数为

$$N(A) = \frac{2k}{\pi}\Big(\arcsin\frac{a}{A} + \frac{a}{A}\sqrt{1 - \Big(\frac{a}{A}\Big)^2}\Big), \quad A \geqslant a \tag{7.12}$$

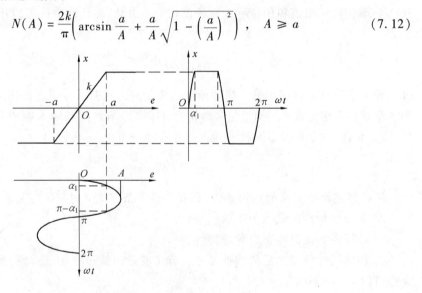

图 7.12 饱和特性及其输出波形

2. 死区

如图 7.13 所示,死区的静特性为

$$x(t) = \begin{cases} 0 & 0 \leqslant \omega t \leqslant \alpha_1 \\ kA\sin(\omega t - a) & \alpha_1 < \omega t < (\pi - \alpha_1) \\ 0 & (\pi - \alpha_1) \leqslant \omega t \leqslant \pi \end{cases} \tag{7.13}$$

式中
$$\alpha_1 = \arcsin \frac{a}{A}$$

理想死区是单值奇对称非线性特性,因此有
$$A_1 = 0$$
$$B_1 = \frac{1}{\pi}\int_0^{2\pi} x(t)\sin\omega t\,\mathrm{d}(\omega t) = \frac{4}{\pi}\int_0^{\pi/2} x(t)\sin\omega t\,\mathrm{d}(\omega t) =$$
$$\frac{4}{\pi}\Big[\int_{\alpha_1}^{\pi/2} kA(\sin\omega t - a)\sin\omega t\,\mathrm{d}(\omega t)\Big] =$$
$$\frac{2kA}{\pi}\Big(\frac{\pi}{2} - \arcsin\frac{a}{A} - \frac{a}{A}\sqrt{1-\Big(\frac{a}{A}\Big)^2}\Big)$$

由此,死区特性的描述函数为
$$N(A) = \frac{2k}{\pi}\Big(\frac{\pi}{2} - \arcsin\frac{a}{A} - \frac{a}{A}\sqrt{1-\Big(\frac{a}{A}\Big)^2}\Big), \quad A \geqslant a \tag{7.14}$$

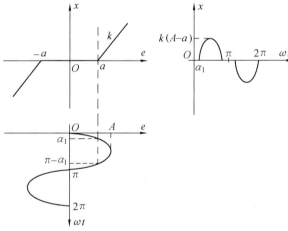

图 7.13 死区特性及其输出波形

3. 继电器

考虑图 7.14 所示继电器非线性特性。静特性定义为
$$x(t) = \begin{cases} 0 & 0 \leqslant \omega t \leqslant \alpha_1 \\ M & \alpha_1 < \omega t < \pi - \alpha_2 \\ 0 & \pi - \alpha_2 \leqslant \omega t \leqslant \pi \end{cases} \tag{7.15}$$

式中
$$\alpha_1 = \arcsin\frac{e_0}{A}, \quad \alpha_2 = \pi - \arcsin\frac{me_0}{A}$$

对于描述函数有
$$A_1 = \frac{2}{\pi}\int_{\alpha_1}^{\alpha_2} x(t)\cos\omega t\,\mathrm{d}(\omega t) = \frac{2}{\pi}\int_{\alpha_1}^{\alpha_2} M\cos\omega t\,\mathrm{d}(\omega t) = \frac{2Me_0}{\pi A}(m-1)$$
$$B_1 = \frac{1}{\pi}\int_0^{2\pi} x(t)\sin\omega t\,\mathrm{d}(\omega t) = \frac{2}{\pi}\int_{\alpha_1}^{\alpha_2} M\sin\omega t\,\mathrm{d}(\omega t) =$$

$$\frac{2M}{\pi}\left(\sqrt{1-\left(\frac{me_0}{A}\right)^2}+\sqrt{1-\left(\frac{e_0}{A}\right)^2}\right)$$

因此得到

$$N(A)=\frac{2M}{\pi A}\left(\sqrt{1-\left(\frac{me_0}{A}\right)^2}+\sqrt{1-\left(\frac{e_0}{A}\right)^2}\right)+j\frac{2Me_0}{\pi A^2}(m-1),\quad A\geqslant e_0 \quad (7.16)$$

注意,在这一情况下描述函数是复数,这当然会产生相位移。

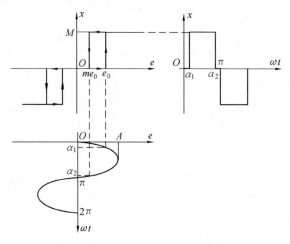

图 7.14 继电器特性及其输出波形

尤其在 $e_0=0$ 的情况下,理想继电器的描述函数为

$$N(A)=\frac{4M}{\pi A},\quad A\geqslant e_0 \quad (7.17)$$

在 $m=1$ 的情况下,具有死区继电器的描述函数为

$$N(A)=\frac{4M}{\pi A}\sqrt{1-\left(\frac{e_0}{A}\right)^2},\quad A\geqslant e_0 \quad (7.18)$$

至于 $m=-1$ 的情况,具有滞环的继电器的描述函数为

$$N(A)=\frac{4M}{\pi A}\sqrt{1-\left(\frac{me_0}{A}\right)^2}+j\frac{2Me_0}{\pi A^2}(m-1),\quad A\geqslant e_0 \quad (7.19)$$

4. 组合非线性特性

在许多系统中,非线性特性发生在系统内不止一个地方。如果非线性和线性环节组合起来时所有非线性特性的输入都大致上是正弦信号,那么对于所研究的情况,每一个这些非线性特性都可以用它们的描述函数代替。

(1) 串联非线性特性

分析串联组合非线性特性的通常做法是,如图 7.15 所示,说明的那样获取总的等效非线性,然后再估算新的等效非线性特性的传递函数。注意,串联组合非线性特性的描述函数不等于各个单独非线性特性的描述函数的乘积,因为由第一个非线性特性产生的谐波对于第二个非线性特性的影响不能忽略,两个非线性特性都不能用描述函数替代。

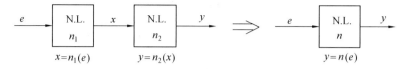

图 7.15　串联组合非线性特性

例如，考虑如图 7.16 所示死区和饱和的组合。等效的非线性特性是一个具有死区的饱和特性。

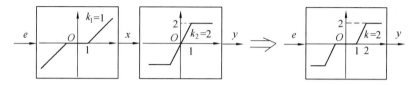

图 7.16　串联非线性组合的例子

(2) 并联非线性特性

两个非线性特性的并联组合见图 7.17。显然，由于两个非线性特性都具有相同的输入 $e(t) = \sin\omega t$，总的输出是各个单独的非线性特性输出的和。因此，如果紧随非线性特性之后的线性部分具有足够好的低通特性，总的描述函数为

$$N(A) = N_1(A) + N_2(A) \quad (7.20)$$

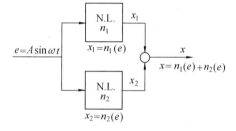

图 7.17　并联非线性组合

尽管上面的非线性结构在实际系统中很少发生，这个概念对于仿真时的非线性特性综合和描述函数的估算非常有用。例如，可以通过死区描述函数的获取图示说明获取描述函数的过程。

由图 7.18 可以看到，在这种情况下死区特性可以由以下特性获取：

① $n_1(e)$ —— 单位幅值的线性增益；

② $-n_2(e)$ —— 线性区单位增益、线性区间为 a 的饱和特性。

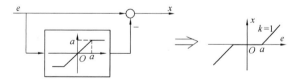

图 7.18　死区的描述函数

因此死区的描述函数为

$$N(A) = 1 - \frac{2}{\pi}\left(\arcsin\frac{a}{A} + \frac{a}{A}\sqrt{1-\left(\frac{a}{A}\right)^2}\right) =$$
$$\frac{2}{\pi}\left(\frac{\pi}{2} - \arcsin\frac{a}{A} - \frac{a}{A}\sqrt{1-\left(\frac{a}{A}\right)^2}\right)$$

表 7.1 给出了刚才推导的一些描述函数，再加上其他一些常用的描述函数。

表 7.1　非线性特性及其描述函数

非线性特性	描述函数 $N(A)$
(饱和特性，斜率 k，线性区 $\pm a$)	$\dfrac{2k}{\pi}\left(\arcsin\dfrac{a}{A}+\dfrac{a}{A}\sqrt{1-\left(\dfrac{a}{A}\right)^2}\right),\ A\geqslant a$
(死区特性，斜率 k，死区 $\pm a$)	$\dfrac{2k}{\pi}\left(\dfrac{\pi}{2}-\arcsin\dfrac{a}{A}-\dfrac{a}{A}\sqrt{1-\left(\dfrac{a}{A}\right)^2}\right),\ A\geqslant a$
(理想继电器，幅值 M)	$\dfrac{4M}{\pi A}$
(带死区继电器，幅值 M，死区 e_0)	$\dfrac{4M}{\pi A}\sqrt{1-\left(\dfrac{e_0}{A}\right)^2},\ A\geqslant e_0$
(带滞环继电器，幅值 M，滞环 e_0)	$\dfrac{4M}{\pi A}\sqrt{1-\left(\dfrac{e_0}{A}\right)^2}-\mathrm{j}\dfrac{4Me_0}{\pi A^2},\ A\geqslant e_0$
(带死区滞环继电器)	$\dfrac{2M}{\pi A}\left(\sqrt{1-\left(\dfrac{me_0}{A}\right)^2}+\sqrt{1-\left(\dfrac{e_0}{A}\right)^2}\right)+\mathrm{j}\dfrac{2Me_0}{\pi A^2}(m-1),\ A\geqslant e_0$
(间隙特性，斜率 k，间隙 ε)	$\dfrac{k}{\pi}\left(\dfrac{\pi}{2}+\arcsin\left(1-\dfrac{2\varepsilon}{A}\right)+2\left(1-\dfrac{2\varepsilon}{A}\right)\sqrt{\dfrac{\varepsilon}{A}\left(1-\dfrac{2\varepsilon}{A}\right)}\right)+$ $\mathrm{j}\dfrac{4k\varepsilon}{\pi A}\left(\dfrac{\varepsilon}{A}-1\right),\ A\geqslant\varepsilon$
(变斜率特性 k_1,k_2，转折点 a)	$k_2+\dfrac{2}{\pi}(k_1-k_2)\left[\arcsin\dfrac{a}{A}+\dfrac{a}{A}\sqrt{1-\left(\dfrac{a}{A}\right)^2}\right],\ A\geqslant a$

7.2.3　非线性系统的稳定性分析

考虑图 7.11 所示系统。假设描述函数分析适用于这一系统，那么得到图 7.19 的等效系统，其中 $N(A)$ 为非线性特性的描述函数，而 $G(s)$ 则为线性部分的传递函数。

该系统的闭环频率响应为

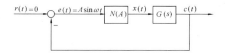

图 7.19　用描述函数表示的非线性系统

$$\Phi(j\omega) = \frac{N(A)G(j\omega)}{1 + N(A)G(j\omega)} \tag{7.21}$$

而相应的特征方程则为 $1 + N(A)G(j\omega) = 0$，即

$$G(j\omega) = -\frac{1}{N(A)} \tag{7.22}$$

式中，$-1/N(A)$ 称为负倒描述函数。这样，非线性特性用其描述函数代替后，非线性系统的稳定性就可以利用任何一种线性方法进行研究，例如极坐标图和尼柯尔斯图。

当频率响应 $G(j\omega)$ 和负倒描述函数 $-1/N(A)$ 一起画在同一个称为奈奎斯特平面的复平面上时，如图 7.20(a) 所示的任一交点都将是式(7.22)的一个周期解，这意味着在控制系统内存在一个对应于该交点的振荡。由于一般情况下 $G(j\omega)$ 是 ω 的复函数，而 $N(A)$ 是 A 的复函数，式 (7.22) 的解同时给出了振荡的频率和幅值。振荡有可能是稳定的，也可能是不稳定的，而稳定的振荡则称为零输入造成的自振荡。本节随后将对振荡的稳定性作进一步的讨论。

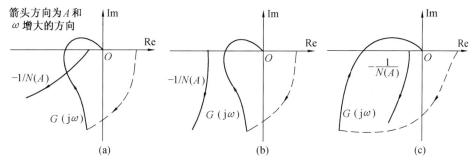

图 7.20　稳定性的图示说明

另一方面，如果 $G(j\omega)$ 和 $-1/N(A)$ 的轨迹不存在交点，系统的稳定性利用一般的奈奎斯特判据关于 $-1/N(A)$ 上的任意一点而不是 $(-1, j0)$ 点进行评估。假设系统的线性部分是最小相位，由此可以得到以下结论，如图 7.20(b) 所示 $G(j\omega)$ 曲线没有包围 $-1/N(A)$ 轨迹则非线性系统是稳定的，如果如图 7.20(c) 所示 $G(j\omega)$ 曲线包围 $-1/N(A)$ 轨迹则非线性系统是不稳定的。

在许多控制系统中，描述函数非常准确地表征了非线性特性的影响。其原因在于控制系统的被控对象，尤其是功率输出器件通常使周期波形的较高次谐波大幅度地衰减。因此较高次谐波的影响很小。但是，另一方面应当记住，描述函数法是一种近似的方法。有时候要确定能否合理地忽略较高次谐波是很困难的，例如当 $-1/N(A)$ 轨迹和 $G(j\omega)$ 曲线在它们相交处差不多相切的情况下就是如此。

7.2.4 振荡的稳定性

当式(7.22)获得一个周期解时,这个解对应于稳定的还是不稳定的振荡条件呢?如果振荡随着振幅有一个小的变化返回原来的解的状态,该振荡称为是稳定的;如果振幅继续远离它的平衡状态,则振荡是不稳定的。不稳定的周期运动是很难观察到的,因为系统中任何扰动都将阻止不稳定的周期运动维持稳态振荡,不稳定的周期运动将或者收敛、或者发散,或者转移到另一个周期运动。

周期运动的稳定性可以应用描述函数法进行研究。例如,考虑图7.21中所示系统可以看到,在 $G(j\omega)$ 和 $-1/N(A)$ 轨迹之间发生有两次相交。预知对于第一个标记为 a 的交点有一个周期运动 $A_a \sin \omega_a t$。但是我们可以注意到,如果工作点移动到 c 点,幅值的增大使得系统变为不稳定,而且由此幅值继续增大;如果工作点移动到 d 点,幅值的减小使得系统变为稳定,而且由此幅值继续减小。因此,该周期运动是不稳定的。至于第二个标记为 b 的交点处预知为 $A_b \sin \omega_b t$ 的周期运动,如果工作点移动到 f 点,幅值的增大使得系统变为稳定,而且由此幅值将减小;如果工作点移动到 e 点,幅值的减小导致系统变为不稳定,而且由此幅值将增大。因此,该周期运动是稳定的。

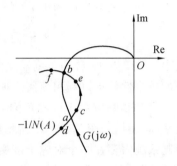

图 7.21 周期运动稳定性的图示说明

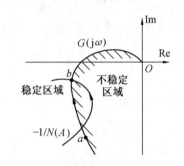

图 7.22 奈奎斯特平面上的稳定区域和不稳定区域

一种类似的图解判别方法在确定周期运动的稳定性时更为方便。如图 7.22 所示奈奎斯特平面可以分为两个部分:稳定区域和不稳定区域。对于像图 7.22 中交点 b 所对应的稳定的周期运动,随着幅值 A 的增大,$-1/N(A)$ 轨迹应当从不稳定区域进入稳定区域。至于像图 7.22 中交点 a 所对应的不稳定的周期运动,随着幅值 A 的增大,$-1/N(A)$ 轨迹应当从稳定区域进入不稳定区域。而且,在 $G(j\omega)$ 和 $-1/N(A)$ 之间没有交点的情况下,假设 $G(j\omega)$ 是最小相位的,如果 $-1/N(A)$ 轨迹位于稳定区域内则非线性系统稳定;如果 $-1/N(A)$ 轨迹位于不稳定区域内则非线性系统不稳定。

例 7.1 分析图 7.23(a) 所示系统的稳定性。

解 对于这一系统,由表 7.1 饱和特性的描述函数为

$$N(A) = \frac{2k}{\pi}\left(\arcsin \frac{a}{A} + \frac{a}{A}\sqrt{1-\left(\frac{a}{A}\right)^2}\right), \quad A \geq a$$

令 $u = a/A$,并取 $N(u) = N(a/A)$ 关于 u 的导数,得到

$$\frac{dN(u)}{du} = \frac{2k}{\pi}\left(\frac{1}{\sqrt{1-u^2}} + \sqrt{1-u^2} - \frac{u^2}{\sqrt{1-u^2}}\right) = \frac{4k}{\pi}\sqrt{1-u^2}$$

注意,当 $A > a$ 时 $u = a/A < 1$,因此 $N(u)$ 是一个关于 u 的递增函数,即 $N(A)$ 是一个

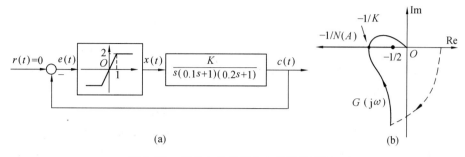

图 7.23 例 7.1 的系统和奈奎斯特平面

关于 A 的递减函数。对于线性运行极限值 $A=1$ 的情况，有
$$-1/N(A) = -1/2$$
而且 $A \to \infty$ 时，有
$$-1/N(A) = -\infty$$
由此可以有图 7.23(b) 所示 $-1/N(A)$ 的轨迹。在该图中还显示有 $G(j\omega)$ 曲线。通过观察可以看到，在 $G(j\omega)$ 和 $-1/N(A)$ 轨迹之间有一个交点，而且该交点对应于一个稳定的周期运动，即该系统存在一个自振荡。

令线性部分的相位穿越频率为 ω_g，可以得到
$$-90° - \arctan 0.1\omega_g - \arctan 0.2\omega_g = -180°$$
即 $\omega_g = \sqrt{50}$ rad/s，注意到
$$|G(j\omega_g)| = \frac{K}{\sqrt{50} \cdot \sqrt{0.5+1} \cdot \sqrt{2+1}} = \frac{K}{15}$$
可以看到，当 $K > 7.5$ 时可以预期有一个稳定的周期运动。

例如，假设给定的 K 值为 15，那么预期的自振荡的频率为 $\omega = \sqrt{50}$ rad/s，而幅值则为 $A = 2.47$。

例 7.2 分析图 7.24(a) 所示系统的稳定性，图中非线性特性是一个死区。

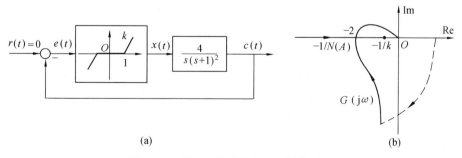

图 7.24 例 7.2 的系统和奈奎斯特平面

解 由表 7.1，该非线性特性的描述函数为
$$N(A) = \frac{2k}{\pi}\left(\frac{\pi}{2} - \arcsin\frac{1}{A} - \frac{1}{A}\sqrt{1-\left(\frac{1}{A}\right)^2}\right), \quad A \geq 1$$
比较饱和特性和死区特性的描述函数很容易看到，现在 $N(A)$ 是一个关于 A 的递增

函数。由于
$$\lim_{A\to 1}[-1/N(A)] = -\infty, \quad \lim_{A\to\infty}[-1/N(A)] = -1/k$$

有 $-1/N(A) \leq -1/k$。$k > 0.5$ 时的描述函数分析见图 7.24，图中 $G(j\omega)$ 的相位穿越频率为 $\omega_g = 1$ rad/s，而且 $|G(j\omega_g)| = 2$。$-1/N(A)$ 和 $G(j\omega)$ 之间的交点对应于一个周期运动。可以看到，幅值 A 的增大使得系统变为不稳定，然后 A 继续增大；幅值 A 的减小使得系统变为稳定，然后 A 继续减小。因此，该周期运动是不稳定的。

这一不稳定的周期运动可以根据物理考虑给予说明。对于 $|e(t)| \leq 1$ 的小信号，没有信号通过死区，系统当然是稳定的。对于稍大一点的信号，非线性特性的等效增益很小，系统仍然是稳定的。对于很大的信号，死区的影响可以忽略不计，非线性特性看上去就像一个线性增益。对于这一线性增益，该系统是不稳定的，系统的响应将无限制地增大。

例 7.3 考虑图 7.25(a) 所示非线性系统。对于以下情况应用描述函数法分析系统的稳定性。

(a) $G_c(s) = 1$

(b) $G_c(s) = \dfrac{0.25s + 1}{8.3(0.03s + 1)}$

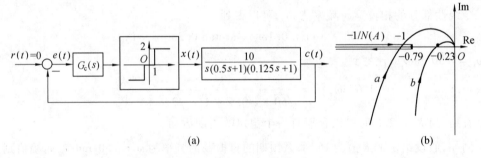

(a)   (b)

图 7.25   例 7.3 的系统

解 由表 7.1，具有死区的继电器的描述函数为
$$N(A) = \frac{4M}{\pi A}\sqrt{1 - \left(\frac{e_0}{A}\right)^2}, \quad A \geq e_0$$

令 $u = e_0/A$，并取 $N(u) = N(e_0/A)$ 关于 u 的导数得到
$$\frac{dN(u)}{du} = \frac{4M}{\pi e_0}\left(\sqrt{1-u^2} - \frac{u^2}{\sqrt{1-u^2}}\right) = \frac{4M}{\pi e_0} \cdot \frac{1-2u^2}{\sqrt{1-u^2}}$$

求解 $d[N(u)]/du = 0$，结果为
$$u_m = \frac{e_0}{A_m} = \frac{1}{\sqrt{2}}$$

由于 $e_0 \leq A \leq A_m$ 时，$d[N(u)]/du > 0$，且 $A > A_m$ 时，$d[N(u)]/du < 0$，$N(A)$ 和 $-1/N(A)$ 两者都在 $A = A_m$ 时有极大值。这样就有
$$-\frac{1}{N(A)} = -\frac{\pi e_0}{2M} = -0.785$$

注意到

$$\lim_{A \to e_0} -\frac{1}{N(A)} = -\infty, \quad \lim_{A \to \infty} -\frac{1}{N(A)} = -\infty$$

然后就可以画出 $-1/N(A)$ 的轨迹如图 7.25(b) 所示。

情况 a $G_c(s) = 1$。线性特性部分的极坐标图见图 7.25(b) 中曲线 a,图中增益穿越频率为 $\omega_g = 4$ rad/s,而相应的幅值则为 $|G(j\omega_g)| = 1$。由观察可知,在 $G(j\omega)$ 和 $-1/N(A)$ 轨迹之间有两个交点。由式(7.22)可以得到周期解

$$\begin{cases} \omega_1 = 4 \text{ rad/s} \\ A_1 = 1.1 \end{cases}, \quad \begin{cases} \omega_2 = 4 \text{ rad/s} \\ A_2 = 2.3 \end{cases}$$

根据稳定性判据,幅值为 A_2 的周期运动是稳定的,而幅值为 A_1 的周期运动则是不稳定的。因此可以得出结论,如果初始条件或者扰动使得 $A < A_1$ 则系统内没有自振荡;如果初始条件或者扰动使得 $A > A_1$,则存在自振荡 $e(t) = 2.3\sin 4t$。

情况 b $G_c(s) = \dfrac{0.25s + 1}{8.3(0.03s + 1)}$。现在线性特性部分的极坐标图见图 7.25(b) 中的曲线 b,图中增益穿越频率为 $\omega_g = 11.97$ rad/s,相应的幅值为 $|G(j\omega_g)| = 0.226$。由观察可知,该系统是稳定的。

7.3 相平面法

7.3.1 引 言

相平面法基本上是一种图解法,主要适用于二阶非线性系统。对于非线性系统 $\ddot{x} = f(x,\dot{x})$,相平面法只是在 $[x,\dot{x}]$ 平面,即所谓的相平面上绘制系统许多不同的运动轨迹。给定一个初始条件后,就可以利用这种方法在相平面上绘制一条轨迹。对于许多初始条件,在相平面上得到的一簇轨迹称之为相轨迹,相轨迹能够提供关于稳定性和是否存在极限环的信息。

例如,图 7.26 是由方程 $\ddot{x} + \omega_n^2 x = 0$ 所描述的二阶系统的相轨迹。箭头指示变量随着时间的增大沿轨迹运动的方向。注意到这一点是很有用的,如果 $\dot{x} > 0$,那么 x 的数值必定递增,也就是说如果轨迹在上半平面,那么这时轨迹运动的结果必定为从左到右。类似地,如果 $\dot{x} < 0$,那么 x 的数值必定递减,也就是说如果轨迹在下半平面,那么轨迹运动的结果必定为从右到左。

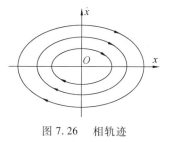

图 7.26 相轨迹

这种方法的主要优点是无需微分方程的解析解就可以概略地绘制出轨迹。它也特别适合于非线性特性为分段直线的系统的分析,因为相平面可以划分为对应于运行在非线性特性特定直线段的一些区域。

7.3.2 绘制相轨迹

1. 解析法

较简单系统的相轨迹方程在数学上处理起来不太困难,常常没有什么问题就可以获得易于处理的解析解。然后就可以如同下面的例子所说明的那样,通过直接积分或者消除中间变量的解析方法绘制相轨迹。

例 7.4 考虑图 7.27(a) 所示的卫星。该卫星一个姿态控制系统的用途是通过触发推进器使姿态角 x 维持在指定的角度上,由此该系统可以描述为

$$\ddot{x} = \frac{\tau}{J} = m, \quad x(0) = x_0, \quad \dot{x}(0) = \dot{x}_0$$

其中 τ 为推进器产生的力矩,J 为卫星的转动惯量,$m(t)$ 则为推力。假设推进器触发时推力为恒值,由此有 $m(t) = \pm M$。

首先考虑 $m(t) = M$ 的情况。由系统的运动方程得到

$$\ddot{x} = \frac{d\dot{x}}{dt} = \frac{d\dot{x}}{dx} \cdot \frac{dx}{dt} = \dot{x}\frac{d\dot{x}}{dx} = M$$

即

$$\dot{x}d\dot{x} = Mdx$$

由于变量 $\dot{x}$ 和 x 是分离的,对该式积分得到相轨迹方程为

$$\frac{\dot{x}^2}{2} = Mx + C_1$$

其中 C_1 为积分常数,由初始条件确定,即

$$C_1 = -Mx_0 + \frac{\dot{x}_0^2}{2}$$

相轨迹方程描述了 $[x,\dot{x}]$ 平面上的一簇抛物线,相轨迹绘制在图 7.27(b) 内。

对于 $m(t) = -M$ 的情况,由相同的方式可以得到相轨迹方程为

$$\frac{\dot{x}^2}{2} = -Mx + C_2$$

其中 C_2 是由初始条件确定的积分常数。这一情况下的相轨迹如图 7.27(c) 所示。

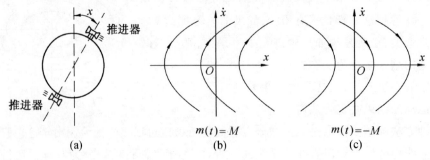

图 7.27 卫星及其相轨迹

绘制相轨迹的另一种解析方法是从 $x(t)$ 和 $\dot{x}(t)$ 的表达式中消去 t 获得轨迹的解析表达式。

2. 等倾线法

获取相平面上运动轨迹的实用作图方法是等倾线法。考虑二阶系统由以下方程描述

$$\ddot{x} = f(x, \dot{x}) \tag{7.23}$$

其中 $f(x, \dot{x})$ 可以是线性的或者非线性的解析函数。在 $[x, \dot{x}]$ 平面的任意一点上,轨迹的斜率为

$$\frac{\mathrm{d}\dot{x}}{\mathrm{d}x} = \frac{\mathrm{d}\dot{x}}{\mathrm{d}t} \bigg/ \frac{\mathrm{d}x}{\mathrm{d}t} = \frac{\ddot{x}}{\dot{x}} = \frac{f(x, \dot{x})}{\dot{x}} \tag{7.24}$$

式(7.24)称为相轨迹的斜率方程,令

$$\frac{\mathrm{d}\dot{x}}{\mathrm{d}x} = \alpha$$

对于相平面上的给定点,α 为通过该点的轨迹的斜率。式(7.24)可以改写为

$$\dot{x} = \frac{f(x, \dot{x})}{\alpha} \tag{7.25}$$

对于给定的 α 值,式(7.25)描述了相平面上的一条曲线。这一曲线称为等倾线,因为当相轨迹与这一曲线相交时总是具有相同的斜率 α。在相平面上构造若干条等倾线,就可以绘制出相轨迹。当然,如果式(7.25)是一簇直线,例如在线性二阶系统的情况下,等倾线的构造一定会比较简单

如图 7.28 所示,应用等倾线法按照以下步骤绘制相轨迹:

① 绘制一些稀疏适当的等倾线。(一般情况下,建议相邻两条等倾线之间的夹角大约为 5° ~ 10°。)

② 从起始点 A 画一条具有斜率 $(\alpha_1 + \alpha_2)/2$ 的直线,并将该直线从对应于斜率为 α_1 的等倾线延伸到对应于斜率为 α_2 的等倾线上的 B 点。

③ 从 B 点出发继续步骤 ② 的过程。

④ 用一条光滑的曲线连接得到的各点。

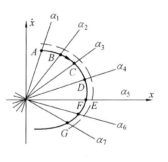

图 7.28 应用等倾线构造一条轨迹

例 7.5 考虑由以下方程描述的线性振荡器

$$\ddot{x} + x = 0$$

斜率方程为

$$\alpha = \frac{\mathrm{d}\dot{x}}{\mathrm{d}x} = -\frac{x}{\dot{x}}$$

而等倾线方程则为

$$\dot{x} = -\frac{x}{\alpha}$$

因此,在这一情况下,等倾线是一簇通过原点的直线。图 7.29 所示为一组等倾线方程、显示

带有等倾线的相平面,以及一条典型的相轨迹。注意,在这一情况下,相轨迹是一个圆。

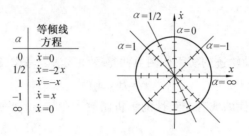

图 7.29　线性振荡器的等倾线法

3. δ 法

和等倾线法一样,δ 法既可应用于线性系统也可应用于非线性系统。但是,它更适合于非线性系统,因为非线性系统相轨迹的等倾线通常不再是直线。在这种方法中,轨迹用一些小圆弧近似表示。

首先,将式(7.23)改写为

$$\ddot{x} + \omega^2 x = \omega^2 \delta(x, \dot{x}) \tag{7.26}$$

其中 $\delta(x, \dot{x})$ 定义为

$$\delta(x, \dot{x}) = \frac{f(x, \dot{x}) + \omega^2 x}{\omega^2} \tag{7.27}$$

选择 $\omega^2 x$ 时,要使得在所考虑的 x 和 $\dot{x}$ 的范围内 $\delta(x, \dot{x})$ 既不太大也不太小。这样式(7.26)就变为

$$\ddot{x} + \omega^2 (x - \bar{\delta}) = 0 \tag{7.28}$$

相轨迹的斜率方程为

$$\frac{d\dot{x}}{dx} = -\frac{\omega^2(x - \bar{\delta})}{\dot{x}} \tag{7.29}$$

然后,对斜率方程积分得到

$$\dot{x}^2 + \omega^2 x^2 - 2\omega^2 \bar{\delta} x = C \tag{7.30}$$

其中 C 为积分常数。重新整理这一方程,得到相轨迹方程

$$\left(\frac{\dot{x}}{\omega}\right)^2 + (x - \bar{\delta})^2 = R^2 \tag{7.31}$$

这一方程是个圆。

图 7.30 给出了这种方法的图示解释。对于给定点 $P_i(x_i, \dot{x}_i)$,相轨迹垂直于 $P_i Q_i$,并可以用一段以 $Q_i(\bar{\delta}_i, 0)$ 为圆心、以 $R = P_i Q_i$ 为半径的小圆弧近似表示。

通过画圆弧的方法得到了轨迹上的下一点 P_{i+1},其中记为 $\Delta\theta$ 的夹角 $\angle P_i Q_i P_{i+1}$ 很小。在点 P_{i+1} 处可以再次计算 $\bar{\delta}_{i+1}$ 的值,在确定新的圆心 Q_{i+1} 后,就可以绘制下一段圆弧。

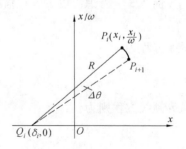

图 7.30　δ 法的图示说明

7.3.3 线性系统基本的相轨迹

相平面法特别适合于包含分段直线非线性特性的系统的分析。另一方面，许多高阶系统常常可以近似为一阶或者二阶的模型。因此，熟悉一阶和二阶线性系统的相轨迹是很重要的。

1. 一阶系统的相轨迹

考虑由以下方程描述的一阶系统

$$\dot{x} + Tx = 0 \tag{7.32}$$

相轨迹方程为

$$\dot{x} = -\frac{1}{T}x \tag{7.33}$$

假设 $x(0) = x_0$，则结果有 $\dot{x}(0) = \dot{x}_0 = -x_0/T$。如图 7.31 所示，在相平面上的轨迹只是一条斜率为 $-1/T$、通过原点的直线。如果 $T > 0$，那么相平面上的轨迹沿着该直线收敛到原点；如果 $T < 0$，那么相平面上的轨迹沿着该直线离原点而去。

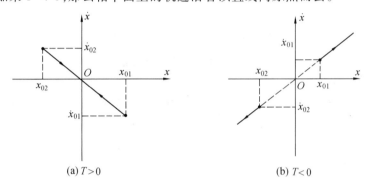

图 7.31 一阶系统的相轨迹

2. 二阶系统的相轨迹

考虑由以下方程描述的二阶系统

$$\ddot{x} + 2\zeta\omega_n\dot{x} + \omega_n^2 x = 0 \tag{7.34}$$

轨迹的斜率方程为

$$\frac{\mathrm{d}\dot{x}}{\mathrm{d}x} = \frac{-2\zeta\omega_n\dot{x} - \omega_n^2 x}{\dot{x}} \tag{7.35}$$

相轨迹可以用解析法或者图解法绘制。

在相轨迹的分析中，对于 $\mathrm{d}\dot{x}/\mathrm{d}x = 0/0$ 的点，即在它们上面轨迹的斜率不确定的点特别感兴趣。这些点称之为奇点。显然，系统在奇点上可以处于一种平衡状态。对于由方程(7.34)给出的线性系统，唯一的奇点位于原点。

二阶系统的行为明显地与其特征根有关。下面将研究方程(7.34)的二阶系统以及其他一些特殊二阶系统在奇点附近的相轨迹。这些奇点将按照在它们附近的相轨迹的图形进行分类。

(1) 两个负实根

在 $\zeta \geq 1$ 且 $\omega_n > 0$ 的情况下，方程(7.34)的两个特征根

$$s_{1,2} = -\zeta\omega_n \pm \omega_n\sqrt{\zeta^2-1}$$

都是负实数。

对于 $\zeta > 1$ 时过阻尼系统的相轨迹如图7.32(a)所示。可以看到，所有的轨迹都以非振荡衰减的形式趋向并终止于原点，也就是奇点。在相轨迹中，有两条特殊的轨迹是斜率分别为 s_1 和 s_2 的直线，它们是其他所有轨迹的渐近线。在靠近原点的地方，各条轨迹与其中一条渐近线相切。当 $\zeta = 1$ 即系统临界稳定时，如图 7.32(b) 所示，除了只有一条渐近线外，相轨迹与(a)中的情况相似。过阻尼响应(和临界阻尼)的奇点称之为稳定节点。

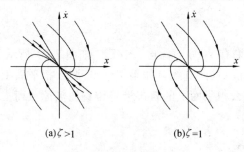

图 7.32　稳定节点的相轨迹

(2) 负实部的复根

在 $0 < \zeta < 1$ 且 $\omega_n > 0$ 的情况下，方程(7.34)的特征根

$$s_{1,2} = -\zeta\omega_n \pm j\omega_n\sqrt{1-\zeta^2} \quad (7.36)$$

是具有负实部的复根。由时域分析可知，零输入响应呈现为有阻尼振荡的形式。

欠阻尼二阶系统的相轨迹如图 7.33 所示。它的特征是所有轨迹都是收敛的对数螺旋线。沿着轨迹的运动随时间的增大以顺时针方向趋向于原点。欠阻尼响应的奇点称之为稳定焦点。

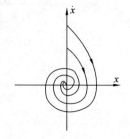

图 7.33　稳定焦点的相轨迹

(3) 虚根

在 $\zeta = 0$ 且 $\omega_n \neq 0$ 的情况下，方程(7.34)的特征根为

$$s_{1,2} = \pm j\omega_n$$

其零输入响应为无阻尼振荡。

相轨迹如图 7.34 所示。在这种情况下，所有的轨迹都是以原点、也就是奇点为中心的椭圆，奇点(0,0)称之为中心点。

(4) 两个正实根

在 $\zeta \leq -1$(负阻尼)且 $\omega_n > 0$ 的情况下，方程(7.34)的两个特征根

$$s_{1,2} = -\zeta\omega_n \pm \omega_n\sqrt{\zeta^2-1}$$

都是具有正实部的根，系统不稳定。

相轨迹见图 7.35。在这一情况下，所有轨迹都离开原点并趋向于无穷远处，这种奇

点称之为不稳定节点。

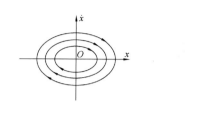

图 7.34 中心点的相轨迹

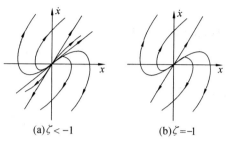

图 7.35 不稳定节点的相轨迹

(5) 具有正实部的复根

对于方程(7.34)中 $-1 < \zeta < 0$ 的负阻尼情况,特征根为

$$s_{1,2} = \zeta\omega_n \pm j\omega_n\sqrt{1-\zeta^2}$$

是具有正实部的共轭复数根,系统不稳定。

相轨迹如图 7.36 所示。在这种情况下,所有的轨迹都是离原点而去发散的对数螺旋线,奇点称为不稳定焦点。

(6) 符号相反的实根

考虑由以下方程描述的二阶系统为

$$\ddot{x} + 2\zeta\omega_n\dot{x} - \omega_n^2 x = 0 \tag{7.37}$$

如图 7.37(a) 所示,特征根为

$$s_{1,2} = -\zeta\omega_n \pm \omega_n\sqrt{\zeta^2+1}$$

是实数,但符号相反。当然,这个系统是不稳定的。

相轨迹如图 7.37(b) 所示。几乎所有的轨迹都是离原点而去的双曲线。在相轨迹中有两条特殊的轨迹是斜率分别为 s_1 和 s_2 的直线,它们是所有其他轨迹的渐近线。这种情况下的奇点称之为鞍点。

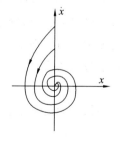

图 7.36 不稳定交点的相轨迹

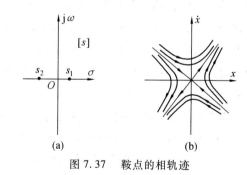

图 7.37 鞍点的相轨迹

(7) 一个零根和一个非零根

考虑由以下方程描述的系统为

$$\ddot{x} + T\dot{x} = 0 \tag{7.38}$$

特征根为

$$s_1 = 0, \quad s_2 = -\frac{1}{T}$$

由相轨迹的斜率方程为

$$\frac{d\dot{x}}{dx} = \frac{\ddot{x}}{\dot{x}} = \frac{-T\dot{x}}{\dot{x}}$$

相轨迹方程为

$$\dot{x}\left(\frac{d\dot{x}}{dx} + T\right) = 0 \tag{7.39}$$

该方程意味着对应于方程 $d\dot{x}/dx = -T$ 的轨迹是一簇斜率为 $-T$ 的直线,而且 x 轴是相轨迹的一部分。相轨迹见图 7.38,图中每一条轨迹都是从起始点 $(x_0, \dot{x}_0)$ 出发,并终止于 x 轴即直线 $\dot{x} = 0$。在这种情况下,x 轴上的每一点都可以看作奇点,或者说,直线 $\dot{x} = 0$ 称之为奇线。

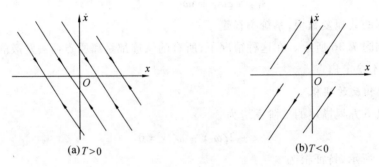

图 7.38 方程 $\ddot{x} + T\dot{x} = 0$ 的相轨迹

7.3.4 相平面分析

控制系统常常可以用二阶微分方程近似表示,然后利用相平面法进行研究。此外,当发生在这样一个系统中的非线性现象可以用分段直线表示其特性时,相平面分析有可能相当简单。分析的过程是把相平面划分成不同区域,每一个区域都对应着非线性特性特定直线段上的运动。这样,相平面各个区域内的运动用不同的线性微分方程描述。而且,根据分区边界上条件相同的关系可以获得对于特定初始值的解。对于一个特定的区域内的轨迹,其奇点有可能位于该区域外,这种情况下的奇点称之为虚奇点。相平面法分析的过程用一些例子说明如下。

例 7.6 现在考虑例 7.3 中所讨论的卫星处于一个反馈结构内,从而使卫星的姿态角 x 维持在 $0°$。这一姿态控制系统见图 7.39。

当姿态角 x 不是 $0°$ 时,相应的推进器将点火迫使 x 变为 $0°$。注意,$m(t)$ 的切换发生在 $x = 0$ 处。因此,直线 $x = 0$ 称为切换线。图 7.40(a) 说明了一条典型的轨迹。这样系统的响应是一个极限环。图 7.40(b) 所示是相轨迹。因此,这一控制系统是不能令人满意的。

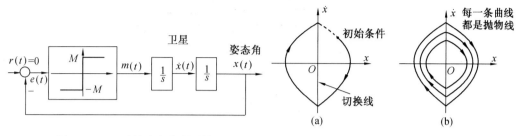

图 7.39 卫星的姿态控制系统　　　图 7.40 卫星的相轨迹

例 7.7 本例还是研究卫星控制系统问题。假设如图 7.41 所示，在控制系统中加入速度反馈，使用一个增益为 β 的速率陀螺仪作为传感装置。

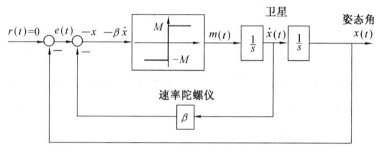

图 7.41 卫星的姿态控制系统

注意，卫星的输入仍然是 $\pm M$，因此响应曲线保持图 7.27（b）和（c）所示的抛物线簇。但是，现在 $m(t)$ 的切换线不一样了。现在发生切换时 $e = 0$，即

$$x + \beta \dot{x} = 0$$

这一方程表明，切换线是一条斜率为 $\mathrm{d}\dot{x}/\mathrm{d}x = -1/\beta$ 的直线。这一系统的典型响应如图 7.42 所示。可以看到，由于加入了速率反馈，极大地改善了系统的响应。而且，原点现在是渐近稳定的。

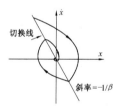

图 7.42 卫星的相轨迹

习　题

题 7.1　非线性系统如题 7.1 图所示，其中 $G(s)$ 为线性部分的传递函数。确定非线性特性 N_1、N_2 和 N_3 的等效非线性特性 N。

题 7.2　确定题 7.2 图所示各系统是否稳定，并分析对应于 $G(\mathrm{j}\omega)H(\mathrm{j}\omega)$ 曲线和 $-N(A)$ 轨迹之间各交点的振荡特性。假设 $G(s)H(s)$ 是最小相位的。

题 7.3　考虑题 7.3 图所示非线性系统。

（a）用描述函数法分析 $k = 10$ 时系统的稳定性。

（b）确定使系统稳定时 k 的临界值。

题 7.4　确定题 7.4 图所示系统的自振荡振幅和频率。

题 7.5　考虑题 7.5 图所示系统。

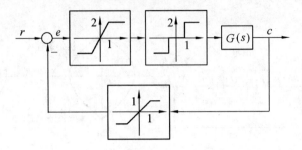

题 7.1 图

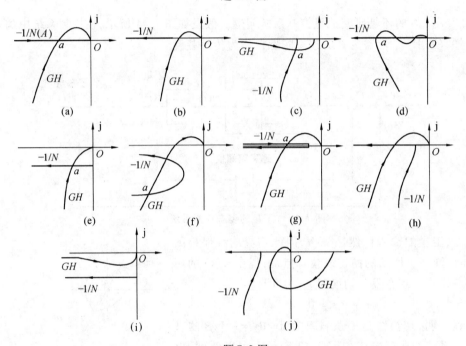

题 7.2 图

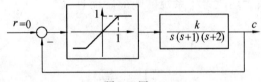

题 7.3 图

（a）用描述函数法分析该系统的稳定性。

（a）确定如何调节参数 e_0 和 M 才能使系统稳定。

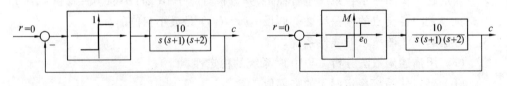

题 7.4 图　　　　　　　　　题 7.5 图

题 7.6　分析题 7.6 图所示系统的稳定性,并确定自振荡的振幅和频率。

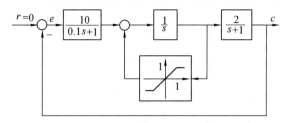

题 7.6 图

题 7.7　找出以下方程的奇点并确定它们的类型。
(a) $2\ddot{x} + \dot{x}^2 + x = 0$；
(b) $\ddot{x} - (1 - x^2)\dot{x} + x = 0$。

参 考 文 献

[1] 多尔夫.现代控制系统(英文影印版)[M].北京:科学出版社,2002.
[2] 多尔西.连续与离散控制系统(英文影印版)[M].北京:电子工业出版社,2002.
[3] 德赖斯.线性控制系统工程(英文影印版)[M].北京:清华大学出版社,2002.
[4] JOHN J D′AZZO, CONSTANTINE H HOUPIS. Linear Control System Analysis and Design [M]. New York:McGraw-Hill,1988.
[5] BENJAMIN C KUO. Automatic Control Systems [M]. New Jersey:Prentice-Hall,1982.
[6] 绪方胜彦.现代控制工程[M].北京:电子工业出版社,2000.
[7] 郑大钟.线性系统理论[M].北京:清华大学出版社,2002.
[8] 李友善.自动控制原理[M].北京:国防工业出版社,1989.
[9] 李文秀.自动控制原理[M].哈尔滨:哈尔滨工程大学出版社,2001.
[10] 胡寿松.自动控制原理[M].北京:科学出版社,2002.
[11] 孙虎章.自动控制原理[M].北京:中央广播电视大学出版社,1994.
[12] 吴麒.自动控制原理[M].北京:清华大学出版社,1991.
[13] 戴忠达.自动控制理论基础[M].北京:清华大学出版社,1991.
[14] 陈启宗.线性系统理论与设计[M].北京:科学出版社,1988.
[15] 王诗宓.自动控制理论例题习题集[M].北京:清华大学出版社,2002.
[16] LAUGHTON M A, SAY M G. Electrical Engineer′s Reference Book[M]. Scotland:Butterworths, 1985.